WJEC
Physics
For AS Level

2nd Edition

Gareth Kelly
Nigel Wood

Illuminate Publishing

Published in 2020 by Illuminate Publishing Ltd, an imprint of Hodder Education, an Hachette UK Company, Carmelite House, 50 Victoria Embankment, London EC4Y 0DZ

Orders: please contact Hachette UK Distribution, Hely Hutchinson Centre, Milton Road, Didcot, Oxfordshire, OX11 7HH. Telephone: +44 (0)1235 827827. Email: education@hachette.co.uk. Lines are open from 9 a.m. to 5 p.m., Monday to Friday. You can also order through our website: www.hoddereducation.co.uk

British Library Cataloguing in Publication Data
A catalogue record for this book is available from the British Library

ISBN 978-1-912820-55-9

Printed by: Ashford Colour Press Ltd.

Impression 4
Year 2024

Hachette UK's policy is to use papers that are natural, renewable and recyclable products and made from wood grown in well-managed forests and other controlled sources. The logging and manufacturing processes are expected to conform to the environment regulations of the country of origin. Every effort has been made to contact copyright holders of material produced in this book. If notified, the publisher will be pleased to rectify any errors or omissions at the earliest opportunity.

This material has been endorsed by WJEC and offers high quality support for the delivery of WJEC qualifications. While this material has been through a WJEC quality assurance process, all responsibility for the content remains with the publisher.

Exam questions are used under licence from WJEC CBAC Ltd. Where there is a solution provided for an example question please be aware that alternative solutions may also be possible, and that solutions have neither been provided nor approved by WJEC.

Editor: Geoff Tuttle
Design: Nigel Harriss
Layout: John Dickinson
Original artwork: John Dickinson and Patricia Briggs

Cover image: © Shutterstock/V.Belov

Acknowledgements

The authors are very grateful to the team at Illuminate Publishing for their professionalism, support and guidance throughout this project. The publisher would like to thank Dawn Booth for her help in sourcing images, and Keith Jones for his help and advice on examination material in particular.

Image credits:

p.1 V.Belov; **p.8** LMSAL; **p.17** NASA, J. P. Harrington (U. Maryland) and K. J. Borkowski (NCSU); **p.20** Eric Isselee; **p.21** (t) BBC News (c) (2015) BBC; **p.21** (b) technotr/iStock; **p.33** (t) Copyright:© 1997 Richard Megna – Fundamental Photographs; **p.33** (b) http://lannyland.blogspot.co.uk/#uds-search-results; **p.35** (t) jflyn98; **p.35** (b) Loren M Winters/Visuals Unlimited; **p.36** Fotalia/salamahin; **p.44** Allstar Picture Library Ltd/Alamy Stock Photo; **p.45** (t) Germanyskydiver; **p.45** (b) Mike Price; **p.50** Daniel White; **p.51** Germanyskydiver; **p.55** (l) Hussain Warraich; **p.55** (m) Red_Shadow'; **p.55** (r) Creative commons; **p.57** Shutterstock/Florian Augustin; **p.61** schther5; **p.64** (l) smikeymikey1; **p.64** (r) ChameleonsEye; **p.67** (t) Creative Commons Attribution 3.0 Unported, http://en.wikipedia.org/wiki/Grain_boundary; **p.67** (b) Gareth Kelly; **p.68** Ductile Test/MaterialScience2000/YouTube; **p.69** Fernando Sanchez Cortes; **p.70** Wrexham Borough Council; **p.75** (t) Creative Commons Attribution-Share Alike 3.0 Unported license, https://commons.wikimedia.org/wiki/File:TaurusCC.jpg; **p.75** (bl) LMSAL; **p.75** (br) LMSAL; **p.76** (t) Courtesy Newport Corporation; **p.76** (m) STSci; **p.76** (b) Voyagerix; **p.79** © ESA–D. Ducros, 2013; **p.80** (l) © 1995 Richard Megna – Fundamental Photographs; **p.80** (r) NASA/Hubble; **p.81** (r) James B. Kaler; **p.81** (t) Hydrogen: Courtesy of the University of Texas Libraries, The University of Texas at Austin; **p.81** (m) Koen van Gorp/NASA; **p.81** (b) Luc Viatour/www.Lucnix.be; **p.82** (t) Public domain; **p.82** (b) Courtesy NASA/JPB-Caltech; **p.83** Image courtesy of NRAO/AUI; **p.87** (t) LAWRENCE BERKELEY LABORATORY/SCIENCE PHOTO LIBRARY; **p.87** (b) SCIENCE PHOTO LIBRARY; **p.98** Pavel L Photo and video; **p.99** Public domain; **p.108** The Life & Experiences of Sir Henry Enfield Roscoe (Macmillan: London and New York), **p.108** Author Henry Roscoe/Public domain; **p.112** Professor Harry Jones, University of Oxford; **p.121** Lotus Overseas & Marketing; **p.123** Evelta; **p.129** Tsunami Hits Jamaican Shore January2020/O This Random/YouTube; **p.139** Colwell, Catharine H. "Single Slit Diffraction." PhysicsLAB.com. Hosted at Mainland High School, Volusia County Public Schools, FL. 2003. Web. 26 Mar 2015; **p.140** Public domain; **p.142** Nigel Wood; **p.154** Colwell, Catharine H. "Single Slit Diffraction." PhysicsLAB.com. Hosted at Mainland High School, Volusia County Public Schools, FL. 2003. Web. 26 Mar 2015; **p.158** (l) Vladimir Wrangel; **p.158** (r) Joe Orman; **p.158** (b) Public domain; **p.167** GNU Free Documentation License; **p.169** Courtesy of the University of Texas Libraries, The University of Texas at Austin; **p.170** (l) Creative Commons 'Attribution-NonCommercial-NonDerivative 3.0 (US)'; **p.170** (r) Rainbow Symphony, Inc.; **p.170** (b) Pi-Lens; **p.171** Hitachi; **p.172** Photo by Robin Dhakal; **p.178** Lawrence Livermore National Laboratory; **p.179** https://farm8.staticflickr.com/7008/6681662889_6606e572b2_b.jpg; **p.180** (t) NASA, J.P. Harrington (University Maryland) and K. J. Borkowski (NCSU); **p.180** (b) NASA, ESA, M. Robberto (Space Telescope Science Institute/ESA) and the Hubble Space Telescope Orion Treasury Project Team; **p.188** lightpoet; **p.201** R. MACKAY PHOTOGRAPHY, LLC

Contents

How to use this book

This book has been written to support the WJEC AS Physics specification and the first half of the A level specification. The layout of the book matches that of Units 1 and 2 respectively of the AS Physics specification. The same material is specified for Units 1 and 2 of the A level.

It provides you with information which covers the content requirements of the course as well as plenty of practice questions to allow you to keep track of your progress.

The main chapters in the book are the AS Units 1 and 2.
- Unit 1 covers Motion, Energy and Matter
- Unit 2 covers Electricity and Light

Additional chapters are
- Chapter 3 – Practical skills
- Chapter 4 – Mathematical skills

Unit 1 covers the seven topics of the Physics specification and Unit 2 has eight. Each has its own section in this book and includes the appropriate **specified practical work**. Chapter 3 covers generic practical skills and Chapter 4 summarises the maths skills you will need. Chapters 3 and 4 both come with their own Knowledge check and Test yourself questions.

Level of coverage
This book contains material which is examined at both AS and A level. It is expected that, whilst some students will take AS Physics only, a large percentage of its students will proceed to take the full A level. Because of this, the level of coverage of both the material and some of the practice questions is higher than required for AS.

Practice questions
As well as the Knowledge check questions in the margin of the main text, each section in Units 1 and 2 ends with a Test yourself exercise. As well as containing material relating to the content of the sections, the exercises contain data-analysis questions around the specified practical work for the unit. Some questions also relate to the content of more than one unit: A level students need to be able to answer such synoptic questions, which bring ideas from different topics in the Physics specification. The solutions to these exercises as well as the Knowledge check questions are to be found at the end of the book.

Caution: Questions which ask for explanations or other working can often be answered in various ways. The solutions given in this book cannot cover all the possibilities.

Exam practice questions
At the end of each of Units 1 and 2, you will find a set of Exam Practice questions. These are all taken from past WJEC examinations and a set of sample answers is given at the end of this book.

Margin features
The margins of each page hold a variety of features to support your learning:

 **Key term**

Quantity: A quantity is represented by a number multiplied by a unit.

Key terms are terms that you need to know how to define. They are highlighted in blue in the body of the text and appear in the Glossary at the back of this book. You will also find other terms in the text in bold type, which are explained in the text, but have not been defined in the margin. You should ensure that you can accurately define these terms or state the physical laws.

 **Study point**

The symbols for quantities are printed in *italics*, e.g. m, T. The symbols for units are printed in plain print, e.g. kg, K.

As you progress through your studies, Study points are provided to help you understand and use the knowledge content. In this feature, factual information may be emphasised, or restated to enhance your understanding.

 **Knowledge check**

Derive the unit of volume by considering a cube.

Knowledge check questions are short questions to check your understanding of the subject, allowing you to apply the knowledge that you have acquired. Many of these questions ask for calculations related directly to the text alongside. The answers to all the Knowledge check questions are in the back of the book.

 **Stretch & challenge**

(c) Use $x = ut + \frac{1}{2}at^2$ to solve part (b) of the example.

Material developed in these is designed to make you think more deeply about the subject. Some S&C boxes contain material which is beyond the specification. You will not be examined on this content but it does provide a background for students who will progress to higher education in physics or engineering. Other S&C boxes contain questions, sometimes requiring more advanced mathematical techniques than most AS questions.

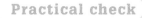

Top tip ▷
Remember that you measure the *diameter* of a wire, not the *radius*!

Maths tip ▷▷
See Chapter 4 for help with finding the gradient of a graph for straight and curved graphs.

Practical check ▷
See Section 3.2 for a suggestion on measuring the diameter of a wire.

Top tips are to help you avoid common unnecessary mistakes.

Maths tips and Practical checks refer to particular techniques and often direct you to Chapters 3 and 4 for a fuller treatment.

◁ **Link** ▷
See Chapter 3 for combining and reducing uncertainties.

Links to other sections of the course are highlighted in the margin, near the relevant text. They are accompanied by a reference to any areas where sections relate to one another. It may be useful for you to use these Links to recap a topic, before beginning to study the current topic.

The AS examination

Examination papers are constructed from questions that require a variety of skills to answer. Some will assess whether you can recall facts, experimental procedures and physical laws, such as Newton's laws. In others you will need to apply your knowledge in verbal or mathematical ways and to analyse experimental results. Some questions require maths; some require experimental skills. There are three basic assessment objectives (AOs), each of which contributes a fixed percentage of the marks in each paper and which include all the other skills. Here is a brief description of each of the AOs, with its examination weighting.

Assessment objective 1 (AO1)

Learners must: Demonstrate knowledge and understanding of scientific ideas, processes, techniques and procedures

35% of the marks for questions set on the examination papers are for AO1. As well as pure recall, such as stating laws and definitions, this includes knowing which equations to use, substituting into equations and describing experimental techniques.

Assessment objective 2 (AO2)

Learners must: Apply knowledge and understanding of scientific ideas, processes, techniques and procedures:
* in a theoretical context
* in a practical context
* when handling qualitative data
* when handling quantitative data.

45% of the marks for questions set on the examination papers are for AO2. Bringing together ideas to explain phenomena, solving mathematical problems and performing calculations using experimental results and graphs are categorised as AO2. Application involves using the skills you have acquired in situations you have not previously encountered, e.g. in synoptic questions.

Assessment objective 3 (AO3)

Learners must: Analyse, interpret and evaluate scientific information, ideas and evidence, including in relation to issues, to:
* make judgements and reach conclusions
* develop and refine practical design and procedures.

20% of the marks for questions set on the examination papers are for AO3. These marks include determining quantities using experimental results and also responding to data to draw conclusions.

AS Physics Units 1 and 2 written papers

The two papers usually consist of 7 or 8 structured questions (see below). You cannot, however, assume that each topic will have a question related to it:

* Some topics have closely related material, e.g. kinematics, dynamics and energy in Unit 1. Hence a question might cover areas of two (or all three) of these topics.
* Questions involving practical skills tend to be longer questions.
* Some topics, e.g. basic physics, have a larger content than the average.

Answering examination questions

Before you start answering any questions, you need to do some checks.

1. Is it the right examination paper?
2. Is the examination paper complete?
3. Do you know where the last question finishes? (Look for the message END OF PAPER.)
4. Do you know when the exam finishes?
5. Do you have a copy of the Data Booklet?

Time and mark allocation

When you answer the questions, if you allow yourself a minute for each mark this will give you ten minutes for checking over your work. Check the number of marks for each section of a question – don't miss out on marks because you have left a question part unanswered, so scan down each question prior to answering it, identifying the parts and the marks allocated to each one.

Mark allocations for question parts

The mark allocation for each question part is given in square brackets, e.g.

State the principle of conservation of momentum. **[2]**

The mark allocation gives a good clue to the detail required in your answer. The [2] is a good clue that there are two aspects to the answer. In this case your answer should include:

1. a statement that the (vector) sum of the momenta remains the same, and
2. the conditions, i.e. in a closed system or if no resultant external force acts.

On some occasions, a question may only have one mark, *but still require both points to be made.*

Question structure

Most questions are structured, having several parts with a linking theme. The way these parts are arranged gives a clue to their relationship to one another. For example, look at question 1 of the Unit 1 Exam practice questions. It has the following structure:

1. (a)
   (b) (i)
       (ii)
   (c)

Part (a) asks you to write down an equation. This shows you the physics ideas you will need to employ in answering the question – in this case, moments.

Part (b) is in the same question, so the hint in part (a) applies. The two subsections (i) and (ii) are closely related, a fact emphasised in this case by the word 'hence' in part (ii).

Part (c) is less directly related than (b)(i) and (ii) but it is still about moments.

Sometimes there is an extra level of subdivision. For example, look at question 1 of the Unit 2 Exam practice questions: Part 1(a)(iii) has two parts, labelled (I) and (II).

All this structure is to guide you through what might otherwise be a lengthy multistage calculation. Without it, you are in danger of starting off in an unhelpful direction, getting lost in the undergrowth, and ending up with an incorrect answer.

Answering QER questions

QER (Quality of Extended Response) questions test how logically you can present a detailed piece of physics. There will be at least 12 answer lines and perhaps a space for a diagram.

Many QER questions ask for a standard piece of knowledge, i.e. it is AO1 and you should know it. For example, question 5 in the Unit 1 Exam practice questions asks for the information which can be gained from the spectrum of a star. Your job is to take what you know and write it down in a logical, structured way.

You should pay attention to both your spelling and grammar as well as the accuracy of physics in your response.

A common QER question is to ask for the method for one of the assessed practicals, e.g.

Describe an experiment to measure the acceleration of a freely falling object. **[6 QER]**

Sometimes QER questions ask about practical work other than the specified practicals. These involve designing the method yourself and so are AO3, e.g.

Explain what properties of light from a laser can be determined using polarisation and interference. Give practical details. **[6 QER]**

Synoptic questions

In Units 3 and 4 of the A level specification, you will encounter questions which use ideas from AS Units 1 and 2. These are called synoptic questions. This means that the material contained in this book will still be useful in your year 13 studies.

Calculation questions

Physics exams always have lots of calculations. In fact a minimum of 40% of the marks for Units 1 and 2 are for maths. In calculations you often have to do several things: selecting equations, substituting values, doing some algebra, calculating, giving an answer with a correct unit.

It is always good practice to give the working when calculating the answer to a question. Look at this example:

A cube, with sides of length 0.30 m has a mass of 120 kg. Calculate its density. **[2]**

The following students have the same wrong answer:

Student 1: $\text{Density} = \dfrac{\text{Mass}}{\text{Volume}} = \dfrac{120 \text{ kg}}{(0.30 \text{ m})^3}$ ✓ $= 1330 \text{ kg m}^{-3}$ ✗

Student 2: $\text{Density} = 1330 \text{ kg/m}^3$ ✗✗

Student 1 gets a mark because she inserted the values correctly into the equation. She squared the 0.30 m instead of cubing it and so loses the second mark. Student 2 has probably done the same things but without the evidence cannot be awarded the first mark.

Practical skills questions

Each unit exam has a **minimum of 15%** of the marks for practical skills. These are for recalling the methods used in the specified practicals (AO1) and the way the results are analysed, for analysing data, drawing conclusions and suggesting improvements (AO2 and AO3).

For example, look at the following questions:

Unit 1 Exam practice questions, Q3: all 10 marks are practical marks.

Unit 2 Exam practice questions, Q4: the 6 marks in Q4(b) are practical marks.

Issues questions

The definition of AO3 refers to issues. This includes ethical judgements, how scientists validate new discoveries, how science informs society's decisions and the costs, risks and benefits of the application of scientific knowledge. Examples of this type of question are:

Unit 1 Exam practice questions, 3(d) – costs and benefits of knowledge

Unit 2 Exam practice questions, 1(c) – benefits, risks and ethical judgement.

You should be aiming to use your scientific knowledge to make reasoned comments.

Error carried forward (ecf)

If you look at a WJEC physics mark scheme, you will come across the letters ecf, which stand for error carried forward. You will find this when a calculation uses a value which you have just calculated in an earlier question part – which might be incorrect. You will not be penalised further if you use this value.

Example: Question 2 of the Unit 1 Exam practice questions.

(b) (i) Calculate the strain in the rubber at point **B**. [1]
(ii) Determine the Young modulus of the rubber in the region **AB**. [3]

The calculation of the Young modulus depends upon the value of the strain, so you can still get 3 marks for part (ii) even if your answer to (i) was wrong.

Command words used in WJEC exam questions

These are words which give you information about what sort of answer is required. There are quite a few command words – here are the most common ones:

State

Give a value or statement without any explanation.

Example: State the Principle of Moments.

Answer: For a body in equilibrium under the action of a number of forces, the sum of the clockwise moments (of the forces) about any point is equal to the sum of the anticlockwise moments about the same point.

This example shows the value of becoming familiar with the WJEC Terms and Definitions booklet.

Describe

Write a short account with no explanation.

Example: Describe the motion shown in the $v-t$ graph between 0 and 30 s.

Answer: 0–10 s: constant velocity of 15 m s^{-1}.
10–30 s: constant acceleration to 35 m s^{-1}

Experimental methods can also be asked for using this command word.

Explain

You need to give a reason or reasons.

Example: Explain why the skydiver falls at a constant velocity.

Answer: Because the upward force of air resistance on her is exactly equal (and opposite) to the downward force of gravity.

Calculate

You should use one or more equations together with data to find the value of an unknown quantity.

Determine

This word is used when the calculation requires more than just the application of an equation.

Example: Unit 1 Exam practice questions, 2(b).

To answer part (ii) you have to relate the graph to the equation $E = \sigma/\varepsilon$.

Discuss and Evaluate

These command words are often used in AO3 questions, e.g. in the context of deciding which (if either) of two statements is likely to be correct. Your aim should be to provide good application of physics together with a final judgement.

Example: Unit 1 Exam practice questions, Q1(c).

These command words are also used in the concluding parts of a practical skills question.

Example: Evaluate whether the data are consistent with the relationship $y = kx$, where k is a constant.

In this case, the evaluation would consist of deciding whether the data were consistent with a straight-line graph through the origin and whether the degree of scatter in the data points made for a confident conclusion.

Suggest

This word indicates that there is no single correct answer. It often occurs as the last part of a structured question and it involves using knowledge gained from the specification and applying it creatively to additional material.

Example: Unit 1 Exam practice questions, Q3(d).

The earlier parts of the question are about the properties of subatomic particles and are testing knowledge and the ability to apply the knowledge. This part includes a degree of speculation. In answering this kind of question, be prepared to have an opinion and to use it.

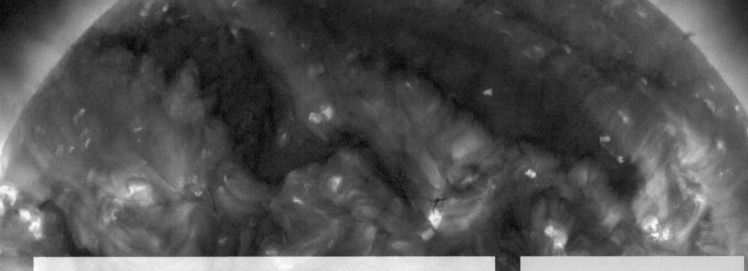

Unit 1

Motion, energy and matter

This foundation unit of the AS Physics course builds upon concepts developed in Key Stage 4 in addition to introducing entirely new material.

- The initial topic, Basic physics, examines the language of physics in terms of quantities and units, which are written in the standard manner of the scientific community using negative indices where appropriate.

- The core of Unit 1 comprises the concepts of motion and energy. These are explored in greater depth than in earlier courses, the vector aspects of motion being examined and the mathematical relationships between quantities of motion investigated.

- Engineers and materials scientists rely on knowledge of the properties of materials to be able to make buildings and machines. These properties are investigated and explained in terms of the behaviour of the constituent molecules.

- Electromagnetic radiation is used to explore the nature of the universe and its constituent parts – stars, galaxies and the cosmic microwave background radiation. Using the whole range of the e-m spectrum allows a much fuller picture of the universe to be obtained than by visible light alone.

- The familiar structures of the material world, atoms and molecules, are seen to be built upon combinations of the fundamental particles of nature, leptons, quarks and antiquarks, interacting by four fundamental forces. The rules of particle interactions are explored.

Content

1.1 Basic physics
1.2 Kinematics
1.3 Dynamics
1.4 Energy concepts
1.5 Solids under stress
1.6 Using radiation to investigate stars
1.7 Particles and nuclear structure

Practical work

Practical work is integral to any physics course. Unit 1 provides a wealth of opportunities for students to hone their practical skills as well as to develop their understanding of the contents.

1.1 Basic physics

Physics is an experimental science. It involves making measurements of quantities, such as pressure, speed, electric current and temperature, and discovering laws, which concern relationships between quantities, and formulating theories, to explain why natural phenomena occur. This topic includes some details of how to handle physical quantities, which are covered in Level 2 courses such as GCSE Physics. Some of its contents will therefore be familiar but they are taken to a higher level.

1.1.1 Quantities and units

In physics, a **quantity** is a physical property of an object or material which can be measured. An example of a quantity is *density*. The value of the density of air at room temperature and pressure is 1.28 kg m^{-3}. Notice that the unit of density (kg m^{-3}) is expressed in terms of two other units, **kg** and **m** (kilograms and metres). This is explored in the next section.

(a) Base quantities and units

In order to measure a quantity, such as length, we need to have a defined standard to compare the length with. In our system of units, *Le système international d'unités* (abbreviated to SI), the defined unit of length is the metre with the abbreviation **m**. What does a reported length of, say, **53.7 m** mean?

$$53.7 \text{ m} = 53.7 \times \text{the defined unit of length;}$$

in other words the distance light can travel in 53.7/ 299 792 458 of a second!

Table 1.1.1 shows the 7 base quantities with their SI units.

Quantity		Unit	
Name of quantity	Symbol	Name of unit	abbreviation
mass	m	kilogram	kg
length	ℓ	metre	m
time	t	second	s
electric current	I	ampère	A
temperature	T	kelvin	K
amount of substance	n	mole	mol
luminous intensity	L	candela	cd

Table 1.1.1 SI quantities and units

Notice that the definition of the metre depends upon another definition (the second) as well as a physical property (the speed of light). The table also has common symbols for the quantities, e.g. t for time and ℓ for length. Other symbols can be used, e.g. x and r for lengths and M for mass.

1.1.1 Knowledge check

Simplify the following:

(a) $6a + 2a$
(b) $6a \times 3a$
(b) $6a \div 3b$
(d) $(6a)^2$

1.1.2 Knowledge check

Derive the unit of volume by considering a cube.

Study point

It is rather tedious to keep writing *unit of*, so we use square brackets to stand for this:

$[\text{length}] = \text{m}$

$[\text{area}] = \text{m}^2$

Study point

A useful symbol for a change of something is Δ (delta). So

Δv = change of velocity.

Top tip

Learn the expressions for N, J and W in terms of kg, m and s, and how to derive them.

1.1.3 Knowledge check

The unit of the coefficient of viscosity, η, is usually written as 'Pa s' (pascal second) where **Pa** is the unit of pressure, defined by

$$\text{pressure} = \frac{\text{force}}{\text{area}}.$$

Show that this unit is the same as that derived in the example.

(b) Derived quantities and units

Most of the time physicists work with quantities other than the base quantities, e.g. area, volume, pressure, power. They use the base units in combination to express these. In order to derive these units we treat them as algebraic letters and remember some simple algebraic rules. To remind yourself of them see Knowledge check 1.1.1.

The easiest way of understanding how to derive a unit is to look at some examples:

1. **Unit of area.** We start with a defining equation:

   Area of a rectangle = length × breadth

   ∴ Unit of area = unit of length × unit of breadth

   But length and breadth are both distances so they both have **m** as their unit.

   ∴ Unit of area = m × m = m².

2. **Unit of change of speed (or change of velocity).**

   The unit of speed (or velocity) is m s⁻¹. If the speed of a car changes from 15 m s^{-1} to 33 m s^{-1} then

   change of speed = final speed − initial speed

   $$= 33 \text{ m s}^{-1} - 15 \text{ m s}^{-1}$$

   $$= 18 \text{ m s}^{-1} \text{ (remember that, in algebra, } 33a - 15a = 18a)$$

   So the unit of a change of speed is the same as the unit of speed.

3. **Unit of acceleration.** Again we start with a defining equation:

   $$\text{acceleration} = \frac{\text{change of velocity}}{\text{time}} \qquad \text{or } a = \frac{\Delta v}{t}$$

   $$\therefore [a] = \frac{[\Delta v]}{[t]} = \frac{\text{m s}^{-1}}{\text{s}} = \text{m s}^{-2}$$

Some derived units are used very frequently and it is useful to learn how to express them in terms of the base SI units.

Example

Express the unit of force, the newton (N), in terms of base SI units.

Answer

Equation: Force (N) = mass (kg) × acceleration (m s⁻²).

Or in symbols: $F = ma$

$$\therefore [F] = [m][a] \text{ so } \text{N} = \text{kg m s}^{-2}$$

Using the result of the example and the equations

$$\text{Work} = \text{Force} \times \text{distance} \quad \text{and} \quad \text{Power} = \frac{\text{Work}}{\text{time}}$$

we can express the units of work (J) and power (W) in terms of base SI units.

Here is another example, this time using an unfamiliar quantity:

Example

The drag force, F_D, on a sphere moving through a fluid is given by Stokes' formula, $F_D = 6\pi\eta av$, where a is the radius of the sphere, v the velocity and η [eta] is the *coefficient of viscosity* of the fluid. Find the unit of η in terms of the base SI units.

Answer

Rearranging the equation $\eta = \dfrac{F_D}{6\pi av}$. Both 6 and π have no units, so $[\eta] = \dfrac{[F_D]}{[a][v]}$

$[F_D] = $ kg m s^{-2}, $[a] = $ m and $[v] = $ m s^{-1} $\therefore [\eta] = \dfrac{\text{kg m s}^{-2}}{\text{m}^2\,\text{s}^{-1}} = $ kg m^{-1} s^{-1}.

> **Maths tip**
>
> See Section 4.2.1 (c) and (d) for SI multipliers and standard form.

(c) Using SI multipliers and standard form

Many problems arise in which the quantities are either much larger or much smaller than the basic units. The data are thus given either in *standard form* or using SI multipliers. This example has data in mixed forms.

Example

Calculate the energy transmitted by a 44 kV power cable in one day if it carries a current of 2.5×10^2 A. [Use $P = IV$ and $E = Pt$]

Answer

From the two equations, $E = IVt$.

$\therefore$ Converting to the base units: $E = 2.5 \times 10^2$ A $\times 44 \times 10^3$ V $\times 86\,400$ s

$= 9.5 \times 10^{11}$ J (2 s.f.)

> **Top tip**
>
> In the example, we could have written 44 kV as 4.4×10^4 V. However, to avoid mistakes it is easier to write it as 44×10^3 and let the calculator handle it!

1.1.2 Checking equations for homogeneity

Consider the equation: $v^2 = u^2 + 2ax$, where u and v are the initial and final velocities, a the acceleration and x the displacement of a uniformly accelerating object. We're going to take this equation apart and look at the units of its various bits.

1. The u^2 term: Now $[u] = $ m s^{-1}, so $[u^2] = ($m s$^{-1})^2 = $ m^2 s^{-2}.

2. The $2ax$ term: $[2ax] = [a] \times [x] = $ m s$^{-2} \times$ m $=$ m^2 s^{-2}

Let's just stop here for a moment: the u^2 term and the $2ax$ term **have the same units!** Why is this important? Because it means that they **can** be added together. See Rule 1 in the margin. This means that the unit of the right-hand side of the equation is m^2 s^{-2}.

3. The v^2 term: $[v] = $ m s^{-1}, so $[v^2] = ($m s$^{-1})^2 = $ m^2 s^{-2}

Notice that **the left-hand side has the same unit as the right-hand side.** Why is this important? Two things can only be equal if they have the same units; 53 V can never be equal to 53 A – similarly 1 day and 1 cm could never be the same!

We say that this equation is **homogeneous** – only terms with the same units are added or subtracted and the units of the two sides are the same. If the 'equation' isn't homogeneous it cannot be right – you must have remembered it incorrectly.

> **Study point**
>
> **Homogeneity – Rule 1**
>
> Two quantities a and b can only be added together if they have the same units – and then the answer has the same units.
>
> The same goes for subtraction.
>
> **Homogeneity – Rule 2**
>
> An equation is homogeneous only if the units of the two sides are the same.

> **Knowledge check**
>
> Show that the equation
>
> $x = ut + \frac{1}{2}at^2$
>
> is homogeneous.
>
> (Remember that $\frac{1}{2}$ has no units.)

> **Study point**
>
> **Warning**
>
> Just because an equation is homogeneous doesn't mean that it is right, e.g. $v^2 = u^2 + 3as$ is homogeneous and incorrect!

1.1.3 Scalar and vector quantities

Some quantities, e.g. mass, are completely specified by their magnitudes. These are called **scalar quantities**. Others, e.g. force, have a direction too. These are called **vector quantities**.

Think about the effects of the two forces on the sledge in Fig. 1.1.1(a) and (b). We need to specify a direction for a vector quantity, e.g.

$$F = 25 \text{ kN horizontally} \quad \text{or} \quad F = 25 \text{ kN due north.}$$

In both cases, the 25 kN is the magnitude, consisting of a number (25) and a unit (kN).

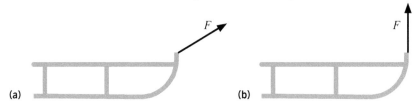

(a)   **(b)**

Fig. 1.1.1 Forces on a sledge

(a) Forces and how to add them

Scalar quantities are easy to add, e.g.

$$3.0 \text{ kg} + 4.0 \text{ kg} = 7.0 \text{ kg}$$

and similarly, subtract

$$4.0 \text{ kg} - 3.0 \text{ kg} = 1.0 \text{ kg.}$$

We just use the normal rules of arithmetic.

Fig. 1.1.2 Adding masses

When we are adding forces, other rules apply.

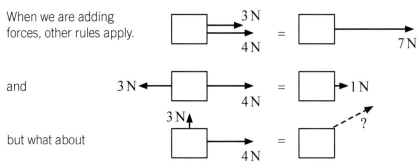

and

but what about

Fig. 1.1.3 Adding forces

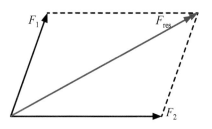

Fig. 1.1.4 The parallelogram law of vector addition

The combined effect of two (or more) forces is called the **resultant force**, F_{res}. We can calculate the sum of two forces using the *parallelogram law of vector addition*, as shown in Fig. 1.1.4. You *could* find the resultant force, ΣF, by drawing a scale diagram but it is more accurate to calculate it using trigonometry, e.g. the cosine rule or (if the forces are at right angles) Pythagoras' theorem.

Note that for the AS exam you only need to be able to add forces at right angles but if you are going on to higher studies you will need to be able to cope with other angles.

Example

Find the resultant of the 3.0 N and 4.0 N forces acting at right angles in Fig. 1.1.3.

Answer

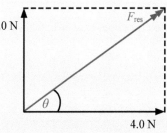

Fig 1.1.5 Adding forces at right angles

Step 1: Draw the parallelogram of the forces, which in this case is a **rectangle**. [Note the angle, θ].

Step 2: Use Pythagoras' theorem to calculate the resultant force.

$F_{res}^2 = 3.0^2 + 4.0^2$, so $F_{res} = \sqrt{3.0^2 + 4.0^2} = 5.0$ N

Step 3: Calculate, θ. $\sin \theta = \dfrac{\text{opp}}{\text{hyp}} = \dfrac{3.0}{5.0} = 0.6$, so $\theta = \sin^{-1} 0.6 = 36.9°$

∴ The resultant = 5.0 N at 36.9° to the 4.0 N force

Note that the direction given in the example answer does not specify the resultant fully, e.g. the angle could be below the horizontal. However, together with the θ in the diagram, this is enough.

Most of the time you will only have to combine two vectors at right angles, which you can do using Pythagoras' theorem and simple trig, such as $\sin \theta = \dfrac{\text{opposite}}{\text{hypotenuse}}$.

Knowledge check 1.1.5 is an example.

(b) Motion scalars and vectors

Like length, distance is a scalar quantity. The question, 'What is the distance between Aberystwyth and Bangor?' doesn't ask about direction. If you know the answer, it doesn't help you to navigate from A to B. However, 'Bangor is 91 km due north of Aberystwyth' would enable a pilot to fly from one to the other. This quantity, which includes direction as well as distance, is called **displacement**. The displacement of Bangor from Aberystwyth is 91 km north. Similarly, Flint is 65 km east of Bangor.

The displacement of a point B from a point A is the shortest distance from A to B together with the direction.

What is the displacement from Aberystwyth to Flint? Fig. 1.1.6 shows how we can add the displacements $\overrightarrow{AB}$ and $\overrightarrow{BF}$ to give the resultant $\overrightarrow{AF}$ displacement (shown in red). You should be able to show that $\overrightarrow{AF} \sim 112$ km at 35.5° E of N.

We calculate speed (strictly, **mean speed**) using speed = $\dfrac{\text{distance}}{\text{time}}$. Distance is a scalar so speed is as well. The vector equivalent of speed is **velocity**, which is defined by:

$$\text{velocity} = \frac{\text{displacement}}{\text{time}}$$

The next example shows the distinction between the two.

Top tip

A force has a direction as well as a magnitude so, if calculating a force, you need to specify the direction. Make sure you make it clear what angle you mean.

Knowledge check 1.1.5

Calculate the resultant force. Remember, magnitude and direction.

Key term

Mean speed: $\dfrac{\text{distance}}{\text{time}}$

Mean velocity: $\dfrac{\text{displacement}}{\text{time}}$

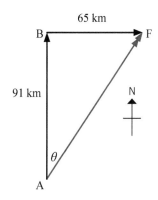

Fig. 1.1.6 Adding displacements

Maths tip ≫

If you look at Figs 1.1.4 and 1.1.6 we seem to have used two different ways of adding vectors.

In fact they give the same answer.

Here is 1.1.6 drawn like 1.1.4:

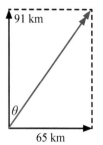

Example

A light aircraft flies from Aberystwyth to Bangor and then to Flint in a time of two hours. Using the data on the previous page, calculate (a) the mean speed and (b) the mean velocity.

Answer

(a) Distance travelled $= AB + BF = 91 + 65 = 156$ km

$$\therefore \text{Mean speed} \quad = \frac{156 \text{ km}}{2 \text{ h}}$$

$$= 78 \text{ km h}^{-1}$$

(b) Displacement $\overrightarrow{AF} \sim 112$ km at 35.5° E of N

$$\therefore \text{Mean velocity} \quad = \frac{112 \text{ km}}{2 \text{ h}}$$

$$= 56 \text{ km h}^{-1} \text{ at } 35.5° \text{ E of N.} \quad \text{[Note: direction!]}$$

(c) Lists of scalar and vector quantities

These lists include most of the scalar and vector quantities that you'll encounter in AS/A level physics. The ones in italics are only in the full A level course.

Scalars – density, mass, volume, area, distance, length, speed, work, energy (all forms), power, time, resistance, temperature, potential (or pd or voltage), electric charge, *capacitance*, *activity*, pressure, refractive index.

Vectors – displacement, velocity, acceleration, force, momentum, *electric field strength*, *magnetic field strength* (or *magnetic flux density*), *gravitational field strength*.

(d) Adding more than two vectors

We have seen how to add two vectors using either the parallelogram method (see Fig. 1.1.4) or the nose-to-tail method (Fig. 1.1.6). Fig. 1.1.7 shows how we can extend the latter way of addition to more than two vectors. An alternative method is given in Section 1.1.4.

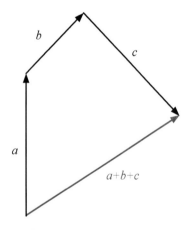

Fig. 1.1.7 Adding more than two vectors

1.1.4 Working with vectors

(a) Subtracting vectors

To calculate an acceleration we need first to find a change in velocity, Δv. If the first velocity is v_1 and the second is v_2, then $\Delta v = v_2 - v_1$. We know how to subtract scalars. How does it work for vectors?

For numbers we know that instead of writing

$53 - 45$

we get the same answer (**8**) if we write

$(-45) + 53$.

We have changed a subtraction into the addition of a negative number. The same procedure works with vectors. Look at Fig. 1.1.8(a).

To find $v_2 - v_1$, we **add** $-v_1$ to v_2. The vector $-v_1$ is the same magnitude as v_1 but has the opposite direction. The dotted lines in (b) show us how we can think of it, using the nose-to-tail way of adding vectors: go backwards along v_1 and forwards along v_2. So $v_2 - v_1$ is the vector from the end of v_1 to the end of v_2.

Diagram (c) shows the same calculation using the parallelogram method. It doesn't matter which method we use: $v_2 - v_1$: the red vector obviously has the same length and direction in (b) and (c).

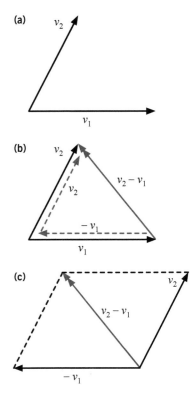

Fig. 1.1.8 Subtracting vectors

Example

A car changes velocity from 25 m s^{-1} due E, to 20 m s^{-1} due N in 8.0 seconds. Calculate the mean acceleration.

Answer

Step 1: Draw the diagram. Take care with the direction of $v_2 - v_1$: Go backwards along the v_1 vector ($- v_1$) then forwards along the v_2 ($+v_2$).

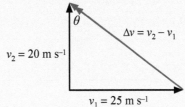

Step 2: Use Pythagoras' theorem to calculate Δv.

$(\Delta v)^2 = 25^2 + 20^2 = 1025.$ $\therefore \Delta v = 32.0$ m s^{-1}.

Step 3: Calculate θ. $\tan \theta = \dfrac{25}{20} = 1.25.$ $\therefore \theta = 51.3°$

Step 4: Calculate a. $a = \dfrac{\Delta v}{\Delta t} = \dfrac{32.0}{8.0} = 4.0$ m s^{-2} at 51.3° W of N. [NB direction!]

(b) Components of vectors

Look back at Fig. 1.1.1(a). How much of the force F is pulling the sledge forwards and how much of it is lifting the sledge? In other words, if a force, F, acts at an angle θ to the horizontal, what are its horizontal and vertical components, F_h and F_v? Fig. 1.1.9 clarifies the question: F_h and F_v are the horizontal and vertical forces which add together to give F as the resultant.

Using elementary trigonometry: $F_h = F \cos \theta$

$$F_v = F \sin \theta$$
$$\text{and} \ \ F = \sqrt{F_h^2 + F_v^2}$$

This process is called **resolving**. Why is this a useful technique? For all sorts of reasons. Here are just two:

1. If the motion is horizontal (like the sledge) the horizontal component of the force multiplied by the distance moved gives the work done, i.e. the energy transferred.

2. When adding several (i.e. more than two) vectors, it is often easier to find the horizontal and vertical components of each and add them.

Sometimes it is useful to find the components in directions other than horizontal and vertical, e.g. for the forces on a car on a slope the sensible directions to calculate components of the forces or velocity are parallel to and at right angles to the slope.

We will meet this sort of situation frequently. The important thing to remember is that the **component of a vector, A, in a direction at angle θ to the direction of the vector is always $A \cos \theta$.**

◀ Top tip

At AS, you will only have to subtract vectors at right angles.

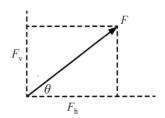

Fig. 1.1.9 Force components

Knowledge check 1.1.6 ◀

In Fig. 1.1.9, $F = 150$ N and $\theta = 30°$. Calculate F_v and F_h.

≫ Key term

Resolving: Finding the component of a force: we **resolve** a force into its horizontal and vertical components.

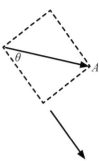

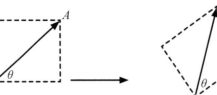

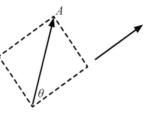

Fig. 1.1.10 Component directions on a slope

Fig. 1.1.11 In all cases the component of A in the direction of the arrow is $A \cos \theta$

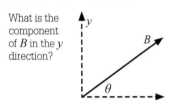
Example

Use components to find the resultant of the forces in Fig. 1.1.12.

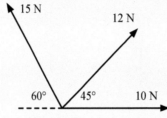

Fig. 1.1.12

Answer

Total horizontal component $= 10 + 12 \cos 45° + 15 \cos 120°$ [see Study point]

(to the right) $= 10 + 8.485 - 7.5$ [Note the '−' sign]

$= 10.99$ N

Total vertical component $= 0 + 12 \sin 45° + 15 \sin 60°$

(upwards) $= 21.48$ N

Combine the two components using Pythagoras' theorem:

$F^2_{res} = 21.48^2 + 10.99^2$

$\therefore F_{res} = 24.1$ N.

And $\theta = \tan^{-1} \dfrac{21.48}{10.99} = 62.9°$

$\therefore$ The resultant force is **24 N** at **63°** to the horizontal (2 s.f.).

1.1.5 Density

For a material of uniform composition, the mass of a sample is directly proportional to its volume. Hence the ratio of mass to volume is a constant, which is characteristic of the material. We call this constant the **density**.

Table 1.1.2 contains the densities of some common materials. You should make sure you can make a reasonable estimate of densities if asked.

Material	ρ / kg m^{-3}	ρ / g cm^{-3}	Material	ρ / kg m^{-3}	ρ / g cm^{-3}
Air*	1.29	0.00129	Steel	7900	7.90
Water	1000	1.00	Aluminium	2800	2.8
Brick	2300	2.30	Mercury	13 600	13.6
Petrol	880	0.88	Gold	19 300	19.3
* At 0°C and atmospheric pressure.					

Table 1.1.2 Densities

The range of densities of materials on the Earth is quite large. The drawings in Fig. 1.1.13 all represent 1 tonne (10^3 **kg**) of material. But this range pales into insignificance compared with the range of densities in the universe. For comparison, Table 1.1.3 gives the size of 1 tonne (1 **t**) cubes of various materials outside the Earth. The range of densities shown in the table is ~10^{33}.

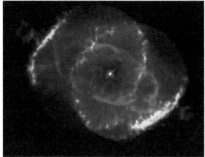

Fig. 1.1.14 The Cat's Eye nebula. The white dot in the middle is a white dwarf star with a density of ~ 10^9 kg m^{-3}.

Material	Width of 1 t cube
Interstellar space	10^6 km
Red giant star	100 m
The Sun	0.89 m
White dwarf	8.9 mm
Neutron star	15 μm

Table 1.1.3 Density in the universe

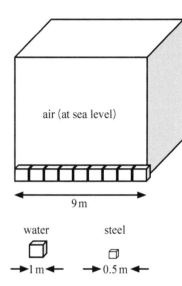

Fig. 1.1.13 Range of densities

Problems involving density will usually require you to convert units. Either the density or the volume often needs to be converted. If the volume is given in **cm**^{3} and the density in **kg m**$^{-3}$ then:

either convert the volume using 1 cm^3 = 1 × 10^{-6} m^{-3}

or convert the density using 1000 kg m^{-3} = 1 g cm^{-3}

Example

A rectangular block of steel, of density 7900 kg m^{-3}, has length 10.0 cm, width 5.0 cm and height 4.0 cm. Calculate its mass.

Answer

Equation first: $\rho = \dfrac{M}{V}$ $\therefore M = \rho V$. We'll use **kg** and **m**^{3} as the units.

Mass $= 7900$ kg m^{-3} $(10.0 \times 10^{-2}$ m $\times 5.0 \times 10^{-2}$ m $\times 4.0 \times 10^{-2}$ m$)$

$= 7900$ kg m$^{-3} \times 2 \times 10^{-4}$ m^3

$= 1.6$ kg (2 s.f.)

1.1.6 Moments of forces

(a) The turning effect of a force

Sometimes forces cause things to accelerate. Sometimes they stretch or compress an object or make it rotate.

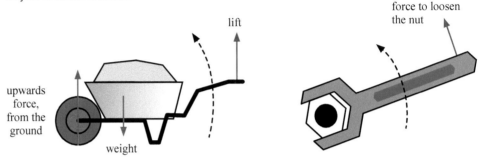

Fig. 1.1.15 Forces causing rotation

The forces (red arrows) in Fig. 1.1.15 cause the wheelbarrow, and the spanner and nut to turn about the pivot. Anyone who has used a spanner knows that the longer the handle, the easier it is to undo a nut. In other words, the turning effect of the force is bigger if it is applied further away from the pivot.

An easy experiment to show this difference in turning effect is for two people to push on either side of a door. It is easy for a child to hold a door shut against an adult – if the adult pushes close to the hinge! (See Fig. 1.1.16)

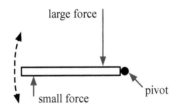

Fig. 1.1.16 Forces on a door

 **Key term**

Moment: The moment of a force about a point is the product of the force and the perpendicular distance from the point to the line of action of the force.

(b) The principle of moments

The turning effect of a force about a point depends upon its direction as well as the distance from a point – see Fig. 1.1.17. We take account of this when we define the **moment** of a force, which is the mathematical expression of its turning effect:

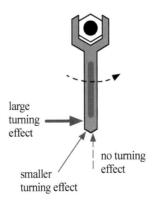

Fig. 1.1.17 Direction matters

This is clarified in Fig. 1.1.18. The force F is applied at a distance x from P. But the perpendicular distance from P to the line of action of F is d. So:

Moment of F about P = Fd.

Looking at Fig. 1.1.18 again we notice that $d = x \cos \theta$.

∴ Moment of F about P = $Fx \cos \theta$.

We can also write this as $(F \cos \theta)x$.

If you look at Fig. 1.1.18, you'll notice that θ is also the angle between the line of action of the force and the vertical grey dotted line. Hence $F \cos \theta$ is the component of the force perpendicular to the line joining P and the point of application of F. So this gives an alternative way of calculating the moment of F about P.

If we look back at Fig. 1.1.16, we see that the two forces are acting in opposite senses: the small force tends to make the door move clockwise about the hinge; the large force, anticlockwise. We say that the small force has a **clockwise moment** (CM) and the large force has an **anticlockwise moment** (ACM).

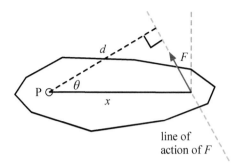

line of action of F

Fig. 1.1.18 Moment of F about P = Fd

Example

Calculate the moment about O of each of the forces in Fig. 1.1.19.

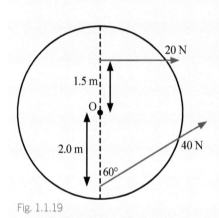

Fig. 1.1.19

Answer

(a) The perpendicular distance of the line of action of the 20 N force is 1.5 m from O.

∴ The clockwise moment of the 20 N force about O = 20 N × 1.5 m = 30 N m.

(b) **Either**: The perpendicular distance of the line of action of the 40 N force is 2.0 m sin 60° from O.

∴ The ACM of the 40 N force about O = 2.0 m sin 60° × 40 N m = 69.3 N m

Or: The perpendicular component of the 40 N force to the 2.0 m displacement is 40 sin 60°.

∴ The ACM of the 40 N force about O = 40 sin 60° × 2.0 N m = 69.3 N m

Knowledge check 1.1.9

Identify the moment of each of the forces in Fig. 1.1.15 as CM (clockwise) or ACM.

Now, if those are the only forces acting, the disc in the example will start turning anticlockwise – the anticlockwise moment is larger than the clockwise moment. There is a **resultant anticlockwise moment** of 69.3 − 30.0 = 39.3 N m. What happens after that is not clear because we don't know how the forces are applied to the disc; will they stay the same in magnitude, direction and position of application? But it does lead us to an important principle:

The principle of moments (PoM)

PoM states that, for a body to be in equilibrium, the sum of the anticlockwise moments about any point is equal to the sum of the clockwise moments about the same point. Some people define a positive direction of moment, either clockwise or anticlockwise, and use an alternative, equally valid, statement of the PoM: the PoM states that, for a body to be in equilibrium the resultant moment about any point is zero. (For the time being, we'll ignore the phrases 'about any point' and 'about the same point' and come back to them in Section 1.1.7.)

With the aid of the PoM, we are in a position to solve a real-life problem!

Example

In Fig. 1.1.20 where must the fat cat sit to balance the other two on the see-saw?

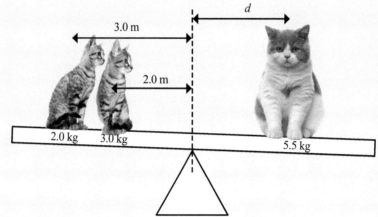

Fig. 1.1.20 Calculate d

Answer

The weights of the cats are (using mg), 19.62 N, 29.43 N and 53.96 N respectively. Using PoM, the resultant moment about the pivot must be zero.

∴ Taking clockwise as positive: $53.96d - 29.43 \times 2.0 - 19.62 \times 3.0 = 0$

∴ Solving this equation $d = 2.18$ m

∴ The fat cat must sit **2.18 m** from the pivot.

(c) The centre of gravity (C of G)

In the last example, we treated the cats (of all sizes) as though they were point masses. This is obviously not true – they are spread out. However, for any object we can identify a point at which we can consider all its weight to act. This is called its **centre of gravity**. In a uniform gravitational field (which will always be the case in AS Physics) the C of G of a symmetric body, of uniform density, will lie on any plane of symmetry. Fig. 1.1.21 has the examples you are likely to meet:

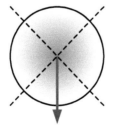 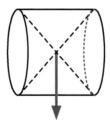 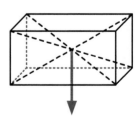

Fig. 1.1.21 Centres of gravity

For a standing object, such as a bus or a racing car, the lower the centre of gravity and the wider the base, the more stable it is. That means that objects with low centres of gravity can be tipped more before they topple over.

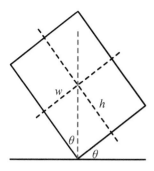

Fig. 1.1.23 Stability of a block

Fig. 1.1.24 Stability testing

Fig. 1.1.23 shows the principle of this using a tall rectangular block. The block is just on the edge of tipping – it could go either way because the C of G is vertically above the balance point. From the geometry,

$$\tan \theta = \frac{w}{h},$$

where h and w are as shown in the diagram.

Racing cars have a very wide wheel base and are very low on the ground, so their centres of gravity are also low. Fig. 1.1.24 shows a tilt test on an F1 car.

Fig. 1.1.22

1.1.7 Conditions for equilibrium

A body is said to be **in equilibrium** if it is moving and rotating at a **constant rate**. In many cases, especially when applied to engineering objects, such as bridges and buildings, this means it is not moving at all. In order for an object to be in equilibrium:

1. The resultant force on the object must be zero, and

2. The resultant moment (about any point) must be zero (the principle of moments).

The object in Fig. 1.1.25 is clearly not in equilibrium: the resultant force is downwards and to the right and the resultant moment is clockwise (about the centre of gravity).

What about the metre rule arrangement in Fig. 1.1.26? The rule weighs $1.0\ \text{N}$ and we can

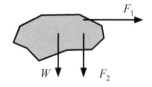

Fig. 1.1.25 This object is not in equilibrium

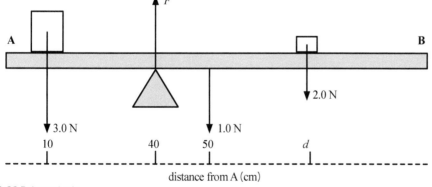

Fig. 1.1.26 Balanced rule

▶▶ Study point

In Fig. 1.1.26, there are two unknown quantities (F and d), so we need two equations to calculate them.

consider this weight to act at the $50\ \text{cm}$ mark, the centre of gravity. Assuming it balances, what are the values of d and F?

Applying condition 1: The resultant force = 0

$\therefore F = 3.0\ \text{N} + 1.0\ \text{N} + 2.0\ \text{N} = 6.0\ \text{N}$,
$\therefore$ The pivot exerts an upward force of $6.0\ \text{N}$ for the ruler to be in equilibrium.

Knowledge check 1.1.11

Find d in Fig. 1.1.26 by assuming $F = 6.0\ \text{N}$, (from applying $\Sigma F = 0$) and taking moments about **B**.

▶1.1.12 **Knowledge check**

Find d and F in Fig. 1.1.26 by taking moments, about **A**, then about **B** and solving the simultaneous equations.

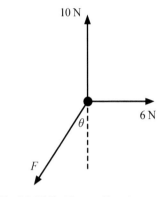

Fig. 1.1.27 Find the equilibrant

Applying condition 2: The resultant moment = 0 (about any point).

Let's take moments about end **A**: The 3, 1 and 2 N forces each have a clockwise moment about **A**; F (= 6 N) has an anticlockwise moment. Taking CM as positive:

∴ $3.0 \text{ N} \times 10 \text{ cm} - 6.0 \text{ N} \times 40 \text{ cm} + 1.0 \text{ N} \times 50 \text{ cm} + 2.0 \text{ N} \times d = 0$
∴ (simplifying) $2d = 160$ cm. ∴ The 2.0 N weight must be at the 80 cm mark.

The last problem we are going to look at is how to find an unknown force if there are forces at different angles. For example, what force F must we apply in Fig. 1.1.27 so that the forces are in equilibrium? We don't need to worry about rotations because all the forces pass through the same point. We have a choice of three techniques – but two of them are essentially the same!

a) Add the 10 N and 6 N. Then F must be equal and opposite to the resultant.

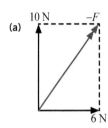

b) Add the 10 N, the 6 N and F as in Fig. 1.1.7, so that the resultant is 0.

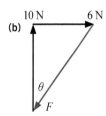

c) Resolve in two directions – horizontally and vertically are the obvious ones:

Vertically: $F \cos \theta = 10$ N [1]   Remember: through the angle ⟶ cos

Horizontally: $F \sin \theta = 6$ N   [2]   Through $90° - \theta$ ⟶ sin

Dividing [2] by [1] and remembering that $\dfrac{\sin \theta}{\cos \theta} = \tan \theta$ ⟶ $\tan \theta = \dfrac{6}{10}$.

This lets us calculate θ and then we can use [1] or [2] to calculate F.

▶1.1.13 **Knowledge check**

Find F and θ in Fig. 1.1.27 using methods (a), (b) and finish off (c).

We finish off with two more difficult examples. The first shows that the principle of moments applies even when there isn't a pivot: if something doesn't start to rotate, the moments of the forces about any point must add to zero, even for a bridge! The second is more difficult because the forces are not parallel.

Example

Fig. 1.1.28 shows a load on a bridge. Calculate the forces, F_1 and F_2, provided by the supports.

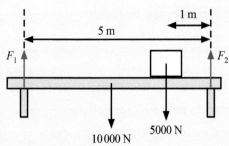

Fig. 1.1.28 Forces on a bridge

Answer

Apply the conditions for equilibrium.

1. Resultant force = 0

$$\therefore F_1 + F_2 = 10\,000 + 5000$$
$$= 15\,000 \text{ N} \qquad [1]$$

2. Resultant moment about any point = 0

Take moments about the left-hand support with clockwise being positive.

$$\therefore 10\,000 \text{ N} \times 2.5 \text{ m} + 5000 \text{ N} \times 4 \text{ m} - F_2 \times 5 \text{ m} = 0$$

$$\therefore 5F_2 = 45\,000 \text{ N}$$

$$\therefore F_2 = 9000 \text{ N} \qquad [2]$$

Substitute the value for F_2 in equation [1] to calculate $F_1 \longrightarrow F_1 = 6000$ N

Knowledge check 1.1.14

Calculate the magnitude and direction of the force F in Fig. 1.1.29.

Example

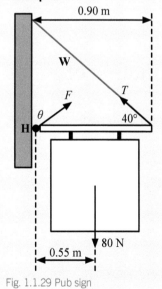

Fig. 1.1.29 Pub sign

A pub sign is supported on a vertical wall by a hinge, H and a wire, W, as shown in Fig. 1.1.29. Calculate the tension, T, in W.

Answer

Take moments about H. ACM positive. Using the principle of moments:

$$T \sin 40° \times 0.90 \text{ m} - 80.0 \text{ N} \times 0.55 \text{ m} = 0$$

$$\therefore T = 76.1 \text{ N}$$

Following on from this example, we can also use the equilibrium conditions to find the force exerted by the hinge on the pub sign bar – F in Fig. 1.1.29. We could do this using the triangle of forces:

The three forces on the sign have a **0** resultant. Using the result for the tension, the triangle looks like this (Fig. 1.1.30).

To find F and θ we use components and the fact that the resultant horizontal and vertical components are zero:

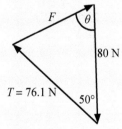

Fig 1.1.30 Triangle of forces

$\therefore$ Horizontally: $F \sin \theta = T \cos 40°$ $\qquad \therefore F \sin \theta = 58.3$ N $\quad$ [1]

and vertically $\quad F \cos \theta + T \cos 50° = 80$ $\quad \therefore F \cos \theta = 31.1$ N $\quad$ [2]

If we then divide equation [1] by equation [2] we get $\tan \theta = \dfrac{58.3}{31.1} = 1.875$.

So we can calculate θ and hence F.

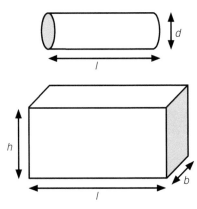

Fig. 1.1.31 Regular solids

‹ Link ›

See Chapter 3 for combining and reducing uncertainties.

Top tip

Remember that you measure the *diameter* of a wire, not the *radius*!

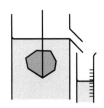

Fig. 1.1.32 Measuring the volume of a rock

Study point

The centre of gravity is not necessarily at the midpoint of the scale. It is a good idea to do a preliminary experiment to find it:

The C of G is above the balance point.

Practical check

The results will be most accurate if both *x* and *y* are as large as possible. Thus, in Fig. 1.1.33, the known mass should be similar to the unknown mass.

1.1.8 Practical work

(a) Measuring the density of solids

Finding the density of a substance involves measuring the mass and the volume and dividing mass by volume. This practical is often used to test understanding of uncertainties and how to combine them. The mass is usually determined from a single reading electronic balance, so the absolute uncertainty is taken as ±1 in the last digit of the reading;

e.g. a reading of 159.73 g would be taken to be (159.73 ± 0.01) g.

The uncertainty in mass is often not as significant as that in the volume.

How the volume is determined depends upon whether the solid object has a regular shape, such as a cuboid (e.g. a microscope slide) or a cylinder (e.g. a wire).

(i) Regular solids

Volume of a cuboid $= \ell bh$; volume of a cylinder $= A\ell = \pi r^2 \ell = \dfrac{\pi d^2 \ell}{4}$.

For lengths up to ~15 cm, digital callipers, with a resolution of 0.01 mm are normally used. It is important when using them to check the zero reading, i.e. close the jaws and take a reading. Any reading should be subtracted from the reading with the object being measured. For lengths > 15 cm a metre rule with a resolution of 1.0 mm is normally used.

The precision can be improved for a set of identical objects by laying them end to end, e.g. laying 10 microscope slides end to end gives a length of ~75 cm; using a mm scale to measure the length gives a % uncertainty of 0.13%.

(ii) Irregular solids

The solid, e.g. a rock, is suspended from a thread and lowered into a measuring cylinder of water until it is completely submerged. The increase in volume reading is the volume of the solid. If the solid is too large for a measuring cylinder, a *displacement can* is used (Fig. 1.1.32) and the water overflow captured in a measuring cylinder. Disadvantages of this method are: (a) the resolution of the measuring cylinder is quite large (typically 1–2 cm³ for a 100 cm³ cylinder) and (b) the volume of water overflowing is not necessarily exactly the same as the volume of the object.

(b) Measuring mass using the principle of moments

In Fig. 1.1.33, the long bar is a $\frac{1}{2}$ metre or metre rule. The triangle is any pivot – it could be as simple as an outstretched finger. The pivot is placed at the centre of gravity of the rule.

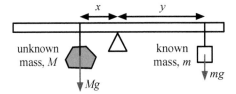

Fig. 1.1.33 Finding an unknown mass

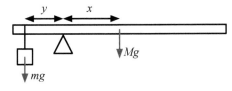

Fig. 1.1.34 'Weighing' a ruler

When the bar balances, the ACM and the CM about the pivot are equal.

$\therefore Mgx = mgy$

$\therefore Mx = my$.

We can also use this technique to find the mass, *M*, of the bar itself. The position of the C of G is found as in the Study point. A known mass, *m*, is hung near one end of the bar. The pivot point is found and distances *x* and *y* measured (see Fig 1.1.34).

Then, as above: $Mx = my$.

Test yourself 1.1

1. A student gave the following definition of the moment of a force:

   Moment is the force multiplied by the distance from the pivot.

   What is wrong with this definition? Give a better one.

2. A constant called Planck's constant has the unit **J s** [joule second]. Express this in terms of the base SI units, **kg, m** and **s**.

3. A student is asked to calculate the acceleration of an object and gives the answer as 7.5 m s^{-2}. Why is this not a complete answer?

4. The diagram shows all the forces acting on a rod. Explain why the rod is not in equilibrium.

   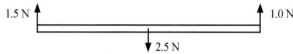

5. A force of magnitude 100 N acts at an angle above the horizontal of $40°$. Calculate (a) the horizontal; and (b) the vertical component of the force.

6. Two forces act on an object of mass 5.0 kg: a horizontal force of 12.0 N and a vertical force of 5.0 N. Calculate (a) the resultant force; and (b) the acceleration of the object.

7. A student determines the density of aluminium by taking measurements on a rectangular block. Using vernier callipers with a resolution of $\pm 0.1 \text{ mm}$, she determines the dimensions of the block to be $10.30 \text{ cm} \times 4.75 \text{ cm} \times 3.21 \text{ cm}$, by taking one measurement of each length. An electronic balance gives the mass to be 427.32 g.

   (a) Use these data to calculate the density of the aluminium.
   (b) State which length reading has the greatest percentage uncertainty. Explain your answer.
   (c) The student assumes that the uncertainty in balance reading can be neglected. Calculate the percentage uncertainty and the absolute uncertainty in her value for the density and hence express the density to an appropriate number of significant figures.
   (d) Evaluate whether the student's assumption in part (c) was a reasonable one.

8. A bacterium has a mass of 0.95 pg. Estimate its volume (in m^3), assuming its density is 1000 kg m^{-3}.

9. The pressure, p, at a depth, d, below the surface of a liquid of density ρ is given by $p = p_A + g\rho d$ where p_A is the air pressure at the liquid surface and g is the acceleration of free fall.

   (a) State the units of p, g and ρ in terms of the base SI units.
   (b) Show that the equation is homogeneous.
   (c) Calculate the pressure 25 m below the surface of the sea. ($p_A = 101 \text{ kPa}$, $g = 9.81 \text{ N kg}^{-1}$, $\rho_{\text{sea water}} = 1030 \text{ kg m}^{-3}$).

10. A uniform rigid plank of weight 30 N and length 4.8 m, with a load of 20 N at one end, is supported between two fulcrums, X and Y, which apply forces F_1 and F_2, as shown.

    (a) Calculate the moments of the 20 N load and the 30 N weight of the plank about fulcrum X.
    (b) Take moments about X and use the principle of moments to calculate the value of F_2.
    (c) Explain why $F_1 = F_2 + 50 \text{ N}$. Hence calculate the value of F_1.

    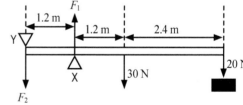

11. Newton's law of gravitation states that two small bodies of masses M_1 and M_2, separated by a distance d attract each other with a force, F, given by: $F = \dfrac{GM_1M_2}{d^2}$, where G is the universal gravitational constant.

    (a) Show that $[G] = \text{N m}^2 \text{ kg}^{-2}$.
    (b) Express $[G]$ in terms of the base SI units, **m, kg** and **s**.

12. A cylinder of length 1.5 m and diameter 60 mm is made from iron of density 7900 kg m^{-3}. Calculate its mass.

13. (a) Find the horizontal and vertical components of each of v_1 and v_2.

    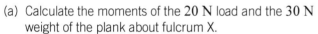

    (b) Hence find the sum of v_1 and v_2, expressing the direction as an angle to the dotted line.
    (c) By a similar method, find the difference ($v_2 - v_1$) of the vectors.

Kinematics is to do with motion and its mathematical description. It is the study of how things move without considering why they move, which is covered in the next topic, Dynamics.

1.2.1 Speed and velocity

(a) Motion in a straight line: displacement–time graphs

Looking at motion in a straight line is not as restrictive as it might appear. Indeed, the kinematic equations for straight-line motion can be applied, with care, to motion along a bendy path. Also, when we consider motion in two (and three) dimensions, we'll often look at the components of the motion: essentially, 3D motion consists of three sets of motion in straight lines!

The displacement–time graph, Fig. 1.2.1 is of a car moving along a road. We'll use it to help understand certain terms. The displacement is the distance along the road measured to the right.

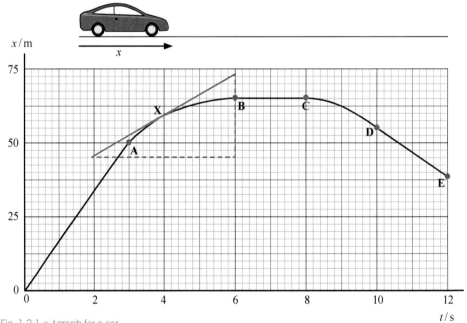

Fig. 1.2.1 x–t graph for a car

1. An upward slope (between $(0,0)$ and **B**) represents an *increase* in x with time, so a movement in the positive x direction; downward slope (**C** to **E**) is moving in the negative direction; horizontal (**B** to **C**) means stationary.

2. The *gradient* of the graph represents the increase in displacement per second, so the gradient of a displacement–time graph is the velocity. If the graph is a straight line, the gradient is constant and so the velocity is constant.

Interrogating Fig. 1.2.1 mathematically:

The *mean velocity* between 3 and 8 seconds (**A** to **C**) $= \dfrac{\Delta x}{\Delta t} = \dfrac{(65-50) \text{ m}}{5.0 \text{ s}} = 3.0 \text{ m s}^{-1}$.

From **D** to **E** the velocity is constant. $v_{DE} = \dfrac{\Delta x}{\Delta t} = \dfrac{(40-55) \text{ m}}{2.0 \text{ s}} = \dfrac{-15}{2.0} = -9.1 \text{ m s}^{-1}$.

The *instantaneous velocity* at **X** is the gradient of the tangent at **X**:

$$v_X = \frac{\Delta s}{\Delta t} = \frac{(73-46) \text{ m}}{4.0 \text{ s}} = \frac{27}{4.0} = 6.8 \text{ m s}^{-1}.$$

The displacement, x, for the whole journey is 37 m to the right but the distance, d, travelled by the car is 93 m. These are made up as follows:

$x_{OB} + x_{BC} + x_{CE} = 65 \text{ m} + 0 \text{ m} + (-25 \text{ m}) = 40 \text{ m}$

$d_{OB} + d_{BC} + d_{CE} = 65 \text{ m} + 0 \text{ m} + 25 \text{ m} = 90 \text{ m}$

So the *mean speed* for the whole journey $= \dfrac{\text{distance travelled}}{\text{time taken}} = \dfrac{90 \text{ m}}{12 \text{ s}} = 7.5 \text{ m s}^{-1}$.

(b) Motion in two dimensions

For motion in two dimensions, we just use the same equations:

$$\text{Mean speed} = \frac{\text{distance travelled}}{\text{time taken}}$$

$$\text{Mean velocity} = \frac{\text{displacement}}{\text{time taken}}$$

For velocity and displacement, we can also use the vector ideas from Sections 1.1.3 and 1.1.4. As no new concepts are involved, we'll go straight to an example.

Example
A drone flies from A to D at a constant speed of 2.5 m s^{-1}, as shown in Fig. 1.2.2.

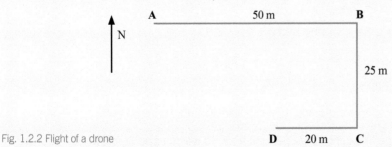

Fig. 1.2.2 Flight of a drone

Calculate: (a) the total flight time, t
(b) the displacement from A to D, d
(c) the mean velocity, v, for the flight.

1.2.5 Knowledge check

For the drone flight in the example, state the velocity between:

(a) A and B

(b) B and C

(c) C and D.

Hint: Remember magnitude and direction.

1.2.6 Knowledge check

For the drone flight in the example, calculate:

(a) the displacement from B to D

(b) the mean velocity from B to D.

1.2.7 Knowledge check

The drone in the example travels straight back from D to A at 2.0 m s^{-1}.

(a) State the velocity.

(b) Calculate the time taken.

Study point

If the velocity changes from v_1 to v_2, then $\Delta v = v_2 - v_1$.

Study point

The unit of acceleration is m s^{-2}.

An acceleration of 5 m s^{-2} means that the velocity increases by 5 m s^{-1} every second, e.g. velocities of $2, 7, 12, 17 \ldots$ m s^{-1} at 1 second intervals.

Study point

There can be an acceleration even if the *speed* stays the same – see Section 1.2.2(c).

Answer

(a) Total distance travelled = AB + BC + CD

$$= 50 \text{ m} + 25 \text{ m} + 20 \text{ m}$$

$$= 95 \text{ m}$$

$$\text{speed} = \frac{\text{distance}}{\text{time taken}}$$

$$\therefore \text{ flight time} = \frac{\text{distance}}{\text{speed}} = \frac{95 \text{ m}}{2.5 \text{ m s}^{-1}} = 38 \text{ s}$$

(b) Looking at Fig. 1.2.3, the point D is 30 m East and 25 m South of A.

So, using Pythagoras' theorem,

$$d = \sqrt{30^2 + 25^2}$$

$$= 39.1 \text{ m}$$

and, using trigonometry

$$\tan \theta = \frac{25}{30} = 0.833$$

$$\therefore \theta = 39.8°$$

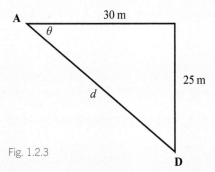

Fig. 1.2.3

So displacement, $d = 39$ m (2 s.f.) on a bearing of 130°.

(c) $\text{Mean velocity} = \dfrac{\text{displacement}}{\text{time taken}}$

$$\therefore \text{ Mean velocity, } v = \frac{39.1 \text{ m}}{38 \text{ s}} = 1.0 \text{ m s}^{-1} \text{ on a bearing of } 130°.$$

1.2.2 Acceleration

(a) Definition of acceleration

If the velocity of a body is changing it is said to be accelerating. The **instantaneous** and **mean acceleration** are defined as follows:

> **Instantaneous acceleration** = rate of change of velocity.
>
> $$\textbf{Mean acceleration} = \frac{\textbf{change in velocity}}{\textbf{time taken}} = \frac{\Delta v}{\Delta t}$$
>
> Unit: m s^{-2}

Note the following things:

- Acceleration is a vector quantity, just like velocity and displacement.

- 'Rate of change' means change per unit time, so we can also define acceleration as change in velocity per second.

Example

During takeoff, the velocity of an aeroplane changes from 10 to 70 m s⁻¹ in 24 s in a constant direction. Calculate the mean acceleration.

Answer

$$\text{Mean acceleration} = \frac{\text{change in velocity}}{\text{time taken}} = \frac{70-10}{24 \text{ s}} = 2.5 \text{ m s}^{-2}$$

Note: Because acceleration is a vector quantity, we should also state, 'in the direction of motion'.

(b) Motion in a straight line: velocity–time graphs

It is often convenient to represent accelerated motion using a graph of velocity against time (a *v–t* graph). Acceleration and displacement can be found from a *v–t* graph as follows:

Acceleration = gradient of the *v–t* graph

Displacement = the area between the *v–t* graph and *t*-axis

As an example, Fig. 1.2.4 is the *v–t* graph of the motion of an underground train between two stations. The train starts off at rest, accelerates uniformly for 5.0 s, travels at a constant velocity for 15.0 s, etc., before finally decelerating to rest after 68.0 s.

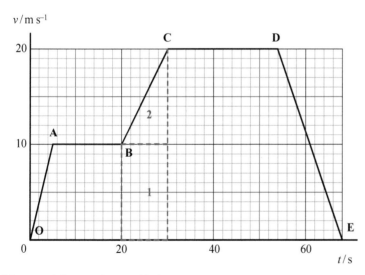

Fig. 1.2.4 *v–t* graph for an underground train

The acceleration of the train in **OA** = $\frac{\Delta v}{\Delta t} = \frac{10 \text{ m s}^{-1}}{5.0 \text{ s}} = 2.0 \text{ m s}^{-2}$.

The acceleration between **D** and **E** = $\frac{\Delta v}{\Delta t} = \frac{(0-20) \text{ m s}^{-1}}{14 \text{ s}} = -1.43 \text{ m s}^{-2}$.

We now use the 'area' between the graph and the *t*-axis to calculate the displacement, e.g. between B and C. From Fig. 1.2.4:

Displacement between **B** and **C** = area between the graph and the *t*-axis

                = area 1 + area 2 [in red on the graph]

                = $10 \text{ s} \times 10 \text{ m s}^{-1} + \frac{1}{2} (10 \text{ s} \times 10 \text{ m s}^{-1})$

                = 150 m

 Study point

People (including scientists) often use *deceleration* to mean the rate of **decrease** of velocity. If the velocity is negative, it can be confusing and it is probably best to avoid this term.

 Study point

An area underneath the *t*-axis is regarded as negative and represents a negative displacement.

 Knowledge check 1.2.8

Describe the journey in Fig. 1.2.4 between 20 and 68 s.

 Knowledge check 1.2.9

In Fig. 1.2.4, calculate the mean acceleration between 10 and 35 seconds.

 Study point

The displacement is often said to be the area *under* the graph, but if the velocity is negative it will be the area under the *t*-axis and above the graph.

The displacement for the whole journey is 945 m (see Knowledge check 1.2.10). From this we can find the mean velocity for the trip:

$$\text{Mean velocity} = \frac{\Delta x}{\Delta t} = \frac{945 \text{ m}}{68 \text{ s}} = 13.9 \text{ m s}^{-1}.$$

Example

The *v–t* graph is of a bullet as it is fired into a tank of water.

Use the graph to calculate:
(a) The deceleration at 2 ms,
(b) The displacement of the bullet in coming to rest.

Answer

(a) Acceleration is the gradient of the tangent at 2 ms.

From the red triangle

$$a = \frac{\Delta v}{\Delta t} = \frac{(10 - 190) \text{ m s}^{-1}}{3.2 \text{ ms}}$$

$$= -56.3 \times 10^3 \text{ m s}^{-2}$$

∴ Deceleration is 56.3 km s⁻².

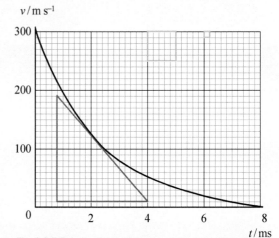

Fig. 1.2.5 Decelerating bullet

(b) Displacement = area under graph. Using the trapezoidal rule (see Maths tip), with $\Delta t = 1.0$ ms:

Area under graph

$$= \tfrac{1}{2} (300 + 390 + 256 + 164 + 104 + 66 + 40 + 20 + 0) \text{ m s}^{-1} \times 0.001 \text{ s}$$

$$= 0.670 \text{ m}$$

∴ Displacement of bullet when coming to rest = 0.67 m (2 s.f.)

▶ 1.2.10 Knowledge check

By dividing the area under the graph in Fig. 1.2.4 suitably, show that the displacement for the journey is 945 m.

Maths tip ▷

See Chapter 4 for various techniques for estimating the area under a non-linear graph, including the *trapezoidal rule*.

Using the square-counting method, the author obtained 0.65 m and 0.69 m respectively when using the large and small green squares.

▶ 1.2.11 Knowledge check

In Fig. 1.2.6, assuming that the acceleration is always downwards and 1.5 m s⁻², calculate the velocity of the ball when it hits the ground 20 s after **C**.

[Hint: calculate Δv using $\Delta v = a\, \Delta t$, and use the horizontal components of velocity at **C**.]

(c) Motion in two dimensions

We'll see in Section 1.2.4 how to study horizontal and vertical components of motion separately. Here we'll have a brief look at how to calculate acceleration when the direction of the motion changes. Consider in Fig. 1.2.6 the motion of the cricket ball in the interplanetary test match on the lunar Tycho base. The Jovian opening bat despatched a full toss to the (rather distant) boundary.

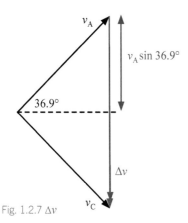

Fig. 1.2.7 Δv

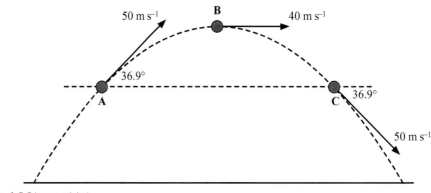

Fig. 1.2.6 Lunar cricket

The ball was at positions **A**, **B** and **C** at times of 20 s, 40 s and 60 s after it was struck. We'll determine the mean acceleration $\langle a \rangle$ between positions **A** and **C**. Fig. 1.2.7 shows how to calculate Δv.

$$\Delta v = 2 \times 50 \sin 36.9° = 60 \text{ m s}^{-1}$$

$\therefore$ Between **A** and **C**, $\langle a \rangle = \dfrac{\Delta v}{\Delta t} = \dfrac{60 \text{ m s}^{-1}}{40 \text{ s}} = 1.5 \text{ m s}^{-2}$ vertically downwards.

1.2.3 Uniform acceleration equations

(a) Deriving the equations

Fig. 1.2.9 *xuvat*

In this section, we'll consider an object which is initially moving with a velocity, u, and accelerates with a constant acceleration, a, for a time, t. In this time, it attains a final velocity, v, and moves through a displacement, x. The motion is along a straight line. We'll derive some relationships between these quantities, x, u, v, a and t.

From the definition of acceleration, $a = \dfrac{v - u}{t}$, $\therefore$ (rearranging)

$$v = u + at \qquad [1]$$

The displacement, x, is the 'area under' the v–t graph – see Fig. 1.2.10 (a). The graph is a straight line because the acceleration is constant. For convenience we've assumed that $a > 0$ [so the gradient is positive] and $u > 0$. The equations we derive will still be valid for either a or u [or both] < 0.

The displacement, x, is the area of the trapezium. From the formula for the area of a trapezium,

$$x = \tfrac{1}{2}(u + v)t \qquad [2]$$

Stretch & challenge

A car takes a 90° bend at 15 m s^{-1}. It completes the bend in 7.0 s. See diagram. Calculate the mean acceleration.

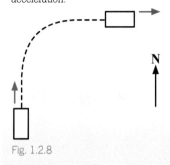

Fig. 1.2.8

》 Study point

These equations are often referred to as *xuvat*.

x = displacement
u = initial velocity
v = final velocity
a = acceleration
t = time

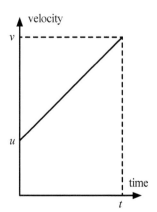

Fig. 1.2.10 (a) *v–t* graph

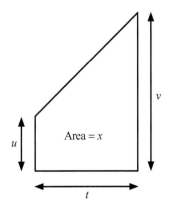

Fig. 1.2.10 (b) Trapezium

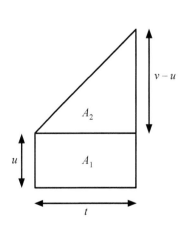

Fig. 1.2.10 (c) Rectangle plus triangle

If we break up the trapezium into a rectangle plus a triangle we obtain Fig. 1.2.10 (c)

Adding the areas, $A_1 + A_2$, we get: $x = ut + \frac{1}{2}(v - u)t$

From equation [1] $v - u = at$ ∴ $x = ut + \frac{1}{2}(u + at - u)t$

$$x = ut + \tfrac{1}{2}at^2 \qquad [3]$$

From equation [1], $t = \dfrac{v - u}{a}$. Substitute for t in [2] we get $x = \frac{1}{2}(u + v)\dfrac{(v - u)}{a}$,

So, $2ax = (v + u)(v - u) = v^2 - u^2$

$$\therefore \quad v^2 = u^2 + 2ax \qquad [4]$$

You should **learn equations [1] – [4]**. Make sure you can derive them.

A fifth equation to complete the set is:

$$x = vt - \tfrac{1}{2}at^2 \qquad [5]$$

(b) Applying the equations

It is important to be systematic when applying these equations.

Start by identifying and writing down which of the quantities, x, a, etc., you already know and which you need to calculate.

Say you know the initial velocity (u), the acceleration (a) and the time (t) and are asked to calculate the displacement (x). The equation that contains these four quantities is $x = ut - \frac{1}{2}at^2$, so this is the equation to apply.

Example

A car, travelling at 26 m s^{-1}, decelerates at 1.2 m s^{-2} to a speed of 10 m s^{-1}. Calculate (a) the distance travelled and (b) the time taken in this process.

Answer

(a) Writing down the quantities. $u = 26$ m s^{-1}; $v = 10$ m s^{-1}; $a = -1.2$ m s^{-2}.

Unknown quantity = x. ∴ Use the equation $v^2 = u^2 + 2ax$. From this we get $x = 240$ m.

(b) Now we know u, v, a and x and need to calculate t. So we can use any of the equations 1, 2 and 3. The easiest is $x = \frac{1}{2}(u + v)t$, which gives $t = 13.3$ s.

Comments on the example:

1. It is possible to answer part (b) of the question before part (a).

   With u, v and a known and t to be calculated, equation [1] is the obvious route.

2. If you use $x = ut + \frac{1}{2}at^2$ to calculate t, then you will usually obtain two possible solutions – see Study point. In this case, $t = 13.3$ s or 30.0 s. To see how this second solution arises, look at the sketch graph of x against t in Fig. 1.2.11. A good reason for avoiding quadratics.

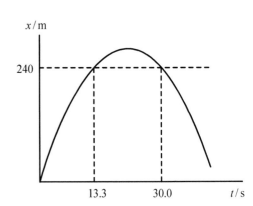

Fig. 1.2.11 x–t graph

(c) Vertical motion under gravity[1]

The image in Fig. 1.2.12 is of a 'freely' falling golf ball (white) and table-tennis ball (blue). They are lit by a strobe which is flashing at regular intervals. The distances between the positions of both spheres increase as they fall showing that they are accelerating. We can use the scale to investigate the acceleration.

The positions of the centre of the golf ball in the four images are approximately (in cm): 3.5, 10.5, 22.0 and 38.0. The distances fallen between the images are (approx.) 7.0, 11.5, 16.0 cm, showing a constant increase and hence a constant acceleration. We cannot measure the acceleration because we do not know the time interval between the flashes. Note that the golf ball is apparently overtaking the table-tennis ball, because the relative effect of air resistance is greater on the lower mass ball.

Around 1590, Galileo famously devised a thought experiment to justify by logic that all **freely** falling objects (i.e. in the absence of air resistance) accelerate to Earth with the same acceleration. We call this acceleration the *acceleration of free fall* or the *acceleration due to gravity*. The symbol for this acceleration is g. See Section 1.3.8 for some more discussion of this and also for the effect of air resistance on falling motion.

The acceleration due to gravity, g, close to the surface of the Earth is almost constant. The value of g at **39 km**, the height from which Felix Baumgartner jumped in October 2012, is only 1.2% less than at ground level. Even at the altitude of the International Space Station (**400 km**) g is only 13% less than at ground level. Unless otherwise told, assume $g = 9.81$ m s^{-2}. When making estimations, it is sensible to use 10 m s^{-2} as the approximate value of g.

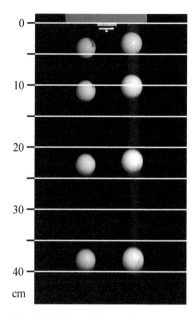

Fig. 1.2.12 Freely falling spheres

Example

A student drops a rock from a tall building. Estimate (a) its speed and (b) its position after 1.0 s, 2.0 s, 3.0 s and 4.0 s, assuming it hasn't hit the ground!

Answer

(a) If the acceleration ~ 10 m s^{-2}, this means that the speed increases by 10 m s^{-1} every second, so the speeds are $\sim 10, 20, 30$ and 40 m s^{-1} respectively.

(b) Applying $x = ut + \frac{1}{2}at^2$ with $u = 0$ and $a = g \sim 10$ m s^{-2}, we get $x = 5t^2$ leading to $x = 5$ m, 20 m, 45 m and 80 m.

Many problems involving motion under gravity involve upwards motion, which decreases and eventually becomes downwards motion. To solve such problems we have to decide which direction is positive.

If we choose *upwards as positive* then $a = -g = -9.81$ m s^{-2}; a downward velocity will be negative and a position below the starting point will be negative. The following example illustrates these points.

Fig. 1.2.13 Galileo's thought experiment

Knowledge check 1.2.13

A passenger in a stationary hot-air balloon at an altitude of 200 m throws a stone downwards at 10 m s^{-1}. Calculate the speed with which it hits the ground.

Study point

We could choose downwards as positive. If we do this, $a = g = +9.81$ m s^{-2}.

In the example, u is -80 m s^{-1} and x is $+250$ m at the ground.

Study point

The equations in the example cannot distinguish between the given question and the situation in which an object is thrown upwards from the ground and achieves a velocity of 80 m s^{-1} at a height of 250 m at time 0. In this second situation, the time and velocity of projection would be -2.7 s and $+106$ m s^{-1} respectively.

1.2.14 Knowledge check

Do the calculations for stages 1 and 2 and show that the total time is 19.0 s.

Why can't we find the total flight time of the rocket from its launch using the equations of motion 1–4?

Stretch & challenge

In the rocket example, use $x = ut + \frac{1}{2}at^2$ for the whole flight from 250 m, to find the time to impact with the ground.

Example

A toy rocket runs out of fuel at an altitude of 250 m and a vertical upwards velocity of 80 m s^{-1}. Calculate the velocity with which it hits the ground. Ignore the effects of air resistance.

Answer

$u = 80$ m s^{-1}, $a = -g = -9.81$ m s^{-2}.

The ground level is 250 m below the point at which our equations start to be applicable, i.e. $x = -250$ m. We need to find v at this point.

$$\therefore v^2 = u^2 + 2ax. \therefore v^2 = 80^2 + 2(-9.81)(-250) = 11305$$

$$\therefore v = \pm 106 \text{ m s}^{-1}$$

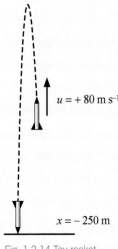

Fig. 1.2.14 Toy rocket

How do we distinguish between the two possible solutions? We could either say, 'Well, it's obvious that +106 m s^{-1} cannot be right, so the answer is -106 m s^{-1}, i.e. 106 m s^{-1} downwards.' This is certainly correct but we could look a little more deeply into it. See the Study point.

A difficult question

A more difficult question is to calculate the time until the rocket reaches the ground. Why is it more difficult? Because it involves the solution of a quadratic equation! (See Stretch & challenge).

An easier solution

Some people find it easier to think of this problem in two parts:

1. Calculate the time and distance to the top of the flight (stage 1), and

2. Calculate the time to fall from the top to the ground (stage 2).

This involves the following sequence:

- For stage 1, $v = 0$. Use $v = u + at$ to calculate the time to the top.

- Use $x = \frac{1}{2}(u + v)t$ or $x = ut + \frac{1}{2}at^2$ to calculate the height gain in stage 1. Add the initial height (250 m) to give the total height.

- For stage 2, use $x = ut + \frac{1}{2}at^2$ to calculate the time taken to fall 576 m from rest. This is a quadratic but the ut term is zero (because $u = 0$). [Hint: it is sensible to take down as positive for this part of the flight.]

- Add the two times together.

Now try Knowledge check 1.2.14.

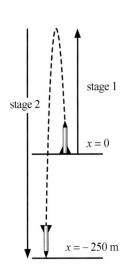

Fig. 1.2.15 Flight time

1.2.4 Projectiles

A *projectile* is an object which is thrown/kicked/made to move obliquely upwards and carries on its path under the influence of gravity, e.g. a rugby ball during a conversion. The study of this kind of motion is called *ballistics* after the Roman siege weapon, the ballista.

Writings in military books in the Middle Ages suggested that the flight path of a cannonball was as shown in Fig. 1.2.17.

Fig. 1.2.16 Parabolic jets of water

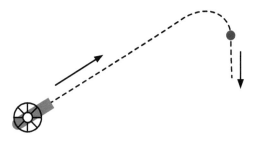

The idea was that the cannonball had a certain 'impulse' which carried it along. When this ran out, gravity took over and caused it to plummet downwards. This is similar to the cartoon character's path when running off a cliff. The image of water jets shows the true parabolic path of projectiles (See Fig. 1.2.16).

Fig. 1.2.17 'Impulse theory' of cannonball flight

To see what is happening, we'll look at a strobe image of two spheres, one of which is moving vertically, the other starting off at the same moment horizontally (Fig. 1.2.18).

The vertical white lines are equally spaced. From this image we see that:

1. The heights of the two spheres are the same at all instants, i.e. they are accelerating downwards at the same rate, which we know is g.

2. The white sphere moves horizontally at a constant velocity.

We therefore conclude that we can treat the horizontal and vertical motions of a projectile independently.

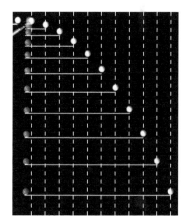

Fig. 1.2.18 Independence of horizontal and vertical motion

Fig. 1.2.19

If we use x, y co-ordinates, with x being horizontal, y being vertically upwards and the projectile stating off from $(0,0)$ with an initial velocity u at θ to the horizontal, the equations of motion become:

Horizontal: $x = u_x t$ and $v_x = u_x$ [i.e. velocity is constant].

Vertical: $v_y = u_y - gt$; $y = \frac{1}{2}(u_y + v_y)t$; $y = u_y t - \frac{1}{2}gt^2$; $v_y^2 = u_y^2 - 2gy$

where $u_x = u \cos\theta$ and $u_y = u \sin\theta$ are the initial horizontal and vertical components of velocity and v_x and v_y the horizontal and vertical components of velocity at time, t. Taking upwards as positive, $a = -g$.

Look at Knowledge check 1.2.15. This type of question can be tackled in a straightforward way by identifying the correct equation at each stage and substituting the data. It is strongly suggested that you do this question and check your answer before proceeding.

Knowledge check 1.2.15

For Fig. 1.2.19, if $u = 30$ m s^{-1} and $\theta = 30°$, calculate (a) the horizontal and vertical components of u, (b) the values of v_x and v_y after 5 seconds, (c) the velocity after 5 seconds and (d) the position after 5 seconds.

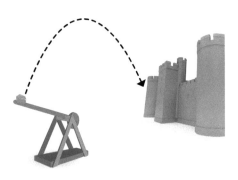

Fig. 1.2.20 A trebuchet

Some problems, such as the example below, involve a multi-stage approach. They often require a time to be calculated before the relevant distance.

Example

A trebuchet (a medieval siege engine; see Fig. 1.2.20) hurls a rock at a speed of 40.0 m s^{-1} and an angle of 30° to the horizontal towards the vertical wall of a castle 100 m away. Calculate how high up the wall of the castle the rock will hit.

Plan

Calculate the time it takes to reach the castle wall, using the horizontal motion and then calculate the height of the rock at this time using the vertical motion.

1.2.5 Measurement of *g* by free fall

To measure the acceleration of free fall, all one needs to do is to measure the time, t, it takes an object to fall from a known height, h.

Then, using $x = ut + \frac{1}{2}at^2$, with $u = 0$, $x = h$ and $a = g$, the equation becomes:

$$h = \frac{1}{2}gt^2 \qquad [1]$$

$$\therefore \quad g = \frac{2h}{t^2} \qquad [2]$$

▶ **1.2.16 Knowledge check**

Use the **plan** in the trebuchet example to answer the question, then use the time to calculate the velocity with which the rock hits the castle wall.

▶ **1.2.17 Knowledge check**

If $h \sim 50$ cm,

(a) Estimate t using $g \sim 10$ m s^{-2}.

(b) Use ± 1 mm and ± 10 ms to estimate the % uncertainty in g.

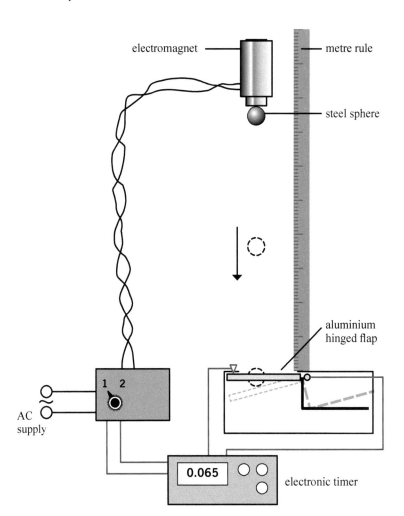

Fig. 1.2.21 *g* by free fall

In principle, one could just drop an object from a high window and use a stopwatch to time its fall. The problem with this method is that the time taken to fall 20 m [quite a high window] is only ~2 s, so the percentage uncertainty in t with a hand-operated stopwatch is quite high and then the equation for g involves t^2 which doubles the uncertainty. For example, if you can measure time with an uncertainty of 0.1 s the percentage uncertainty, p, in 2 s is:

$$p = \frac{0.1}{2} \times 100 = 5\%$$

So the uncertainty in g will be 10%.

In the apparatus in Fig. 1.2.21, the steel sphere is held in position by an electromagnet, the current for which is from the AC supply. Moving the switch from 1 to 2 simultaneously switches off the current to the electromagnet (releasing the sphere) and triggers the electronic timer to start. The sphere hits the aluminium flap, causing it to swing downwards breaking the red circuit, stopping the timer.

Measurements:

- Drop height, h, to ±1 mm using the metre rule. Typically h is up to 75 cm.

- Time, t, using the timer. The scale might be to 1 ms or 10 ms but the measurements of t typically vary with an uncertainty of 10 ms.

Analysis of results

Commonly t is measured for a range of values of h up to ~ 75 cm, and a graph of h against t^2 plotted. From equation 1, the gradient is $\frac{1}{2}g$, so g is double the gradient.

Systematic error

A problem with this technique is that there is often a small delay before the sphere is released because it takes a short time for the magnetisation in the electromagnet and/or the steel sphere to decay. This adds an unknown time – so the true time for the fall is less by an unknown time, τ. The effect of this τ is to produce a curved graph for h against t^2. See the Study point for a technique for dealing with this.

>> **Study point**

With the time delay of τ the true relationship is:

$h = \frac{1}{2}g(t - \tau)^2$

Taking the square root gives:

$\sqrt{h} = \sqrt{\frac{1}{2}g}\,t - \sqrt{\frac{1}{2}g}\,\tau,$

so if we plot a graph of

$\sqrt{h}$ against t we should get a straight line with gradient $\sqrt{\frac{1}{2}g}$.

◄**Stretch & challenge**

Referring to the Study point above:

(a) What is the intercept on the $\sqrt{h}$ axis?

(b) How can τ be found more easily?

Test yourself 1.2

1. A ball is dropped from a height of 5.0 m. It hits the ground at 10.0 m s^{-1} after 1 s. It rebounds at 8.9 m s^{-1} and reaches a maximum height of 4.0 m in a time of 0.90 s. Calculate:

   (a) the mean speed in the descent,
   (b) the mean speed in the ascent,
   (c) the mean speed for the whole 1.9 s,
   (d) the mean velocity for the whole 1.9 s,
   (e) the change in velocity at the bounce.

2. The Earth orbits the Sun at a mean distance of 1.496×10^{11} m in one year (365.25 days). Calculate the mean orbital speed to an appropriate number of significant figures, expressing your answer in km s^{-1}. (You will need to calculate the number of seconds in a year.)

3. A car, travelling at a steady speed of 20 m s^{-1}, changes its direction from due North to due East in 5 seconds. Calculate the mean acceleration.

4. A car travels at a steady speed of 15 m s⁻¹ around a semicircular bend of radius 60 m, as shown in the diagram. Calculate the:

(a) displacement AC [Hint: magnitude and direction]
(b) time taken to travel from A to C
(c) mean velocity, $\overline{v_{AC}}$, between A and C
(d) change in velocity between A and C
(e) mean acceleration between A and C
(f) mean velocity between A and B
(g) mean acceleration between A and B.

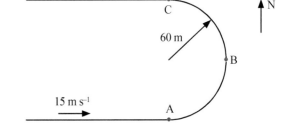

5. Use the v–t graph to determine:

(a) the acceleration between 0 and 10 seconds;
(b) the distance travelled in the first 20 s;
(c) the acceleration at 30 s;
(d) the mean velocity over the 40 s.

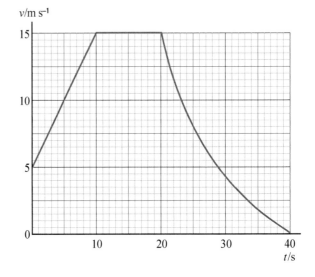

6. Explain why the equation $v = u + at$ is homogeneous but that $x = u + at^2$ is not and therefore cannot be correct.

7. A train accelerates from rest at 0.50 m s⁻² for a period of 90 s. It travels at a steady velocity for 4.5 km before decelerating to rest in a further 1800 m. Calculate the mean velocity for the journey. [Hint: Calculate the total displacement and the total time taken.]

8. A ball is thrown vertically upwards from ground level at 35 m s⁻¹. Taking upwards as positive and using the approximation, g = 10 m s⁻², calculate the: (a) velocity after 1.0 s, 2.0 s, 5.0 s; (b) time at the highest point [i.e. when the velocity = 0]; (c) time taken to reach the ground again (assuming it is not intercepted by a passing gull);

9. A stone is catapulted horizontally from the top of a cliff with a speed of 30 m s⁻¹. It hits the sea 5.0 s later. Calculate: (a) the height of the cliff; (b) the vertical component of its velocity when it hits the water; (c) the velocity with which it hits the water.

10. An arrow is shot at 50 m s⁻¹ at an angle of 30° to the horizontal. Ignoring air resistance and other aerodynamic effects, calculate: (a) the maximum height reached by the arrow; (b) the time to the maximum height; (c) the velocity at the maximum height; and, (d) the horizontal range of the arrow.

11. An experiment is undertaken to determine a value for the acceleration due to gravity using the apparatus illustrated in Fig. 1.2.21. A centisecond timer was used. The measured times for the steel sphere to drop are given in the table:

Distance, x / cm	10.0	20.0	30.0	40.0	50.0	60.0	70.0
Time, t / ms	180	250	280	330	350	390	420

There is a small systematic error due to the magnet. Compare graphically the two methods of analysis given in the text and determine the delay time before the magnet drops the sphere.

1.3 Dynamics

Section 1.2 was concerned with the mathematical language of motion – describing uniform and accelerated motion in terms of equations. Dynamics is concerned with the causes of motion and its changes. Before the 17th century, the natural state of motion of objects was considered to be at rest. Some agency was considered necessary for an object to be moving: it is difficult to force a log to move at all and the moment you stop dragging it, the log ceases motion; even a rolling ball stops in quite a short time; the Earth was 'known' to be at rest at the centre of the universe.

Different rules were held to apply to objects outside the Earth, i.e. the Moon and beyond. They were not made of normal stuff (earth, air, fire and water) but *quintessence*, a fifth substance whose natural state of motion was to circle the Earth. Section 1.3 presents the results of the revolution in the 16th and 17th centuries in which Isaac Newton built upon the work of such famous people as Copernicus, Kepler, Galileo and Descartes. (The effect of Newton on our understanding of the way that objects move cannot be overstated. In the words of Alexander Pope's epitaph for him, 'Nature and nature's laws lay hid in night; God said "Let Newton be!" and all was light.')

Newton's three laws of motion may be stated as:

N1. A body's velocity will be constant unless a resultant force acts upon it.
N2. The rate of change of momentum of a body is directly proportional to the resultant force acting upon it.
N3. If a body **A** exerts a force on body **B**, then **B** exerts an equal and opposite force on **A**.

These laws are often referred to as N1, N2 and N3.

The principle of **conservation of momentum** states:

The vector sum of the momenta of the bodies in a system is constant provided there is no resultant external force.

Alternative statement:

The vector sum of the momenta of the bodies in an isolated system is constant (see Study point).

In fact these four statements are not independent; the third law (N3) can be derived from N2 and conservation of momentum. Notice that these laws are framed in terms of velocity, momentum and force. We shall start with momentum.

1.3.1 Momentum

The momentum, p, of a body was defined by Newton as

$$p = mv,$$

where v is the velocity of the body and m is the mass.

Mass is a scalar quantity which is a measure of the the body's **inertia**. Inertial mass is assumed to be independent of velocity.

Momentum is a vector quantity, like displacement and velocity, i.e. it possesses magnitude and direction. Hence:

1. If asked to find the momentum of a body we should always specify the direction (see Study point).

2. We add or subtract momenta in the ways we met in Sections 1.1.3 and 1.1.4.

Study point

A car approaching a bend on an icy road could well fall victim to N1. Without the frictional grip of the road on the tyres it will plough straight on!

Study point

An **isolated system** is one on which no external forces act and no particles enter or leave.

Study point

From the definition

$[p] = [m][v]$.

So $[p]$ = kg m s^{-1}. N s is also a correct unit. See Knowledge check 1.3.6.

Study point

If the motion is along a single straight line, we use + or – to indicate the direction. See Section 1.3.1(b).

The **unit of momentum** is kg m s^{-1}, which we can also write as N s (see Section 1.3.3).

Example

A boat of mass 10 000 kg is sailing at 8.0 m s^{-1} on a bearing of 60°. Calculate:

(a) its momentum

(b) the northerly component of its momentum.

Answer

(a) $p = mv = 10\,000 \times 8.0$

$= 80\,000$ kg m s^{-1} on a bearing of 60°.

(b) Northerly component $p_N = 80\,000 \cos 60°$

$= 40\,000$ kg m s^{-1}

In the laboratory, we often investigate momentum changes using riders on air tracks. A rider sits on a cushion of air so that:

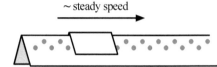
~ steady speed

- There is zero resultant vertical force because the upward force on the rider due to air pressure is equal and opposite to the downward gravitational force and
- There is zero frictional force on the rider and the force due to air resistance is very small.

Fig. 1.3.1 Air track

Hence we can consider a collection of riders (usually two) on the air track to constitute an almost isolated system. A single rider moving at low speed is seen to travel at very close to constant velocity, in line with Newton's 1st law (N1).

a) When colliding objects stick together

Fig. 1.3.2 shows two collisions on the air track. These collisions, in which the objects combine on impact, are called **inelastic collisions**. For illustration purposes the velocities are deliberately idealised! The riders are identical, so their masses are the same – say 0.15 kg.

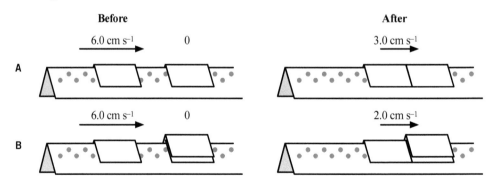

Before **After**

6.0 cm s^{-1} 0 3.0 cm s^{-1}

A

6.0 cm s^{-1} 0 2.0 cm s^{-1}

B

Fig. 1.3.2 Inelastic collisions

Assuming the riders each have a mass of 0.15 kg, the **momenta** before and after the collisions, p_1 and p_2, are for collision **A**:

$p_1 = 0.15$ kg $\times$ 6.0 cm s^{-1} + 0.15 kg $\times$ 0 = 0.90 kg cm s^{-1}

$p_2 = 0.30$ kg $\times$ 3.0 cm s^{-1} = 0.90 kg cm s^{-1}.

The momenta before and after are the same. Of course, the actual mass of the riders is irrelevant as long as they are all the same.

b) When colliding objects bounce apart

If we mounted repelling magnets on the riders we might observe collision **C** in Fig. 1.3.3. There are two differences from **A** and **B**:

- There is movement in both directions.
- The riders move apart after the collision.

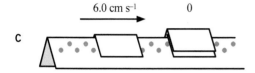 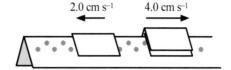

Fig. 1.3.3 Elastic collision

Just as with motion under gravity (Section 1.2) we need to define a positive direction. Let's choose right as the positive direction.

Then: $p_1 = 0.15 \text{ kg} \times 6.0 \text{ cm s}^{-1} + 0.30 \text{ kg} \times 0 = 0.90 \text{ kg cm s}^{-1}$

$p_2 = 0.15 \text{ kg} \times (-2.0) \text{ cm s}^{-1} + 0.30 \text{ kg} \times 4.0 \text{ cm s}^{-1} = 0.90 \text{ kg cm s}^{-1}$.

The example shows how to apply the principle to calculate a final velocity.

Example

An empty wagon of mass 1000 kg travelling at 6.0 m s^{-1} to the right collides with a full wagon of mass 4000 kg travelling at 2.0 m s^{-1} in the opposite direction. If they couple on impact, calculate their common velocity after the collision.

Answer

Step 1: Diagram:

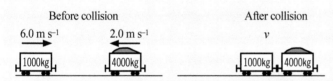

Fig. 1.3.4

Step 2: CoM equation: sum of initial momenta = sum of final momenta

Step 3: Taking right as positive: $1000 \times 6.0 + 4000 \times (-2.0) = 5000v$

Step 4 Solving this $\therefore v = -0.40 \text{ m s}^{-1}$

$\therefore$ The common velocity is 0.40 m s^{-1} to the left.

1.3.5 Knowledge check

Show that the initial and final kinetic energies in collision **C** (Fig. 1.3.3) are both 2.7×10^{-4} J.

1.3.2 Types of momentum conservation problems

a) Elastic and inelastic collisions

Three of the collisions in Section 1.3.1 were **inelastic**: collisions **A**, **B** and the example. These collisions result in a loss of *kinetic* energy. Note the word kinetic. Energy is always conserved but it can be transferred from one object to another or one form into another. You should be able to show that the collision in the example above results in a loss of 25.6 kJ of kinetic energy (from 26.0 kJ to 0.4 kJ).

If you've done Knowledge check 1.3.5 you'll have shown that no kinetic energy is lost in collision **C** in Fig. 1.3.3. This kind of collision is called an **elastic** (or sometimes a **perfectly elastic**) collision. An elastic collision is one in which there is no change in the total kinetic energy. An inelastic collision is one in which kinetic energy is lost.

Contact collisions between macroscopic objects, i.e. objects you can see, such as tennis balls and cars, always involve some loss of kinetic energy – it is transferred to the vibrational energy of the molecules of the objects. Low energy collisions between subatomic particles or between molecules of monatomic gases at room temperature are usually elastic. Collisions between hard spheres such as snooker balls will typically preserve 90% of the kinetic energy. Hence in the macroscopic world, such collisions are rarely perfectly elastic.

Elastic collisions are more difficult to analyse than inelastic ones because there are two unknown velocities. We need to solve simultaneous equations.

Study point

Conservation of momentum doesn't only apply to collisions. It also applies to situations where one object ejects another, such as a nucleus emitting an α particle or a gun firing a bullet. The momentum is zero before the event, so the total momentum is zero afterwards.

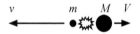

So: $mv = MV$. If we know the total energy, E then we can write

$\frac{1}{2}mv^2 + \frac{1}{2}MV^2 = E$

and solve the equations for v and V.

Example

The collision in Fig. 1.3.5 is elastic. Find v_1 and v_2.

Before   After

$u = 12$ m s^{-1} 0 v_1 v_2

2 kg   4 kg

Fig. 1.3.5

Answer

Momentum before the collision $= 2 \times 12 + 4 \times 0 = 24$ N s

$\therefore$ Using the principle of conservation of momentum: $2v_1 + 4v_2 = 24$ [1]

Kinetic energy before the collision $= \frac{1}{2} \times 2 \times 12^2 + \frac{1}{2} \times 4 \times 0 = 144$ J

$\therefore$ Using conservation of energy $\frac{1}{2} \times 2 \times v_1^2 + \frac{1}{2} \times 4 \times v_2^2 = 144$

$\therefore$ Simplifying $v_1^2 + 2v_2^2 = 144$ [2]

We now solve equations [1] and [2] for v_1 and v_2 as follows:

Dividing equation [1] by 2 and rearranging: $v_1 = 12 - 2v_2$

Substituting in equation [2] for v_1: $(12 - 2v_2)^2 + 2v_2^2 = 144$

Expanding $144 - 48v_2 + 4v_2^2 + 2v_2^2 = 144$

Simplifying 2 $-48v_2 + 6v_2^2 = 0$

Factorising $6v_2 - (8 + v_2) = 0$

So $v_2 = 0$ or 8 m s^{-1}.

Solving equations 1 and 2 for v_2, gives $v_2 = 0$ or 8 m s^{-1}. We can ignore the 0 because that clearly represents a 'no collision' (i.e. the 2 kg ball has missed!) so $v_2 = 8$ m s^{-1}. Substituting in [1] gives $v_1 = -4$ m s^{-1}, i.e. 4 m s^{-1} to the left: the lower mass ball has bounced back (as we might have expected).

Notice that in this collision the balls separate at the same rate (12 m s^{-1}) after the collision as that with which they approached each other before colliding. This is always true in elastic collisions and we can use this fact to simplify the calculation of v_1 and v_2. See Stretch & challenge.

(b) 'Explosion' problems

Conservation of momentum can be applied not only to collisions but to any interactions in which external forces can be neglected. If part of a single body is thrown off in one direction, the remaining part will move in the opposite direction. We see this in the recoil of a gun when a bullet is fired, in the acceleration of a rocket when exhaust gases are ejected at the rear or the recoil of an atomic nucleus when an α-particle is emitted.

We'll look at the general problem. Fig. 1.3.6 shows a body splitting into two parts with masses m and M. These parts move apart with velocities v and V respectively. Assuming that the original body (of mass $m + M$) was at rest before splitting up, the sum of the momenta of the two parts must be zero after the explosion, assuming that external forces can be ignored.

We'll look at the principles involved using the following example.

Example

A stationary astronaut of mass 75 kg in the International Space Station gently throws a ball of mass 1.5 kg forwards at a speed of 2.0 m s^{-1}. What is the effect on the motion of the astronaut?

Answer

The initial momentum is zero. Momentum must be conserved so the astronaut must move backwards to give a total momentum of zero. Let the recoil velocity of the astronaut be v.

Then, applying CoM: $\qquad 1.5 \times 2.0 + 75v = 0$

Rearranging: $\qquad v = \dfrac{-1.5 \times 2.0}{75}$

$\qquad\qquad\qquad = -0.040$ m s^{-1}

Hence the astronaut moves backwards at 0.040 m s^{-1}.

You may encounter a more difficult question on this topic, in which you are told the total kinetic energy, E_{total}, released and the masses of the two objects, e.g. uranium-238 decays by the emission of an α particle with a total energy release of 4.2 MeV (6.7×10^{-13} J). From this we can calculate the kinetic energy acquired by the α particle and the resulting nucleus (thorium-234). It can be shown that the energy, E_α, of the alpha particle is given by

$$E_\alpha = \frac{E_{total}}{\left(1 + \dfrac{m_\alpha}{m_{Th}}\right)}$$

where m_α and m_{Th} are the α-particle and thorium nuclear mass respectively.

Stretch & challenge

If $v_2 - v_1 = 12$ m s^{-1} (see text underneath the example), we can use this as equation 2 instead. This makes the solution easier. Check this by solving:

$2v_1 + 4v_2 = 24$ $\quad$ [1]

$v_2 - v_1 = 12$ $\qquad$ [2]

Why is there only one solution this time?

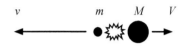

Fig. 1.3.6 CoM in an explosion

Knowledge check 1.3.6

In the astronaut example, calculate:

(a) the kinetic energy of the ball

(b) the kinetic energy of the astronaut

(c) the fraction of the total kinetic energy gained by the ball.

Knowledge check 1.3.7

Use the equation

$$E_\alpha = \frac{E_{total}}{\left(1 + \dfrac{m_\alpha}{m_{Th}}\right)}$$

and the data to find the fraction of the total kinetic energy acquired by the α particle.

(Hint: There is no need to convert m_α and m_{Th} to kg.)

Fig. 1.3.7

1.3.3 Force and momentum

The cricketer in Fig. 1.3.7 is playing a hook shot. The ball, which was originally moving rapidly towards her, is suddenly moving sideways even more rapidly. The change of momentum, Δp, is illustrated in Fig. 1.3.8.

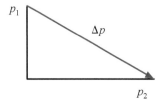

Fig. 1.3.8

We know (Newton's 1st law – N1) that an object's velocity and therefore its momentum stays constant in the absence of a force. So this change of momentum occurs because something, in this case the bat, applies a **force** to the ball.

Newton's 2nd law (N2) says that the resultant force applied is 'directly proportional to the rate of change of momentum'. In SI, we **define** the rate of change of momentum as equal to the force expressed in newtons, i.e.

$$F_{res} = \frac{\Delta p}{\Delta t} \text{ or } \Sigma F = \frac{\Delta p}{\Delta t}$$

where Δt is the duration of the force.

Example

A cricket ball has a mass of **0.16 kg**. Its speed hitting the bat is **30 m s⁻¹**, it turns through a right angle, leaves the bat at **40 m s⁻¹** and the duration of the impact is **1.5 ms**.

Calculate the magnitude of the mean force exerted by the bat on the ball.

Answer

Using Fig. 1.3.8 above: $p_1 = 0.16 \times 30 = 4.8 \text{ N s}$

$p_2 = 0.16 \times 40 = 6.4 \text{ N s}$

Using Pythagoras' theorem $\Delta p = \sqrt{p_1^2 + p_2^2} = 8.0 \text{ N s}$

∴ The mean force, $F = \frac{\Delta p}{\Delta t} = \frac{8.0}{1.5 \times 10^{-3}} = 5300 \text{ N (2 s.f.)}$

Just as the gradient of a *v–t graph* is the acceleration of an object, the gradient of the momentum–time graph is the resultant force upon it.

This is because $F = \frac{\Delta p}{\Delta t}$ and $\frac{\Delta p}{\Delta t}$ is the gradient of the *p–t* graph.

Study the *p–t* graph for a car moving east along a straight road and braking to rest (Fig. 1.3.9). Now answer Knowledge check 1.3.9.

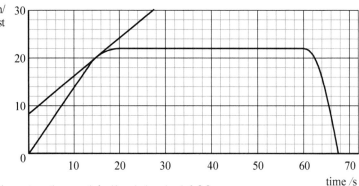

Fig. 1.3.9 Momentum–time graph for Knowledge check 1.3.9

1.3.4 Momentum and Newton's 3rd law of motion

Fig. 1.3.10 Body A exerts a force on B

As promised in the introduction, we are going to show that N3 is a consequence of the principle of conservation of momentum and the definition of a force from N2. Fig. 1.3.10 shows two isolated bodies, **A** and **B**. **A** exerts a force on **B**, which we'll call F_{AB}. This is drawn as a repulsive force but it could act in any direction.

Consider a small time interval Δt. From N2, body **B** suffers a change in momentum, Δp_B given by:

$$\Delta p_B = F_{AB}\,\Delta t.$$

But, if the bodies are isolated, their total momentum, $p_A + p_B$ is constant so body **A** must suffer an equal and opposite momentum change, Δp_A, i.e.

$$\Delta p_A = -F_{AB}\,\Delta t.$$

Because the momentum of **A** has changed it must have a force on it, which can only be exerted by **B** (because the bodies are isolated). Then F_{BA}, the force exerted by **B** on **A**, is given by:

$$F_{BA} = \frac{\Delta p_A}{\Delta t} = -F_{AB} \text{ (from above).}$$

In other words, the force exerted by body **B** upon body **A** is equal and oppositely directed to the force exerted by body **A** upon body **B**. Let's consider a falling skydiver. The forces acting on the skydiver at a particular instant are shown in Fig. 1.3.11.

We analyse this in terms of N3 and look for the equal and opposite force to each force in the diagram.

1. The $800\ \text{N}$ weight: This is the gravitational force that the Earth exerts on the skydiver. So the skydiver exerts an equal and opposite gravitational force on the Earth. The Earth is pulled (upwards) by an $800\ \text{N}$ force! [The mass of the Earth is $6 \times 10^{24}\ \text{kg}$ so its acceleration is rather small!]

2. The $600\ \text{N}$ air resistance: This is the drag of the air molecules on the clothing as the skydiver plummets down. So (the clothing of) the skydiver exerts an equal and opposite drag force (i.e. $600\ \text{N}$ downwards) on the molecules of air, which doubtless causes some air turbulence.

> ## Study point
>
> N3 is a bit tricky so it's important to understand what it is saying:
>
> All forces arise from interactions between two bodies. In the interaction, both bodies experience a force. These two forces are equal and opposite.

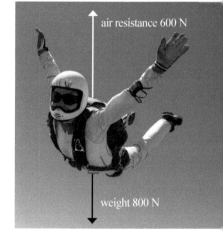

Fig. 1.3.11 The skydiver

The N3 trap: not all equal and opposite forces arising are N3 pairs!

Fig. 1.3.12 shows a sea lion holding up a ball. The ball is in equilibrium under the action of the two labelled equal and opposite forces.

Why aren't they an N3 pair? There are two main reasons:

1. **They act on the same body** (the ball). In an N3 pair, there are two bodies: one force acts on one body; the other force acts on … the other!

2. **The forces are not of the same kind.** The force exerted by the Earth on the ball is gravitational, so the other member of its N3 partner must also be gravitational (it is, as with the skydiver, the gravitational force of the ball on the Earth).

Fig. 1.3.12 Equal and opposite forces but not N3!

3. [Additional reason] N3 is a universal law. There must always be an equal and opposite force. Imagine the sea lion quickly whipping its head out from under the ball: the downward force is still there but the upward force has disappeared – so they can't have been an N3 pair!

Think about reason 2. What 'kinds' of forces are there? Physicists recognise four fundamental forces: the strong nuclear force; the weak force; the electromagnetic force and the gravitational force. The first two are only of significance for subatomic particle interactions so all other forces are gravity and electromagnetic. The forces between the molecules of the ball and sea lion's nose are electromagnetic – they are caused by repulsion of the outer electrons in the molecules in the ball and nose which are in close proximity.

Don't worry about this last point – any legitimate way of describing a force must also apply to its N3 partner, e.g. the sea lion exerts an inter-atomic force on the ball, so the ball exerts an inter-atomic force on the sea lion.

Example – A difficult example of N2 and N3

Two thousand years ago, Hero of Alexandria described an engine which consisted of a water-filled sphere, which was heated over a fire. The steam from the boiling water escaped along two bent pipes, which caused the whole to rotate.

A model of Hero's engine – see Fig. 1.3.13 – ejects 0.10 g of steam per second from each of two nozzles at a speed of 12 m s^{-1}.

(a) Explain why the engine starts to rotate clockwise.

(b) Calculate the moment exerted on the engine by the escaping steam.

Answer

(a) The steam gains momentum to the left at the top and the right at the bottom. So the engine exerts a force on the steam, left at the top. So the steam exerts equal and opposite forces on the engine, which is made to rotate to the right at the top, i.e. clockwise.

(b) At the top:

Momentum change per second $= 1.0 \times 10^{-4}$ kg s$^{-1} \times 12$ m s^{-1}

Force = rate of change of momentum

∴ Force to the left on the steam $= 1.2 \times 10^{-3}$ N

∴ By N3, force to right on engine $= 1.2 \times 10^{-3}$ N

∴ CW moment on engine $= 1.2 \times 10^{-3}$ N $\times 0.15$ m $= 1.8 \times 10^{-4}$ N m

The steam exiting from the nozzle at the bottom produces an equal CW moment, so the total moment $= 3.6 \times 10^{-4}$ N m.

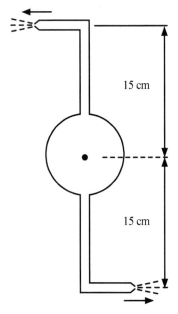

15 cm

15 cm

Fig. 1.3.13 Hero's engine

1.3.5 Forces between materials in contact

Forces arise between objects due to molecular interactions. We shall consider some of these forces: the normal force, friction and air resistance.

(a) The normal contact force, F_N

If an object rests against a surface, the surface exerts a force on the object. This force arises because the molecules in the two bodies are placed in close contact. If molecules are close together, the electrons in the outer shells repel one another and so, in this case, the bodies experience a force at right angles to the surface. The word *normal* is used to mean 'at 90°'.

(b) Friction

The box on the slope in Fig. 1.3.14 will remain stationary on the slope as long as the gradient is not too great. This can only be because a force acts on the box up the slope which counteracts the component of the weight, $W \sin \theta$, down the slope, see Fig. 1.3.15. This force is called *static friction*, F_R, or *grip*.

This force acts to stop the two surfaces sliding over each other, i.e. it opposes relative motion. For the stationary box, $F_R = W \sin \theta$, i.e. the friction is just big enough to stop motion. F_R has a maximum value, often called *limiting friction*. In the case of the box on the slope, if the slope is gradually made steeper, the value of F_R will increase up to this limiting value; at greater angles, $W \sin \theta > F_R$ and the box will start to accelerate down the slope. Grip arises from bonds between molecules of the two surfaces in contact.

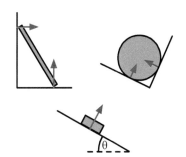
Fig. 1.3.14 Normal contact force examples

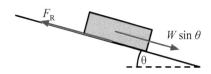
Fig. 1.3.15 Static friction

The force which opposes the relative motion when one surface slides over another is also called friction – in this case *dynamic friction*. It arises from temporary bonds which form as molecules in the surfaces move past one another. When they stretch and break, the stored energy in the bonds is converted to vibrational energy of the molecules, i.e the temperatures of the bodies increase. The value of dynamic friction is typically smaller than the limiting value of grip. This means that once an object starts to slide, it usually accelerates rather than just moving very slowly.

(c) Air resistance

Air resistance is one example of drag. As with friction it opposes relative motion between the object and the fluid [= liquid or gas] through which the object is moving. It also exists when the fluid flows past a stationary object, e.g. the wind on a building. The mechanism of drag is complicated but a simplification is that the molecules of a fluid bounce off a moving object slightly faster than they hit it: they gain velocity in the direction that the object is moving. So there is a momentum transfer to the fluid – and momentum transfer means a force on the fluid in the direction of the object's motion. By N3 the fluid exerts an equal and opposite force on the body.

 Study point

Avoid referring to the normal contact force as the normal reaction. This sounds too much like part of a poor statement of Newton's 3rd law!

Link

This section links to Section 1.4.5.

Study point

Static and dynamic friction oppose relative motion between surfaces. If a surface would slide to the right, the frictional force on it is to the left.

Knowledge check 1.3.10

Sand is fed vertically onto a horizontal conveyor belt which is moving towards the right. In which direction does the friction act on (a) the sand and (b) the belt? Explain your answers.

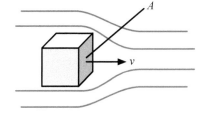
Fig. 1.3.16 Drag

 Link

Stokes' formula on page 11 is applicable only at low speeds.

Show that c_d has no units (i.e. it is dimensionless).

In many situations, the drag force, F_d, is given by the following equation:
$F_d = \frac{1}{2}\rho v^2 c_d A$, where ρ is the density of the fluid, c_d is a dimensionless quantity called the drag coefficient, which depends upon the shape of the object. You do not need to know this equation but you should know that F_d increases with the area A of the object, the velocity v and the density ρ of the fluid.

1.3.6 Free-body diagrams

Study point

An effect of air resistance is to transfer energy from the moving body to the molecules of air, as kinetic energy.

Drawing a free-body diagram is useful in identifying the forces on interacting objects.

Consider the house brick lying on the sloping plane in Fig. 1.3.17(a). We can identify three forces on the brick. Isolating the brick from the slope in Fig.1.3.17(b) makes these forces easier to identify. They are:

1. the gravitational force, mg, of the Earth on the brick – vertically downwards

2. the normal contact force, N, of the plane on the brick – at right angles to the slope

3. the frictional force, F, of the plane on the brick – up the slope.

Fig. 1.3.17 Forces on a brick resting on a slope

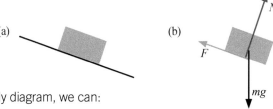

Identify the N3 partner to each force, on the box in Fig. 1.3.17.

From the free body diagram, we can:

(a) calculate the resultant force if all the forces are known

(b) calculate any unknown force on an object in equilibrium if all the other forces are known

(c) identify the N3 partner forces on any interacting object, e.g. on the plane in Fig. 1.3.17.

1.3.7 Force and acceleration

When Newton's 2nd law of motion is applied to a single body with constant mass, acted on by a number of forces, the equation is often written in the form:

$$\Sigma F = ma.$$

Study point

Note that a and ΣF are vectors and m is a scalar (> 0) so the direction of the acceleration is the same as that of the resultant force.

We can derive this from $\Sigma F = \dfrac{\Delta p}{\Delta t}$ as follows:

By definition, $p = mv$, $\therefore \Delta p = \Delta(mv) = m\Delta v$ because m is constant.

$\therefore \Sigma F = \dfrac{\Delta p}{\Delta t} = m\dfrac{\Delta v}{\Delta t}$. But $\dfrac{\Delta v}{\Delta t} = a$, $\therefore \Sigma F = ma$

We can rearrange this as $a = \dfrac{\Sigma F}{m}$.

So, in words, the acceleration of an object is the resultant force acting on it divided by its inertial mass.

Example

Calculate the acceleration of the body in Fig. 1.3.18

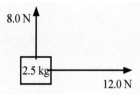

Fig. 1.3.18 $\Sigma F = ma$

Answer

Step 1: Calculate the resultant force:

From the force diagram (Fig. 1.3.19)

$$F_{res} = \sqrt{8^2 + 12^2} = 14.4 \text{ N} \quad \text{and} \quad \theta = \tan^{-1}\frac{8.0}{12.0} = 33.7°$$

Step 2: Calculate the acceleration: $a = \dfrac{F_{res}}{m} = \dfrac{14.4}{2.5} = 5.8 \text{ m s}^{-2}$ (2 s.f.).

Answer: The acceleration is 5.8 m s^{-2} at $33.7°$ to the 12 N force (see Top tip).

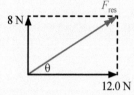

Fig. 1.3.19

1.3.8 Gravitational force

It is an experimental fact that if an object falls freely, i.e. in the absence of air resistance, its acceleration is independent of its mass, density or shape. The classic 'guinea and feather' illustration of this is shown in Fig. 1.3.20. This is a modern version of Galileo's famous thought experiment, in which he imagined dropping two cannonballs of different masses from the top of the Leaning Tower of Pisa. US Apollo 15 astronaut, David Scott, repeated this experiment live on the Moon in 1971 using a hammer and a feather; more recently Brian Cox used a NASA facility to show the same experiment – you can find videos of these using a search engine.

The acceleration of free fall is given the symbol g and is also known as the 'acceleration due to gravity'. Its value on the surface of the Earth is approximately 9.81 m s^{-2}, though this depends on the location because of the shape of the Earth and height of the terrain.

Consider a falling object of mass m. Its acceleration is g, so N2 tells us that the gravitational force on it, which we call its *weight*, W, is given by

$$W = mg.$$

The following example illustrates the use of $W = mg$ in conjunction with other concepts, which we have already met.

Example

Fig. 1.3.21 shows a box accelerating down a slope. F, the frictional force has a value which is $0.30C$, where C is the normal contact force. Calculate:

(a) the minimum value of θ for the box to accelerate down the slope,

(b) the acceleration of the box if $\theta = 20°$. [Take $g = 9.81 \text{ m s}^{-2}$]

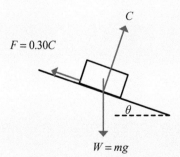

Fig. 1.3.21 Accelerating box on a slope

⟨ Link ⟩

This Section links to Section 1.2.3 (c).

⟨ Top tip

Acceleration is a vector so, if asked for an acceleration, you should give a direction as well as a magnitude.

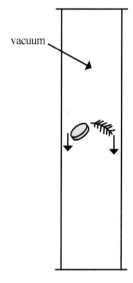

Fig. 1.3.20 Guinea and feather

≫ Study point

We can rearrange the equation

$W = mg$, as $g = \dfrac{W}{m}$, so N kg^{-1}

can be used as the unit of g. If N kg^{-1} is used, g is referred to as the *gravitational field strength* – it gives the force per unit mass on a body placed in the gravitational field.

1.3.13 Knowledge check

Calculate the acceleration of a cyclist who is freewheeling down a 10° slope. The mass of the cyclist and bike is 75 kg; the resistance force is 50 N.

< Link >

This section links to Section 1.2.3(c).

Answer

(a) There is no motion perpendicular to the slope so the resultant force in this direction is 0.

∴ (Resolving perpendicular to the slope): $C = mg \cos \theta$

But $F = 0.30C$, ∴ $F = 0.3\, mg \cos \theta$

The component of W down the slope $= W \sin \theta = mg \sin \theta$

∴ The resultant force, ΣF, down the slope $= mg \sin \theta - 0.3\, mg \cos \theta$

$= mg\, (\sin \theta - 0.3 \cos \theta)$

The box can only accelerate if $\Sigma F > 0$. ∴ $\sin \theta > 0.3 \cos \theta$

$\dfrac{\sin \theta}{\cos \theta} = \tan \theta$ ∴ $\tan \theta > 0.3$ ∴ θ must be at least $16.7°$ (0.29 rad)

(b) At 20°, ΣF down the slope $= mg\, (\sin 20° - 0.3 \cos 20°) = 0.060mg$.

∴ Using $\dfrac{\Sigma F}{m}$, the acceleration is $0.060g = 0.590$ m s^{-2}.

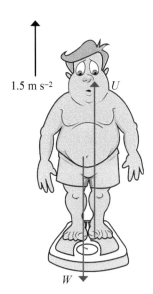

1.5 m s^{-2} U

W

Fig. 1.3.22 How to put on weight.

Your *sensation* of weight is not the same as the force of gravity on you. It arises from the compression of your body between the force of gravity, which is distributed through your body, and the upward contact force on you from the ground. Hence an astronaut in the International Space Station has no sensation of weight. When you are in a lift, your apparent weight depends upon the motion of the lift. Our final example illustrates this effect.

Example

A man of mass **85 kg** is standing on a bathroom scales whilst in a lift, which is accelerating upwards at 1.5 m s^{-2}, as shown in Fig. 1.3.22.

(a)   Calculate the upward force, U, exerted by the scales on the man.

(b)   The scales measure this upward force but their reading, R, is in **kg**, which relates to U by:

$U = Rg$.

What is the reading on the scales?

Answer

(a) $W = mg = 85 \times 9.81 = 833.9$ N

From N2: $\Sigma F = ma$. ∴ $U - 833.9 = 85 \times 1.5 = 127.5$ N

∴ $U = 127.5 + 833.9 = 961.4$ N [= 961 N to 3 s.f.]

(b) $R = \dfrac{U}{g} = \dfrac{961.4}{9.81} = 98.0$ kg

1.3.14 Knowledge check

What would be the reading on the scales if the lift were accelerating downwards at 1.5 m s^{-2}?

What would be the result in this lift on the Moon ($g = 1.5$ m s^{-2})?

Motion under gravity and air resistance

If the 'guinea and feather' in Fig. 1.3.20 are dropped in air, the former reaches the ground first. Whereas the feather appears to drift downwards at a constant speed, the coin accelerates all the way down.

The acceleration of falling objects, such as the guinea, the feather or a skydiver is determined by the resultant force on them. The two significant forces are:

- The weight, W, which is a constant for objects close to the Earth.

- Air resistance, F_d, which varies according to $F_d = \frac{1}{2}\rho v^2 c_d A$ – see Section 1.3.5(c).

Fig. 1.3.24 shows the effect of this combination of forces for a skydiver.

Fig. 1.3.23 A guinea tied to a toy parachute would drift downwards after an initial acceleration.

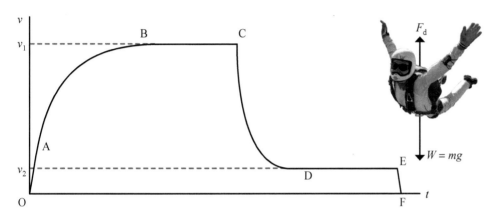

Fig. 1.3.24 The effect of air resistance on a skydiver

Knowledge check 1.3.15

Assuming that $c_d = 1$, estimate the terminal velocity, v_1, of a skydiver. Take $\rho_{air} = 1$ kg m^{-3}. State your assumptions for other quantities.

If we assume, to a first approximation, that ρ, the air density, is constant, the only significant variables which affect F_d are v and A, so the resultant force (F_{res} or ΣF) is given by:

$$F_{res} = mg - \frac{1}{2}\rho v^2 c_d A$$

Acceleration, $a = \dfrac{F_{res}}{m}$ ∴ $a = g - kAv^2$ (where $k = \dfrac{\rho c_d}{2m}$, a constant)

So, to take the descent in stages:

Knowledge check 1.3.16

Explain why a skydiver falls faster if she falls head first rather than spread-eagled (as shown in Fig. 1.3.24).

OB At **O**, $v = 0$ so the acceleration, $a = g$. As the skydiver speeds up, F_d increases so F_{res} and a decrease. At **B**, $F_d = mg$, the two forces balance, so $F_{res} = 0$ and $a = 0$. The velocity is constant $= v_1$ in Fig. 1.3.24. This is called a *terminal velocity*.

BC Assuming the density of the air stays constant, F_d is constant so the skydiver continues to fall at v_1.

C The parachute opens: A increases (massively); F_d increases so $F_{res} \ll 0$ and so $a \ll 0$, i.e. the skydiver very rapidly decelerates.

CD As the speed decreases, F_d decreases ($\propto v^2$) until once again $F_{res} = 0$ and so $a = 0$. This is the second terminal velocity, v_2.

DE As BC (at velocity v_2 this time).

EF Contact with ground. The ground exerts a large upward force on the skydiver. F_{res} is now large and upwards. It quickly reduces the speed to 0.

Detail

The rider on an air track consists of two flanges at 90° to each other on either side of the track. A blower pumps air into the track. This escapes through holes and the rider sits on this cushion of air.

50 g masses can be added to spigots on each flange.

Fig. 1.3.25

Increasing the accuracy

If a broader light shield is used, the time for the shield to pass through the light gate is increased. The timer is usually a millisecond timer. The greater the time, t, the lower the uncertainty in v.

1.3.17 Knowledge check

Using one light gate, the following results were taken:

$x = 60$ cm,

$\Delta x = 2.5$ cm

$\Delta t = 51$ ms

Calculate:

(a) the speed at 60 cm

(b) the acceleration.

▶▶▶ **Study point**

If two light gates are used: start the rider to the left of LG1, measure the speeds at LG1 (u) and LG2 (v) and calculate a using $v^2 = u^2 + 2ax$.

1.3.9 Specified practical work: Investigating Newton's 2nd law of motion

In A level Physics, investigating Newton's 2nd law of motion involves demonstrating that experimental results are consistent with $F = ma$, which means that

- For constant mass: $a \propto F$
- For constant force: $a \propto \dfrac{1}{m}$

The apparatus normally used in school is the air track, a hollow, triangular-section tube with air holes. Air is pumped in and metal riders are held above the track by a cushion of air which escapes from the air holes (see Fig. 1.3.25). Fig. 1.3.26 shows the setup.

The accelerating force is provided by the gravitational force on the 'low mass object' and causes the rider to accelerate.

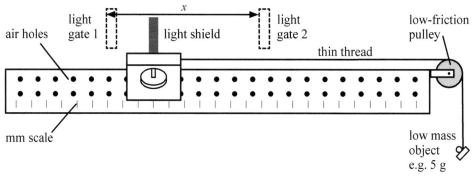

Fig. 1.3.26 Air track used for $F = ma$

The set-up in Fig. 1.3.26 has two light gates separated by a known distance, x. With this arrangement, the rider is released to the left of light gate 1 (LG1). Alternatively, just one light gate (LG2) can be used, in which case the rider is released a distance x before the light gate. The descriptions below are for one light gate. The variations with two light gates are given in the Study point.

(a) Showing that acceleration, a, is proportional to ΣF (for constant M)

The important thing with this experiment is to keep the total accelerating mass constant. This mass is the sum of the masses of the rider and the object on the end of the thread. A good way of doing this is to use a set of (say) five, 10 g masses on a hanger. These can be transferred individually between the end of the thread and the rider, so the total mass is constant.

- Start with a low mass, e.g. 10 g, on the thread (and the other masses on the rider). Calculate the accelerating force using $F = mg$.

- Release the rider from rest and use the light gate (LG2), to measure the time Δt, taken for the light shield of width Δx to cut off the light after the rider has accelerated through a distance x.

- Use the equation $v = \dfrac{\Delta x}{\Delta t}$ to calculate the velocity of the rider as it passes the light gate.

- Use the equation $v^2 = u^2 + 2ax$, with $u = 0$, to calculate the acceleration,

  i.e. $a = \dfrac{v^2}{2x}$.

- Repeat with a series of different accelerating forces by transferring low mass objects from the rider to the end of the thread.

- Plot a graph of a against F. The graph should be a straight line, through the origin, of gradient $\frac{1}{M}$ where M is the total mass of the rider and additional masses.

(b) Showing that acceleration, a, is proportional to $\frac{1}{M}$ (for constant F)

As above but add a series of additional masses to the rider, keeping the small mass on the end of the thread constant. Plot a graph of a against $\frac{1}{M}$, which should be a straight line, through the origin, of gradient $F (= mg)$.

Knowledge check 1.3.18

In experiment (a), identify the dependent, the independent, and the controlled variables.

Experiment (a) analysis

Calculate the acceleration for each of the masses, m, on the thread and plot a graph of a against m. It should be as follows:

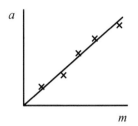

Experiment (b) analysis

Calculate the acceleration for each value of the total mass, M, remembering to include the mass on the thread. Plot a graph of a against $\frac{1}{M}$. It should be as follows:

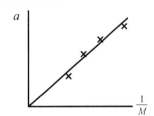

Knowledge check 1.3.19

In experiment (b), explain how the controlled variable is kept constant.

Test yourself 1.3

1. Two objects, A and B, have masses, $20\ kg$ and $30\ kg$, and speeds, $25\ m\ s^{-1}$ and $10\ m\ s^{-1}$ respectively. Calculate the total momentum of the objects if: (a) A and B are both travelling to the right; (b) A is travelling to the right and B is travelling to the left; (c) A is travelling due North and B is travelling due East.

2. Calculate the total kinetic energy for the bodies in question 1.

3. On an air track, two riders together travelling at $6\ cm\ s^{-1}$ collide inelastically with a third identical rider at rest. The common velocity after the collision is $4\ cm\ s^{-1}$. Show that this is as predicted by the principle of conservation of momentum.

4. Calculate the fraction of the initial kinetic energy which is lost in the collision in question 3.

5. A car of mass $1300\ kg$ travelling at $40\ m\ s^{-1}$ collides head on with a second car of mass $1200\ kg$ travelling at $20\ m\ s^{-1}$ and they stick together. Stating an assumption, calculate (a) the total momentum before the collision and, (b) their velocity immediately after the collision. What have you assumed?

6. A neutron travelling at $1200\ m\ s^{-1}$ collides with a stationary U-238 nucleus and is absorbed. Estimate the speed of the resulting U-239 nucleus. You may assume that protons and neutrons have the same mass.

7. Samarium-147 undergoes α decay. The α particles of mass $6.68 \times 10^{-27}\ kg$ are ejected at $1.00 \times 10^{7}\ m\ s^{-1}$, leaving $^{143}_{60}Nd$ nuclei of mass $2.39 \times 10^{-25}\ kg$. Calculate (a) the recoil velocity of the nuclei and, (b) the kinetic energy released in the decay.

8. Find the resulting velocities if the objects in question 1 collide elastically head on.

9. A football of mass $450\ g$ is kicked at $30\ m\ s^{-1}$ against a wall at right angles. It rebounds at $25\ m\ s^{-1}$. If the duration of the collision is $0.04\ s$ calculate the mean force which the ball exerts on the wall. Explain your answer clearly in terms of N2 and N3.

10. Two forces of magnitude 5 N and 12 N are simultaneously applied to a body of mass 1.55 kg. The directions at which the forces act can be varied.

(a) Calculate the magnitudes of the maximum and minimum accelerations.
(b) Calculate the acceleration of the object if the two forces act at right angles to each other.

11. The tension, T, in the longbow string is 4700 N. The mass of the arrow is 0.065 kg.
(a) Calculate the acceleration of the arrow. (b) The arrow leaves contact with the string after 70 cm. Estimate the speed with which it leaves the bow. State your assumption.

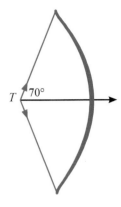

12. (a) 'Momentum is a vector quantity.' Explain what is meant by this statement.
(b) State Newton's 2nd law of motion in terms of momentum.

13. Fig. 1.3.24 includes a free-body diagram of the skydiver. For each of the forces shown, identify the 'equal and opposite' N3 force, by stating the body on which it acts, its direction and its nature.

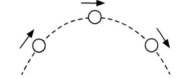

14. The diagram shows the flight of a thrown ball.

Draw free-body diagrams for the ball in each of the 3 positions: (a) without air resistance, (b) with air resistance.

15. A student obtained a set of results to demonstrate that a constant force produces a constant acceleration, using the two light gate technique (see Study point in Section 1.3.9).

x/cm	10.0	20.0	30.0	40.0	50.0	60.0
t/s	0.45	0.74	1.00	1.25	1.38	1.55

The suspended mass, m, was 5.0 g. By drawing a suitable graph, (a) show that the acceleration is constant; (b) calculate the acceleration, a; and, (c) to determine the mass, M, of the rider.

16. An aeroplane is climbing with a constant velocity. Assuming that the 'lift' acts at right angles to the direction of motion and the drag acts directly against the motion of the plane, draw and label a free-body diagram.

17. Copy the v–t graph for the skydiver in Fig. 1.3.24. Add a second v–t graph for the same skydiver making the same jump but carrying an additional load. Explain how the two graphs are related.

◀Stretch & challenge

The diagram shows an air track set up. The chord joining the 40 g mass and the rider has a tension, T.

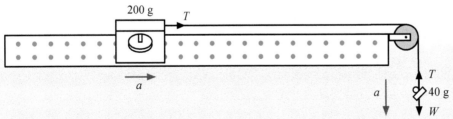

(a) Calculate the weight, W, of the small mass.

(b) Write equations in terms of T for the acceleration of
　(i) the 40 g mass and
　(ii) the 200 g rider.

(c) Solve the two equations and hence determine the acceleration, a.

1.4 Energy concepts

The concept of **energy** is one which physicists developed over a long period, between the late 17th and the early 20th centuries. Unlike momentum, which is a vector quantity and only conserved if there are no external forces, energy is a scalar quantity. The first type of energy to be identified was *kinetic energy* – the energy due to motion. An object that possesses energy can cause events to happen, which means that it can make things move. For example, an asteroid hitting the Earth can cause a large crater to be formed – moving huge quantities of material (and finishing off the dinosaurs in the process). We shall refine these ideas in this section of the book and start by considering the **work** done by a force.

1.4.1 Work and energy

If a force moves something, we say that the force *does work*. For example, the following forces are doing work:

- The force exerted by an electric winch which pulls a car out of a ditch.
- The force exerted by the wind which turns a turbine connected to a dynamo.
- The force exerted by an archer's muscles when he draws a bow.
- The force exerted by the archer's bow when it straightens up and shoots the arrow.

You'll recognise that all these examples need something to drive them, unlike, for example, the force exerted by a table in holding up a book or the force exerted by a nail which holds up a shelf. In these last two examples there is no motion, so no work is done – and there is no need of any 'input' to keep the objects in place. We use the concept of work to define energy.

The work done by a force is defined as follows:

Work done (J)	=	Force (N)	×	Distance moved in the direction of the force (m)

or, in symbols: $W = Fx$.

Applying this to Fig. 1.4.1:

The work done by the 100 N force, $W = 100 \text{ N} \times 50 \text{ m} = 5\ 000 \text{ J}$ (or 5 kJ)

We define **energy** in such a way that the quantity of energy transferred is equal to the work done by the force. How is this energy transferred? This depends upon the details. Various possibilities are shown in Figs 1.4.2 to 1.4.4.

>>> **Key terms**

Work: Done when a force moves its point of application.

Energy: The energy of a body or system is the amount of work it can do.

The unit of work is the **joule** (J).

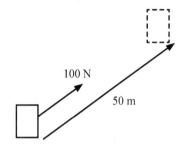

Fig. 1.4.1 Work done

>>> **Study point**

Work done	=	Energy transfer

So, if a force does 5 kJ of work, 5 kJ of energy is transferred.

Fig. 1.4.2 Kinetic energy

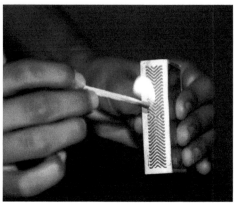

Fig. 1.4.3 Internal energy

Fig. 1.4.4 Gravitational potential energy

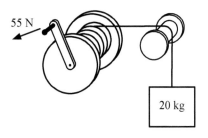

55 N

20 kg

Fig. 1.4.5 Work done using a winch

1.4.1 Knowledge check

What is the excess energy in the example transferred to? (Hint: not *heat* or *sound*)

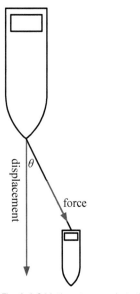

displacement

θ

force

Fig. 1.4.6 Motion at an angle to the force

1.4.2 Knowledge check

Calculate W given the following data:

$F = 500$ kN:

$x = 1.5$ mm: $\theta = 35°$

The captions in Figs 1.4.2 to 1.4.4 give the form the energy takes following the transfer. The crossbow stored energy as elastic potential energy before it is transferred to the quarrel or bolt; in Fig. 1.4.3 the energy was previously stored in the muscles by the compound ATP.

Example

A **20 kg** load is raised by **5.0 m** using a winch as shown in Fig. 1.4.5. The handle of the winch is **0.60 m** long and the radius of the drum on which the rope is wound is **0.15 m**. A force of **55 N** is required to turn the handle. Calculate:

(a) the work done on the **20 kg** block and

(b) the work done by the **55 N** force

Answer

(a) The force needed to raise the load is $mg = 20 \times 9.81 = 196.2$ N

The work done, W, on the **20 kg** block $= Fx$

$\therefore W = 196.2 \times 5 = 980$ J (2. s.f.)

(b) We need to calculate the distance moved by the **55 N** force. The length of the handle is 4× the radius of the drum, so the distance moved is 4×5 m $= 20$ m.

$\therefore$ The work done by the **55 N** force $= 55 \times 20 = 1100$ J

Note that the work done by the person winding the crank is more than the work done raising the load. We'll return to this in Section 1.4.5.

1.4.2 Directions of the force and the displacement

In the examples in the last section the force and displacement were in the same direction. What if this is not the case?

The tug in Fig. 1.4.6 is pulling at an angle θ to the direction of motion of the ship.

The vector diagram (Fig. 1.4.7) shows this more clearly. The distance moved in the direction of the force is $AB = Fx \cos \theta$.

$$\therefore W = Fx \cos \theta.$$

The expression $x \cos \theta$ is the component of x in the direction of F. Equivalently $F \cos \theta$ is the component of the force in the direction of the displacement.

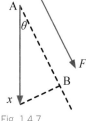

Fig. 1.4.7

So the definition of work done can be written:

Either: Work done = Force × Component of displacement in the direction of the force

or: Work done = Component of force in the direction of displacement × Displacement

We can see that these two ways of writing the work equation give the same answer if we consider the cyclist on a slope in Fig. 1.4.8.

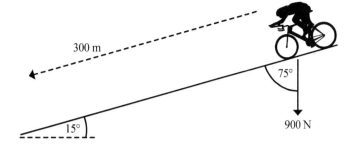

Fig. 1.4.8 Cyclist freewheeling downhill

Using the data in the diagram:

The vertical component of the displacement in **m** is $300 \sin 15°$; the component of the weight in **N** down the slope is $900 \cos 75°$. So applying the either/or from above:

Either: $W = (900 \text{ N}) \times (300 \sin 15° \text{ m}) = 270\,000 \sin 15° \text{ J} = 70 \text{ kJ}$ (2 s.f.)

Or: $W = (900 \cos 75° \text{ N}) \times (300 \text{ m}) = 270\,000 \cos 75° \text{ J} = 70 \text{ kJ}$

The angle between the force and the displacement in Fig. 1.4.8 was 75°. Here are some examples where the angle is 90° or more:

1. A satellite in a circular orbit (Fig. 1.4.9). The gravitational force is at right angles to the movement of the satellite in its orbit. Hence, $\theta = 90°$, so $W = 0$. The gravitational force does no work! This fits in with the fact that the energy of the satellite is constant – there is no change in either the potential or kinetic energies. **Conclusion**: If $\theta = 90°$, no work is done; no energy is transferred.

2. A bullet fired into a tree (Fig. 1.4.10). The diagram shows the frictional force exerted on the bullet by the tree. The angle between the motion and the force is 180°. But $\cos 180° = -1$, so the work done by the force is negative. This negative work means that a negative energy is transferred to the bullet – hence its kinetic energy decreases, i.e. it slows down. **Conclusion**: If $\theta > 90°$, the work is negative and energy is transferred **from** the object.

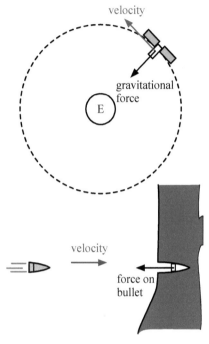

Fig. 1.4.9 & 1.4.10 Large values of θ

1.4.3 Energy conservation: kinetic and potential energy

(a) The principle of conservation of energy

Consider the box sliding to a halt on a surface in Fig. 1.4.11. Free-body diagrams for these two objects are shown in Fig. 1.4.12. Only the forces of the frictional interaction are shown.

Fig. 1.4.11 Interacting objects

force exerted by surface on box

force exerted by box on surface

Fig. 1.4.12 Free-body diagrams

From N3, we know that the two objects exert equal and opposite forces, F, on each other. In a short time Δt, the box slides $v\Delta t$ to the right. In this time, the surface does an amount of work $-Fv\Delta t$ on the box, reducing its kinetic energy by $Fv\Delta t$. At the same time the box does an amount of work $+Fv\Delta t$ transferring this quantity of energy (as internal energy, shared between the surface and the box).

Hence the total quantity of energy in the system is constant; i.e.

$$\Delta E = Fv\Delta t - Fv\Delta t = 0$$

This illustrates the **principle of conservation of energy**.

We shall now use the principle that:

Work done = energy transfer

to derive expressions for kinetic and potential energy.

(b) Kinetic energy

A moving object can do work on other objects in coming to rest. Thus it possesses energy because of its motion. We call this **kinetic energy**. Its symbol is E_k, but it is often just abbreviated to KE.

Consider an object of mass m moving with velocity u subject to a force F over a displacement x as shown in Fig. 1.4.13:

F and x are in the same direction,

$\therefore W = F\,x$.

Fig. 1.4.13 Deriving the formula for KE

But $F = ma$ $\therefore W = max$

The 4th kinematic equation for constant acceleration is $v^2 = u^2 + 2ax$.

$\therefore$ Rearranging, $ax = \frac{1}{2}v^2 - \frac{1}{2}u^2$

So substituting for ax above $\therefore W = \frac{1}{2}mv^2 - \frac{1}{2}mu^2$

From this equation we see that the work done is the change in the value of the quantity $\frac{1}{2}$ mass $\times$ velocity^2. So, because work done is the same thing as energy transfer, we conclude that $\frac{1}{2}mv^2$ is the kinetic energy of a body of mass m travelling with velocity v.

Example
A car of mass 800 kg, travelling at 15 m s^{-1} is accelerated by a force of 1200 N over a distance of 250 m. Calculate the final velocity.

Answer
Work done = change in kinetic energy

$\therefore 1200 \times 250 = \frac{1}{2} \times 800(v^2 - 15^2) \therefore 300\,000 = 400(v^2 - 225)$

$\therefore$ Dividing by 400 and rearranging: $v^2 = 750 + 225 = 975 \longrightarrow v = 31.2$ m s^{-1}.

(c) Gravitational potential energy

An object in a raised position can do work on other objects as it descends. Thus it possesses energy because of its height. We call this **gravitational potential energy**. The symbol is E_p, but it is often just abbreviated to GPE or PE. Strictly speaking, the energy is possessed not by the body on its own but by the Earth–body system: the GPE depends upon their separation.

> **Key term**

The **principle of conservation of energy**: The total energy of an isolated system is constant though it can be transferred within the system.

> **Key term**

Kinetic energy: The energy possessed by a body by virtue of its motion.

> **Study point**

You do not need to know the derivation of E_k (or KE) $= \frac{1}{2}mv^2$.

> **Top tip**

The change in kinetic energy is $\frac{1}{2}mv^2 - \frac{1}{2}mu^2$, which you can write as $\frac{1}{2}m(v^2 - u^2)$ This is **not** the same as $\frac{1}{2}m(v - u)^2$!

1.4.3 Knowledge check

The example can be solved by using $F = ma$ to find a and then using $v^2 = u^2 + 2ax$ to find v.

Do this and compare the methods.

> **Key term**

Gravitational potential energy: The energy possessed by an object by virtue of its position.

As with the kinetic energy, we'll imagine doing work on an object of mass m in such a way as to increase its GPE by raising it a distance Δh.

This is shown in Fig. 1.4.14. This will be carried out at constant speed, so there is no change in kinetic energy.

If the body is not accelerating, $\Sigma F = 0$ [Newton's 1st law]

$$\therefore F = mg$$

Because only potential energy is changing, no other kinds of energy are involved.

$\therefore$ By definition, ΔE_p = work done by $F = F\Delta h$

$$\therefore \Delta E_p = mg\Delta h$$

Warning. This equation for ΔE_p is only valid for small changes in height. That is, small compared with the distance from the centre of the Earth. Because the radius of the Earth is approximately 6400 km this restriction is not a problem for changes in height within the atmosphere, even up to Felix Baumgartner's free fall jump from 39 km!

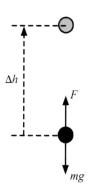

Fig. 1.4.14 Deriving the formula for GPE

>> **Study point**

Unlike kinetic energy, gravitational potential energy has no obvious place where $E_p = 0$.

(d) Elastic potential energy

Elastic objects (e.g. rubber bands, springs, rulers) which are deformed (stretched, compressed or bent) are able to do work on other objects when they return to their normal shape. Thus they possess energy because of their deformation. We call this **elastic potential energy**. The symbol, E_p, is used, as for gravitational potential energy.

When an object is stretched, squashed or bent, the extent of the deformation depends upon the applied force. Many objects obey Hooke's law, at least for small deformations. (See Section 1.5.1.)

>> **Key term**

Elastic potential energy: The energy stored in a body by virtue of its deformation.

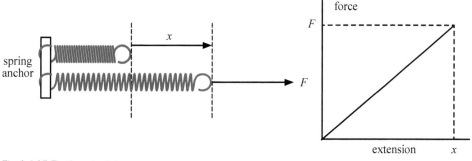

Fig. 1.4.15 Elastic potential energy

>> **Study point**

Particle physicists express energy in *electron volts* (eV). 1 eV is the energy transfer when an electron moves through a potential difference of 1 V:

1 eV $= 1.6 \times 10^{-19}$ J

>> **Study point**

A stationary object can only be stretched or squashed if **two** forces are applied in opposite directions.

The spring constant, k, is defined by $F = kx$, where F is applied force and x the deformation (e.g. stretch) as shown in Fig. 1.4.15. Calculating the work done by the varying force is analogous to calculating the displacement from a velocity–time graph: the work done is the area under the graph.

$\therefore$ The work done in stretching the spring, $\quad W = \frac{1}{2}Fx$.

$\therefore$ Substituting for F from $F = kx$, $\qquad W = \frac{1}{2}kx^2$,

If the spring is allowed to relax, it does work as it relaxes equal to the work done in

stretching it, so the elastic potential energy is given by $E_p = \frac{1}{2}kx^2$.

◀ **Stretch & challenge**

The work done, W, in stretching the spring is given by:

$W = \int F(x)\,\mathrm{d}x$. Given that $F = kx$, show that $W = \frac{1}{2}kx^2$.

NB Don't just ignore the constant of integration!

1.4.4 Knowledge check

Calculate the kinetic energy of: (a) a car of mass 1200 kg travelling at 30 m s^{-1}; and, (b) a bullet of mass 0.04 kg travelling at 500 m s^{-1}.

1.4.5 Knowledge check

A cyclist and cycle, of joint mass 85 kg, freewheel from rest down a sloping road of height 20 m and length (measured along the slope) of 200 m.

(a) Calculate the loss in gravitational potential energy.
(b) Calculate the speed the cyclist would attain in the absence of resistive forces.
(c) The cyclist attains a speed of 10 m s^{-1}. Calculate the mean resistive force acting.

Key term

Power: The work done per unit time or the energy transferred per unit time.

Its SI unit is the watt $(\text{W}) = \text{J s}^{-1}$.

1.4.6 Knowledge check

By considering a power of 1 kW and a time of 1 hour, find the number of joules in 1 kW h.

1.4.7 Knowledge check

Calculate the energy transferred by a 2.5 kW electric kettle in 3 minutes.

Give your answer in (a) J and (b) kW h.

Study point

The annual electrical output of the British national grid is sensibly expressed in TW h.

Example

A 200 g mass is hung on a spring of constant 15 N m^{-1}, as in Fig. 1.4.16, and dropped. Calculate:

(a) The distance, x, the mass falls before it comes to rest.

(b) The speed of the mass when it has fallen half this distance.

Answer

(a) As the mass falls, it loses gravitational potential energy. To start with, the system gains kinetic energy and elastic potential energy. As the spring tightens, the mass slows down until it momentarily comes to rest. Applying the principle of conservation of energy:

Fig. 1.4.16

> loss in GPE = kinetic energy + elastic potential energy

When the mass comes to rest momentarily, the kinetic energy is zero so, at this point, the conservation of energy equation becomes:

> elastic potential energy = loss in GPE

Then $\frac{1}{2}kx^2 = mgx$, $\therefore x = \dfrac{2mg}{k} = \dfrac{2 \times 0.2 \times 9.81}{15} = 0.262 \text{ m}$

(b) When $x = 0.131 \text{ m}$, Kinetic energy = Loss in gravitational PE − gain in elastic PE

$\therefore \frac{1}{2}mv^2 = mgx - \frac{1}{2}kx^2$

Putting in the values and rearranging leads to $v^2 = 1.28$, $\therefore v = 1.13 \text{ m s}^{-1}$

1.4.4 Energy and power

Power, P, is calculated using Power $= \dfrac{\text{energy transfer}}{\text{time}}$ or, in symbols, $P = \dfrac{E}{t}$.

It is not restricted to energy transfer in the context of mechanical work. For example, a 15 W light bulb transfers 15 J into electromagnetic radiation and internal energy every second. For many purposes, the watt is a rather small unit, epecially in the context of heating. A domestic electric kettle typically has a power of 2–3 kW, a small windfarm might have an installed power of ~10 MW and a typical thermal power station has a power of 1–2 GW. A similar approach could be taken with energy (using kJ, MJ, GJ, etc.) but often a different approach is taken:

Rearranging $P = \dfrac{E}{t}$, Energy transfer = Power × time

If we express power in kW and the time in hours, the unit of energy transfer is the kilowatt-hour (kW h). This non-SI unit is a much more convenient size for many purposes.

Example

A nuclear power reactor has an electrical power output of 1.2 GW. Estimate electrical energy output in 1 year.

Answer

No. of hours in 1 year = 365.25 days × 24 hours / day = **8766 hours**.

$\therefore$ Electrical energy output = 1.2 GW × 8766 h = 11 000 GW h = **11 TW h** (2 s.f.)

If the energy transfer arises from mechanical work, then we can re-write the power equation as:

$$\text{Power} = \frac{\text{Work done}}{\text{time}} \text{, i.e. } P = \frac{W}{t}$$

Consider a force F applied to an object moving at velocity v at an angle θ to F (see Fig. 1.4.17). In time Δt, the work done is given by: $W = Fv\Delta t \cos \theta$

$\therefore$ Dividing by Δt $\qquad P = \dfrac{W}{\Delta t} = Fv \cos \theta$

If F and v are in the same direction, i.e. $\theta = 0$, then $P = Fv$

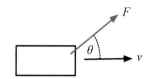

Fig. 1.4.17 Power and velocity

Knowledge check 1.4.8

Express the electrical energy output of the nuclear reactor in J.

1.4.5 Dissipative forces and energy

In Section 1.4.3 we derived the formulae for various forms of energy by considering the work done against external forces:

- The increase in gravitational potential energy is equal to the work done against the force of gravity.

- The increase in elastic potential energy is equal to the work done against the tension within the object when we stretch it.

The reason that we can do this is that these processes are *reversible*. If we release the systems, they will naturally return to their previous states – the energy can transfer in the opposite direction.

However, if a force does work by moving an object against a frictional force or aerodynamic drag, the energy that is transferred cannot be recovered in the same way. This is because the energy is then possessed as an increase in the disordered motion of the molecules of the system (the internal energy) and generally results in a rise in temperature. Reversing the process and turning some of the energy of random motion once more into ordered motion is possible but the efficiency is limited by the 2nd law of thermodynamics, which is beyond the scope of this course.

The familiar pattern in Fig. 1.4.18 is seen in water after an object, e.g. a pebble, is dropped in. A very small fraction of the energy is retained as ordered motion in the rising of the small water droplets and the spreading ripples. This energy is soon converted into random motion of water molecules by the action of viscous drag forces.

Looking back to the example in Section 1.4.1 and to Fig. 1.4.5, the work input into the system is 1100 J but the increase in GPE is only 980 J. This is the *useful* energy transfer. The remaining transfer, 120 J, represents a loss of useful energy. We define the **efficiency** of the system as the fraction of the energy input which is transferred usefully by the system. This is often expressed as a percentage, i.e.

$$\text{Efficiency} = \frac{\text{useful energy transfer}}{\text{total energy input}} \times 100\%$$

In the example this becomes: Efficiency $= \dfrac{980}{1100} \times 100\% = 89\%$

In calculations, efficiency is most conveniently expressed as a number between 0 and 1 (i.e. 0.89 for the winch), with the % figure being reserved for a final communication. In a chain of energy transfers, the total efficiency is the product of the efficiencies at each stage.

We can write the efficiency equation in terms of power instead of energy, i.e.

$$\text{Efficiency} = \frac{\text{useful energy transfer}}{\text{total energy input}} \times 100\%$$

>> **Study point**

The formula $P = VI$ is equivalent to $P = Fv$. The pd, V, drives the current, I, similarly to the way F drives v.

Knowledge check 1.4.9

The drag, F_d, on a car is given by

$F_d = 0.3\rho_{air}v^2$.

(a) Show that the unit of the 0.3 factor is m^2.

(b) Calculate the power developed by the engine at a constant speed of 25 m s^{-1}.

[$\rho_{air} = 1.3$ kg m^{-3}]

Fig. 1.4.18 Energy dissipation

>> **Study point**

What constitutes *useful* transfer depends on the context. The rise in temperature produced by rubbing your hands together could be desired output! The 'waste heat' from a car engine can be used to warm the passengers.

Example

A gas turbine power station, with an electrical power output of 1.0 GW has an efficiency of 60%. It is connected to consumers via a step up transformer (98% efficient), the national grid (97%) and local distribution network (95%).

Calculate the overall efficiency achieved.

Answer

Overall efficiency = $0.60 \times 0.98 \times 0.97 \times 0.95$

$$= 0.54 = 54\%$$

‹ Link ›

See also Section 1.3.5(b) and (c).

Test yourself 1.4

1. A car of mass 1600 kg, travelling at a speed of 25 m s^{-1}, brakes uniformly to rest in a distance of 100 m. Calculate: (a) the initial kinetic energy of the car; and, (b) the braking force.

2. A steel sphere of mass 4 **g** is pushed against a spring with a force of 5 N, compressing the spring by 20 cm. Calculate: (a) The elastic potential energy when the spring is compressed. (b) The speed the sphere attains when the spring is released.
State your assumptions.

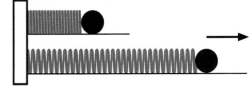

3. A small heavy sphere is suspended from the ceiling by a 1.00 m long thread. A horizontal pin, **P**, is placed 50 cm below the suspension point. The pendulum is pulled aside to the left by $30°$ and released. Calculate: (a) The speed of the sphere at the lowest point. (b) The greatest angle attained by the pendulum thread on the right.

Show your reasoning and your working clearly.

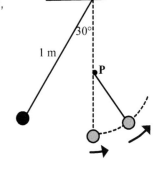

4. A student gives the following definition of work: *Work is force × distance.* Explain why this is not a very good definition and give a better one.

5. A rolling ball possesses kinetic energy in its rotation as well as its translation (forward movement). The rotational kinetic energy is 28.6% of the total. A ball rolls down a slope of height 1.00 m which becomes horizontal before ending in a step of height 1.00 m. Calculate the horizontal distance from the step to the point where the ball hits the ground.

[Note: You can take $g = 9.81$ m s^{-2} but, in fact, the answer is independent of the value of g! You should check this algebraically.]

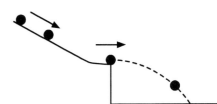

6. A 60 g arrow is nocked onto a long bow which is drawn a distance of 0.70 m with a maximum draw force of 650 N. The arrow is fired into the air at an angle of 30° to the horizontal:

   (a) Assuming the bow transfers 95% of its potential energy to the arrow and that the draw force is proportional to the draw distance, calculate the speed given to the arrow.
   (b) Neglecting aerodynamic and air resistance effects, calculate the maximum height attained by the arrow and its range (assume that the initial height = 0).

7. A space satellite has a kinetic energy of 3.50 MJ. Its motion is changed by the firing of one of its thrusters, which exerts a force of 50 N. During the 'burn', the satellite moves a distance of 12 km.

   (a) What additional information is needed to calculate the new kinetic energy of the satellite?
   Assume there is no change in the gravitational potential energy.
   (b) Calculate the maximum and minimum possible values of the new kinetic energy. Explain how these arise.

8. A ball of mass 2.0 kg is dropped from a cliff of height 40 m.

   (a) Taking the gravitational potential energy of the ball at the base of the cliff to be 0, calculate the GPE of the ball at heights of:
   (i) 40 m
   (ii) 25 m
   (iii) 10 m
   (iv) 0 m.

   (b) Calculate the KE of the ball at heights of:
   (i) 40 m
   (ii) 25 m
   (iii) 10 m
   (iv) 0 m.
   State your assumption.

   (c) Use one of your answers to part (b) to calculate the speed of the ball when it hits the ground.
   Repeat your calculation by a different method. Explain why this second method requires the same assumption as in part (b).

9. (a) State what is meant by the spring constant, k.

   (b) Show that the unit of k can be written as $kg\ s^{-2}$.

10. Object **A** is travelling due north. Its momentum is 25 N s and its kinetic energy is 40 J. Object **B** is travelling due south. Its momentum is 10 N s and its kinetic energy is 50 J.

    (a) Starting from the equations

    $$E_k = \tfrac{1}{2}mv^2 \text{ and } p = mv, \text{ show that } E_k = \frac{p^2}{2m}.$$

    (b) Calculate the masses and velocities of objects **A** and **B**.

11. A ball of mass 0.20 kg is dropped to the ground from a height of 30 m. Its initial velocity is zero. For this question, approximate the gravitational field strength to be 10 N kg^{-1} and take the gravitational potential to be zero at ground level. Consider only the downward motion.

    Neglecting the effects of air resistance, sketch graphs, on the same axes, of:
    • gravitational potential energy, E_p, against distance fallen
    • kinetic energy, E_k, against distance fallen
    • total energy, $E_p + E_k$, against distance fallen.

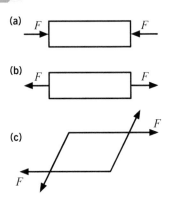

(a) F ← ☐ → F

(b) ← F ☐ → F

(c) F

Fig. 1.5.1 Different types of deforming forces

< Link >

You should refer to Section 1.4.3(d) for further treatment of these concepts.

If equal and opposite forces are applied to the opposite ends of an object, its particles (molecules / atoms / ions) will be forced into new equilibrium positions with respect to one another. The forces can be: (a) **compressive**, (b) **tensile** or (c) **shear** (see Fig. 1.5.1). The objects are said to be under **stress**. The forces shown are those applied externally to the object; by Newton's 3rd law the object exerts equal and opposite forces on the external objects.

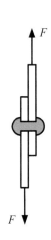

Fig. 1.5.2 Solids under stress

The material of Nelson's column is clearly under compression from the weight of the material above, the bungee cord is subject to tensile forces and the rivet is under shear. The response of different materials to stress is very variable: gases cannot be put under tension because the molecules are not bound together; liquids and gases cannot withstand a shear because they have no rigidity to their shape; many engineering materials, e.g. masonry, fracture easily under tension whereas steel is very strong. This section of the book deals with how materials behave under (mainly) tensile forces.

1.5.1 Hooke's law

Fig. 1.5.1(b) shows tensile forces being applied to an object, which, in physics examples, is often a wire or a spring. Newton's 3rd law applies (it always does!) so the object must exert equal and opposite forces on whatever things are applying the labelled forces. If we imagine cutting the object across, we would need to apply the same size force to hold the two halves together. So this force, F, is transmitted through the object: it is called the **tension**. If an object has a tension, it stretches. The increase in length is called the **extension** and is given the symbol $\Delta \ell$. In many circumstances, **Hooke's law** applies.

Note that tension is normally plotted on the vertical axis and extension on the horizontal. Plotted as shown in Fig. 1.5.3

- the gradient gives the stiffness of the object, which is called the **spring constant**, k, in the case of a spring

- the area underneath the graph is the work done in stretching.

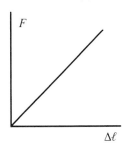

Fig. 1.5.3 Hookean behaviour

1.5.2 Mechanical properties of materials

(a) Stress, strain and the Young modulus

The tension, F, extension, $\Delta\ell$ and spring constant, k, relate to an object: a particular spring, piece of wire, block of concrete, etc. It is more useful for engineers to have quantities which relate to materials and which can be used to predict the properties of many different objects.

The bar in Fig. 1.5.4 of length ℓ and cross-sectional area (c.s.a.) A, is stretched by $\Delta\ell$, which requires a force F. If we imagine two such bars side by side, making a total c.s.a. of $2A$, then each will require F to stretch it by $\Delta\ell$, so the total tension is $2F$.

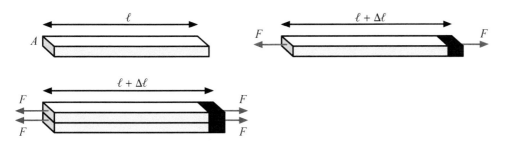

Fig. 1.5.4 Tension ∝ c.s.a

Conclusion: two bars of the same composition, with the same length, will be stretched by the same amount if the ratio $\dfrac{F}{A}$ is the same. This quantity is called the **(tensile) stress**, σ.

We now imagine the two bars being welded end to end, making a total length 2ℓ, and the same stretching force applied. The tension has the same value, F, in each half of the composite bar, so each half extends by $\Delta\ell$ making the total extension $2\Delta\ell$, as in Fig. 1.5.5.

Fig. 1.5.5 Extension ∝ original length

Conclusion: for two bars of the same composition, with the same c.s.a. and tension, the ratio $\dfrac{\Delta\ell}{\ell}$ is the same. This quantity is called the **(tensile) strain**, ε.

If the bar obeys Hooke's law, then $F \propto \Delta\ell$, so from the definitions of σ and ε, it must be true that $\sigma \propto \varepsilon$ and we define the Young modulus, E, *of the material* by:

$E = \dfrac{\sigma}{\varepsilon}$, which we can rewrite as $E = \dfrac{F\ell}{A\Delta\ell}$, from the definitions of σ and ε.

Typical values of E, σ and ε

The newton is, in engineering terms, a rather small unit of force. The m^2 on the other hand is quite a large unit of area. Engineering materials, such as steel and concrete, are very resistant to deformation. Taking these points together it is not surprising that values of stress are very large (in the 100 MPa range) and strains are typically very small: a strain of 0.001 or less for a *Hookean material* (i.e. one that obeys Hooke's law). Hence, the **Young moduli** tend to have values in the 100 GPa range.

Material	$E/$ GPa
Mild steel	210
Copper	117
Aluminium	69
Human long bone	14
Concrete	14–30
Oak (along grain)	11
Glass	50–90
Rubber (small strain)	~0.1
Diamond	1220

Table 1.5.1 Miscellaneous values of the Young modulus

 Knowledge check

A 1 cm² c.s.a. bar of length 2.0 m is subject to a tension of 1 kN. The Young modulus is 100 GPa.

Calculate the extension and give your answer in μm.

 **Key terms**

Elastic strain: The strain that disappears when the stress is removed, that is the specimen returns to its original size and shape.

Plastic strain: The strain that decreases only slightly when the stress is removed.

Elastic limit: The point at which deformation ceases to be elastic.

Study point

We can refer to the **elastic limit** of an object or a material:

For an object (e.g. a spring) the elastic limit is a **load**; for a material (e.g. copper) it is a **stress**.

Study point

Sometimes the symbol x is used for extension and sometimes $\Delta\ell$, so the formula for the work done in stretching can be written:

$W = \frac{1}{2}Fx$ or $W = \frac{1}{2}F\Delta\ell$

Because $F = kx$ [or $k\Delta\ell$] we can also write:

$W = \frac{1}{2}kx^2$ or $W = \frac{1}{2}k(\Delta\ell)^2$ and

$W = \frac{1}{2}\frac{F^2}{k}$ or $W = \frac{1}{2}\frac{F^2}{\Delta\ell}$

Table 1.5.1 gives E for some common materials.

Working with E, σ and ε requires care with the use of SI multipliers and standard form. Study the following example.

Example

Calculate the extension of a 100 m long steel wire of diameter 1.0 mm if it is placed under tension by suspending a 10.0 kg mass from it. [$E_{steel} = 210$ GPa.]

Answer

Tension $= mg = 10.0 \times 9.81 = 98.1$ N; c.s.a. $= \pi(0.5 \times 10^{-3})^2 = 7.85 \times 10^{-7}$ m².

$$\therefore \text{ stress, } \sigma = \frac{F}{A} = \frac{98.1 \text{ N}}{7.85 \times 10^{-7} \text{ m}^2} = 1.249 \times 10^8 \text{ Pa } [125 \text{ MPa}]$$

$$E = \frac{\sigma}{\varepsilon} \therefore \varepsilon = \frac{\sigma}{E} = \frac{1.250 \times 10^8 \text{ Pa}}{210 \times 10^9 \text{ Pa}} = 0.000595$$

$$\therefore \Delta\ell = \varepsilon\,\ell = 0.000595 \times 100 \text{ m} = 0.0595 \text{ m} = 6.0 \text{ cm } (2 \text{ s.f.})$$

Alternatively, we could just start from $E = \dfrac{F\ell}{A\Delta\ell}$ and substitute for F and A.

The approach to take is a matter of choice.

(b) Elastic and plastic strain

For small values of strain, materials return to their original size and shape if the stress is removed. This behaviour is called elastic, and the strain is the **elastic strain**.

Many materials exhibit plastic (or inelastic) strain when they are stretched beyond a certain point called the **elastic limit**. When the stress is removed, they contract slightly but not to their original size – they exhibit permanent deformation, and the strain is called **plastic strain**.

1.5.3 Work of deformation and strain energy

As we showed in Section 1.4.3(d), the work done in stretching a Hookean object is given by $\frac{1}{2}F\Delta\ell$, this being the area under the force–extension graph. Refer to the Study point for alternative ways of writing the formulae for W. Releasing the tension and allowing the object to contract allows it to do work in its turn and, because the graph for relaxation is the same as for tensioning (for Hookean materials), the work done by the object in relaxing is the same as the work done on the object during deformation. Hence the same expressions – $\frac{1}{2}F\Delta\ell$, etc. – give the energy stored in a body by virtue of its deformation. This energy is variously referred to as **strain energy** or elastic **potential energy**.

If the extension of a rubber band is measured when it is loaded (e.g. by hanging 100 g masses) and then unloaded, a load–extension curve similar to that in Fig. 1.5.6 is obtained. The unloading curve is below the loading curve. This

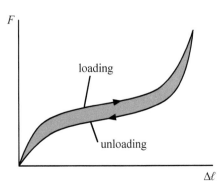

Fig. 1.5.6 Hysteresis in rubber

phenomenon is called **hysteresis** and is responsible for rolling resistance in car tyres. The work done on the rubber in extension is the area under the loading curve; the work done by the rubber in contracting is the area under the unloading curve.

The area between the curves represents the mechanical energy loss in the cycle: it is transferred to internal energy in the rubber and then lost as heat. The elasticity of rubber is dealt with in Section 1.5.6.

1.5.4 Stress and strain in ductile metals

The mechanical properties of solids depend on their structure, that is the way their particles are arranged and the nature of the forces between them. The next three subsections discuss this for different classes of solids.

(a) Structure

Many metals, such as steel, aluminium and copper, are **ductile**, which means that they can be drawn out into wires. Ductile materials are also **malleable**, especially when hot. These metals are crystalline: they have a periodic structure called a lattice. The lattice particles in metals are positive ions – atoms that have lost one or more electrons. They are held together by a 'sea' of negatively charged *delocalised* electrons which are free to move between the ions.

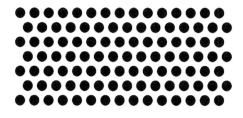
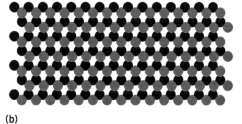

(a)   (b)

Fig. 1.5.7 Hexagonal packing in metals

Because metal ions are spheres, they usually pack with the least potential energy into planes with the hexagonal arrangement shown in Fig. 1.5.7(a), with ions in the plane above nestling in the gaps, the red spheres in (b).

Gas-turbine blades have been developed consisting of single crystals of a 'superalloy' of nickel. Most metal samples, however, are **polycrystalline**. When they solidify from the molten state, crystallisation occurs at many points independently.

This results in a large number of very small interlocking crystals (grains) as shown in Fig. 1.5.8, which represents a ~100 μm square region of a polished and etched surface of a sample of titanium-aluminium alloy. A theoretical arrangement of the metal ions is shown in Fig. 1.5.9. The orientation of the crystal planes is random from one grain to the next. Note that the lattice ions are not shown to the same scale and, in reality, a typical grain will have ~10^5 lattice planes. The grain boundaries have a large component of impurity atoms which are forced out of the grains during crystallisation.

Another important feature of the structure of ductile metals is the presence of irregularities within the lattice. An **edge dislocation** is where an additional $\frac{1}{2}$-plane of ions is present and a **point defect** is where a lattice ion is missing or a 'foreign' atom or just an additional ion is present.

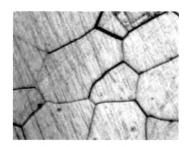
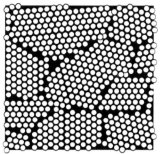

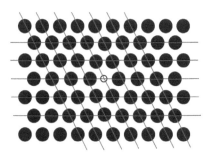

Fig. 1.5.10 Edge dislocation   Fig. 1.5.11 Point defect in a metal lattice

It is the combination of regular lattice, grain boundaries and dislocations that is responsible for the mechanical properties of polycrystalline metals.

(b) Stress–strain graphs

Samples of ductile metals typically have stress–strain curves similar to that in Fig. 1.5.12, in which we can identify the following features:

- A linear portion **OP**. **P** is called the **limit of proportionality**. The gradient of this section is the **Young modulus** of the material.

- Point **E** is the **elastic limit**. Only strains up to **E** are **elastic**; beyond **E** they are **plastic**.

- The **yield point**, **Y**, at which the material shows a large increase in strain for little or no increase in stress. The stress here is called the **yield stress**, σ_Y.

- An extensive **plastic** region, **YX**. The maximum stress is called the **breaking stress** or **ultimate tensile strength**, σ_X. X on the graph marks the breaking point.

- The largest strain region of the σ–ε curve typically bends downwards. In this region, the sample exhibits **necking**, which is a narrowing of the region where it will eventually break. (The 'true' stress goes on rising – see the Study point.)

Fig. 1.5.12 σ–ε curve for a ductile metal

The precise shape of the σ–ε curve varies with material and also the history of the material (e.g. heat or working treatment).

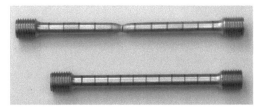

Fig. 1.5.13 Ductile metal specimen before and after destructive testing

The 'before and after' image in Fig. 1.5.13 shows the effect of necking and plastic deformation in a specimen of mild steel. A clip of this tensile test is available on YouTube.

(c) Structure and properties

When a material is put under a low tension, i.e. so that $\sigma < \sigma_E$, the separation between the lattice particles (ions) is increased. This is elastic deformation because the forces between the particles pull them back to their initial position when the tension is removed. This is illustrated in Fig. 1.5.14 – a cubic lattice arrangement has been used as the effect is easier to see – and it is easier to draw!

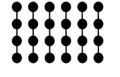

Fig. 1.5.14 Elastic strain

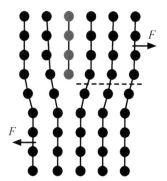
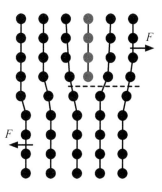
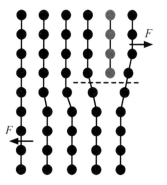
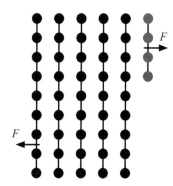

Fig. 1.5.15 Dislocation movement produces plastic strain

Plastic deformation is caused by an irreversible rearrangement of particles. This is made possible by the presence of edge dislocations. The dislocation shown in Fig. 1.5.15 moves to the right under the influence of the applied forces. The individual ions only move slightly; the ions above the dotted line drop into a lower potential energy position in the next plane, so that the extra $\frac{1}{2}$-plane moves to the right until it reaches the grain boundary; the crystal becomes more elongated. The stress at which this happens is the **yield stress**. The dislocation does not move back when the stress is removed, so this elongation is plastic.

The details of what happens next depend on several factors:

1. Edge locations can get entangled (see Knowledge check 1.5.5) limiting their movement.
2. The size of the grains – the smaller the grains, the less the freedom of movement of the dislocations.
3. The presence of point dislocations: foreign atoms can inhibit the movement of edge dislocations; a void in the lattice spawns more edge dislocations.

For different metals, especially alloys such as steel, changing the composition can affect all these factors. Heating and quenching regimes, depending on the metal, can make the metal more or less ductile and cold working generally makes a metal stiffer and less ductile because it causes dislocation entanglement.

1.5.5 Stress and strain in brittle materials

Brittle materials, such as cast iron, ceramics and masonry are totally elastic and generally Hookean, up to the breaking stress (Fig. 1.5.16). Many brittle non-metals have an amorphous (non-crystalline) structure. Figs 1.5.17 and 1.5.18 show the distinction between crystalline (quartz) and amorphous (glass) structures for silicon dioxide, SiO_2, which is covalently bonded.

Glasses form when the SiO_2 cools down too rapidly from the molten state for the molecules to arrange into the crystalline form.

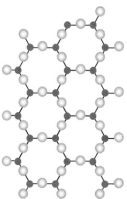
Fig. 1.5.17 Crystalline silicon dioxide

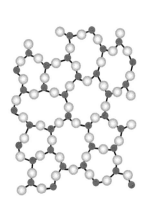
Fig. 1.5.18 Vitreous (glass) silicon dioxide

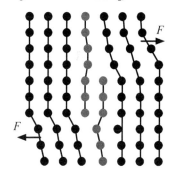

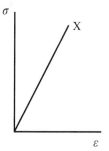

Fig. 1.5.16 σ–ε graph for a brittle material

Fig. 1.5.19 Snowflake obsidian – a natural glass.

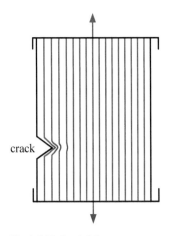

Fig. 1.5.20 Crack failure

Amorphous materials are brittle because the absence of a crystal structure means that there can be no dislocations to move and produce plastic deformation. Cast iron is crystalline, but the crystals are very small and the presence of a large fraction of impurity atoms means that dislocations are pinned down.

Brittle fracture

Brittle materials are weak under tension, i.e. their breaking stress is low. Unlike in ductile fracture, the broken pieces can be fitted back together – a broken cup can be glued. This is because of the absence of plastic deformation.

The failure under tension is due to crack propagation. This is shown in Fig. 1.5.20. The so-called stress lines are shown in red. These show how the tension links the atoms in the material. The interatomic forces cannot cross the gap because the atoms are too far apart, so the forces must be transmitted around the tip of the crack, leading to a greatly magnified stress. The brittle material starts to break at the crack tip: the crack extends which increases the stress at the tip and so the crack propagates (at the speed of sound in the material) resulting in catastrophic failure.

Brittle materials can be used in load-bearing structures if the brittle member is:

* always under compression by the design of the structure, as in the Pontcysyllte brickwork pillars which hold up the Llangollen canal (Fig. 1.5.21),
* or compressively pre-stressed in manufacture.

The bottom of the concrete beam in Fig. 1.5.22 is under tension and is in danger of failure by crack propagation. The diagrams in Fig. 1.5.23 show how a steel pre-tensioning tendon, T, is inserted into a concrete beam:

Fig. 1.5.21 Pontcysyllte aqueduct

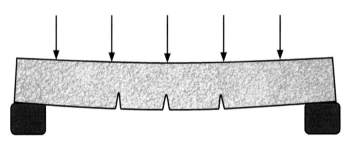

Fig. 1.5.22 Crack failure in a beam

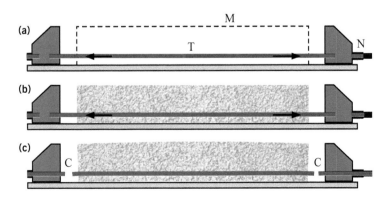

Fig. 1.5.23 Pre-stressing

(a) The tendon, **T**, is placed under tension by tightening the nut, **N**.
(b) The concrete to be cast is poured into the mould, **M**, and allowed to set.
(c) The tendon is then cut and contracts, putting the lower surface of the concrete under compression (and flexing it slightly as shown).

If the beam is subsequently loaded at the top as in Fig. 1.5.22, the lower surface of the beam will not be under tension unless a very large load is applied.

1.5.6 Polymers

(a) What are polymers?

Rubber, polythene, melamine and nylon are **polymers**, i.e their molecules are long chains of repeat units.

The simplest in terms of composition is polythene, which is made from the **monomer** ethene.

Fig. 1.5.24 Ethene, and polythene repeat unit

A significant feature of the C–C bond is that it can rotate, so the polymer molecule can assume a huge number of random shapes, which is significant in the stress–strain properties of rubber.

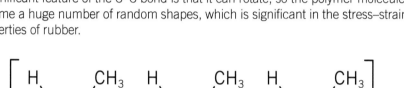

Fig. 1.5.26 Rubber: (a) the polymer; (b) the monomer

(b) Stress, strain and Young modulus of rubber

The load–extension / stress–strain characteristics are as follows:

1. The stress–strain graph is non-linear: rubber is stiff for low extensions, becomes less stiff and finally much stiffer. See Fig. 1.5.27.

2. Loading and unloading curves are different: called *elastic hysteresis*.

3. It exhibits large strains: strains of up to 5, i.e. the final length is ~6× the original length, are possible in some types of rubber.

4. The stress values are much lower than for most engineering materials. The breaking stress is ~16 MPa, compared to ~80 MPa for glass and 400 MPa for mild steel.

Because of the non-linearity we need to be careful when referring to the Young modulus of rubber. Depending on the context it could refer to:

- The gradient of the tangent of the σ–ε curve at the origin.

- The value of $\dfrac{\sigma}{\varepsilon}$ for a particular stress, e.g. halfway along the almost linear region.

These are the gradients of the red lines in Fig. 1.5.27. Then there is the problem of hysteresis… Whichever we use, the value of E is something like 10–20 MPa; much smaller than the 200 GPa for steel. In practice, the Young modulus is only a useful concept in the low strain, linear, region of the σ–ε graph.

Key terms

Polymer: A substance whose molecules consist of long chains of identical sections called repeat units.

Monomer: A molecule with a double bond which is broken open to form the repeat unit of a polymer.

Study point

An artist's impression of an ethene molecule showing the electron orbitals.

Fig. 1.5.25

Knowledge check 1.5.6

When rubber stretches, its volume stays roughly constant in spite of its huge extensions. Estimate the change in thickness of a rubber band which is stretched to 4 times its original length.

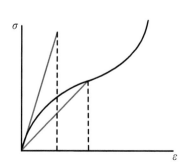

Fig. 1.5.27 Which Young modulus for rubber?

A rubber sample has the same dimensions as a steel wire. The same tension is applied to each. Estimate the ratio of the strains of the two objects.

Fig. 1.5.28 Tangled rubber molecule.

 Stretch & challenge

Statistical random walk theory predicts that, if a route is made of n steps of length $\Delta\ell$, in random directions, the total displacement $\sim\Delta\ell\sqrt{n}$. A rubber molecule consists of 10^6 repeat units of length $\sim10^{-9}$ m. Estimate the extent of a rubber molecule.

 Study point

Materials scientists often introduce additional S–S cross linkages into rubber in order to make it stiffer. The process is called *vulcanisation*.

Study point

The Young modulus is given by the equation $E = \dfrac{F\ell_0}{A\Delta\ell}$. $\Delta\ell$ is made large enough to be measurable with a reasonable precision by making ℓ_0 as long as possible, e.g. the length of a lab bench (perhaps 4 m).

Study point

The wire is clamped using the wooden blocks. This prevents damage to the wire at the point of clamping.

Practical check ⟩

See Section 3.2 for a suggestion on measuring the diameter of a wire.

(c) Structure and properties of rubber

At first sight, the structure of a rubber molecule, Fig. 1.5.26, seems as if it should be linear. However, because the single C–C bonds can rotate freely, a rubber polymer forms in a randomly tangled state – see Fig.1.5.28 (only the C–C bonds are shown).

If a rubber molecule is placed under tension, it responds by straightening out:

Fig. 1.5.29 Straightening a rubber molecule

The rubber (band) can stretch to several times its original length. The force required is much less than when stretching crystalline or amorphous materials because bonds are not being stretched – just rotated. The presence of **cross-linkages** between molecules or just the entanglement of different molecules limits the total extension possible. The thermal motions of the atoms in the molecules, which tend to randomise the shape, provide the opposition to the extension, i.e. the stiffness.

When the tension is removed, random molecular motions of the atoms within the molecules re-randomise the shape of the molecules, leading to the rubber contracting. Some energy is converted by intermolecular collisions into random kinetic energy of the molecules, so not as much work is done in contracting, leading to the hysteresis effect.

1.5.7 Specified practical work

(a) Determination of the Young modulus of the material of a metal wire

Fig. 1.5.30 shows a common set-up used in A level Physics classes for the determination of the Young modulus of metals, such as copper and mild steel, in the form of wires.

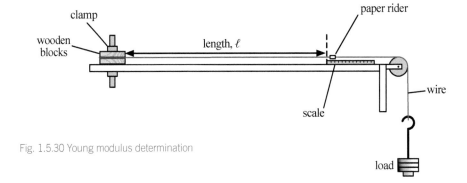

Fig. 1.5.30 Young modulus determination

Procedure:

1. Measure the diameter, d, of the wire using a micrometer or vernier calliper of resolution 0.01 mm. Calculate the c.s.a., A, from $A = \pi\left(\dfrac{d}{2}\right)^2$.

2. Attach a small load to provide original tension to straighten out the wire.

3. Measure the original length, ℓ_0, of the wire, using a metre rule with resolution 1 mm, from the blocks to the paper rider.

4. Add a known mass, m, such as a 10 g / 50 g / 100 g mass, to the hanger.

5. Measure the extension, $\Delta\ell$, from zero load.

6. Repeat 4 and 5 to obtain a series of values of $\Delta\ell$ and m.

7. Repeat 4 and 5 with decreasing values of m.

8. Plot a graph of tension, T, calculated from $T = mg$, against $\Delta\ell$.

9. Measure the maximum and minimum gradient of the straight line graph.

10. Calculate E and its uncertainty using $E = \dfrac{\ell_0}{A} \times \text{gradient}$.

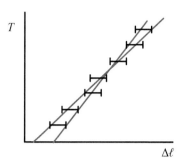

Fig. 1.5.31 Typical T–$\Delta\ell$ graph

Notes:

1. The best fit graph might not go through the origin. This is because the wire often has some slight kinks which need straightening.
2. The accuracy of the $\Delta\ell$ measurements can be improved by using a hand lens or (better) a travelling microscope.

(b) Investigation of the force–extension relationship for rubber

This is simply a case of loading and unloading a rubber band and measuring the extensions from the original length (actually the length with a very small load so that the band is initially straight). The band should be loaded until its additional extension with additional loads is very small. In this experiment, to reveal the hysteresis, the loading and unloading extensions are plotted separately rather than averaged.

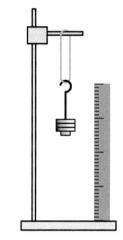

Fig. 1.5.32 Investigating load and extension for rubber

Test yourself 1.5

1. A spring extends by 8.0 cm when a load of mass 200 g is hung from it.
Calculate: (a) the spring constant, k; (b) the elastic potential energy in the spring with the load; (c) the reduction in the gravitational potential energy of the load when it extends the spring by 8.0 cm.

2. The load in Q1 is hung from another spring which has a spring constant half that in Q1. Without repeating the calculations, write down the extension of the spring and the answers to (a), (b) and (c).

3. The load is placed on the unstretched spring in Q1 and dropped. Calculate: (a) the speed of the load when the extension is 8.0 cm; (b) the maximum extension of the spring; (c) the upward acceleration of the load at the point of maximum extension.

4. The constant, k, of a spring is 24 N m^{-1}. Giving your reasoning clearly, determine the constant of two such springs joined: (a) end-to-end; and, (b) side-to-side [sometimes called 'in series' and 'in parallel' respectively].

5. A cylindrical rod of length 50 cm diameter 5 mm is made out of glass with an ultimate tensile strength of 33 MPa and Young modulus 60 GPa. Calculate the maximum tension the rod can take and its increase in length at a tension of 50% of this maximum.

6. A hawser (a rope) is made from nylon with a Young modulus of 3.0 GPa; it is designed to be used safely at stresses of up to 30 MPa. The diameter of the hawser is 5.0 cm and its length is 1 km (Assume Hooke's law is obeyed up to 30 MPa): (a) Calculate the maximum load to which the hawser should be subjected; (b) Calculate the energy stored in the hawser with this load; (c) The hawser is attached to a boat of mass 20 tonnes [1 t = 10^3 kg] which is drifting away at 4 m s^{-1}, to bring it to a halt. Calculate how far the hawser stretches before stopping the boat; (d) Determine the shortest length of hawser which would stop the boat without exceeding its maximum safe stress.

7. Pre-stressed glass, also known as *sight glass*, is manufactured in sheets in such a way that its outside layers are under compression and the middle is under tension. Explain why this glass is more resistant to breaking than ordinary glass.

8. Two cylinders of identical length and diameter are joined firmly end to end. The Young modulus of the material of one cylinder (**A**) is twice that of the other (**B**). The composite cylinder is put under tension. Compare the values of the following quantities for the two sections: the tension; the extension: the stress; the strain; the energy stored.

9. Starting from the definitions of stress (σ), strain (ε), and Young modulus (E), (a) Explain why E and σ have the same unit and state this unit; (b) Express the unit of stress in terms of the base SI units only.

10. The diagram shows a metal rod, composed of two different metals, under tension. The cross-sectional area at **X** is double that at **Y**. The Young modulus of the metal at **X** is $1.5\times$ that of the metal at **Y**. The length of section **Y** is $1.5\times$ that of section **X**.

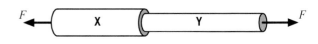

Compare the values of the following quantities for **X** and **Y**: tension; stress; strain; elastic energy stored per unit volume; total energy stored.

11. Some students were asked to measure the Young modulus of the material of a bronze wire as accurately as possible. They were told that the elastic limit of the bronze was more than 100 MPa.

(a) They measured the diameter of the wire in several places using a digital micrometer, with resolution ± 1 μm, each measurement being repeated at right angles. Their readings, in mm, were as follows:
0.273, 0.275, 0.275, 0.273, 0.285, 0.275, 0.273, 0.277

   (i) The students decided to ignore the 0.285 mm reading. Suggest why they did this.
   (ii) Why did they repeat the diameter readings at right angles?
   (iii) They calculated the cross-sectional area of the wire as 0.0591 ± 0.009 mm². Justify this result.

(b) They set up the wire as in Fig. 1.5.30 and measured the length, ℓ, using a metre rule, with a resolution of 1 mm, as 4.365 m. Keeping the total load below 5 N, they placed an additional mass of 0.250 kg ($\pm 1\%$) on the wire. Using a travelling microscope, they measured the increase in length to be 1.73 ± 0.02 mm.

   (i) Explain why they kept the total load below 5 N.
   (ii) Calculate the Young modulus of the bronze together with its absolute uncertainty.

(c) One of the students said that using a shorter wire wouldn't affect the uncertainty because the percentage uncertainty in the length was insignificant compared to the other uncertainties. Evaluate this statement.

(d) Another student suggested that using a thinner wire would give a lower uncertainty in the value of the Young modulus because the extension would be greater. Discuss this suggestion.

12. A student, Nia, measures the Young modulus of the material of a wire. Nia measures the extension of the wire when a load is applied, making sure she keeps within the elastic limit.

(a) State what is meant by 'the elastic limit' and explain how Nia would know whether the wire had exceeded it.
(b) State what other measurements Nia would need to make and how she would use them to calculate a value for the Young modulus.
(c) Alex, another student, says that Nia should have measured the extensions for a series of different loads and plotted a graph. Explain why Alex's idea is a good one and state how Alex could use the graph to obtain a value for the Young modulus.

13. The diagram is a load–extension graph for a rubber band which is taken around one cycle of loading and unloading – as shown by the arrowheads.

(a) Explain in molecular terms why the gradient of the extension graph at **B** is less than at **C**.
(b) What is the name given to the effect by which the contraction curve is below the extension curve?
(c) Explain why the temperature of a rubber band, which is taken repeatedly around the loading–unloading cycle, rises.

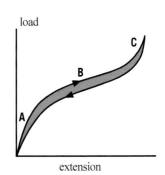

1.6 Using radiation to investigate stars

Almost all the information we have about the universe comes from electromagnetic radiation. The image of the constellation of Taurus in Fig. 1.6.1 was taken using visible radiation, i.e. with wavelengths between ~400 and 700 nm. Until recently this was the only wavelength range available to us because the Earth's atmosphere is opaque to most of the electromagnetic spectrum. However, space telescopes such as Spitzer and Chandra, have allowed us to 'see' the universe across the complete range from radio waves to gamma rays. Multi-wavelength astronomy gives us a much more complete understanding of the processes in the universe.

The images of the Sun in Fig. 1.6.2 were taken in (a) visible light (450 nm) and (b) ultraviolet (17.1 nm) from the Solar Dynamics Observatory on 10 October 2014.

Fig. 1.6.1 Taurus

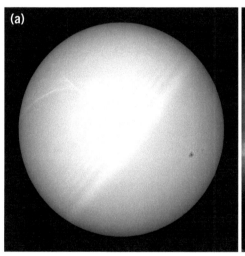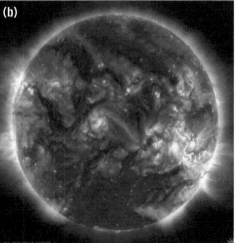

Fig. 1.6.2 The Sun in (a) visible and (b) UV light

Section 1.6.5 considers multi-wavelength astronomy in more detail.

Scientists analyse the light from stars by separating out the different wavelengths, e.g. by passing it through a prism or a **diffraction grating** (see Section 2.5.4), and then either making an image or plotting the energy density at different wavelengths. Examples of both of these spectra for our nearest star, the Sun, are shown on the next page in Fig. 1.6.3.

The graphical spectrum of the Sun, published by the World Meteorological Organisation (WMO), is the continuous graph in the upper part of the figure. The wavelength range is ~ 200–1500 nm. We'll ignore the dashed curve for the moment. The colour image is the appearance of the visible solar spectrum. This is what you can actually see (or image) if sunlight is dispersed though a prism or passed through a diffraction grating. The wavelength range is approximately 400–700 nm and its relationship with the spectrum graph is also shown.

The spectrum of the Sun consists of two parts:

- A **continuous spectrum** (the bright band).
- A **line spectrum** (the dark lines, also visible on the graph).

Study point

The sites of the sunspots which are visible on the 450 nm image (a) are also noticeably sites where violent processes are seen on the 17.1 nm image (b).

Link

Diffraction grating – see Section 2.5.4.

Key terms

Continuous spectrum: Consists of all wavelengths within a range.

Line spectrum: Consists of a series of individual wavelengths (or, more accurately, a series of very narrow wavelength bands).

Study point

The dark lines in the solar spectrum are called *Fraunhofer* lines after the German scientist who noticed them in 1814. The English chemist, William Hyde Wollaston, had discovered them in 1802.

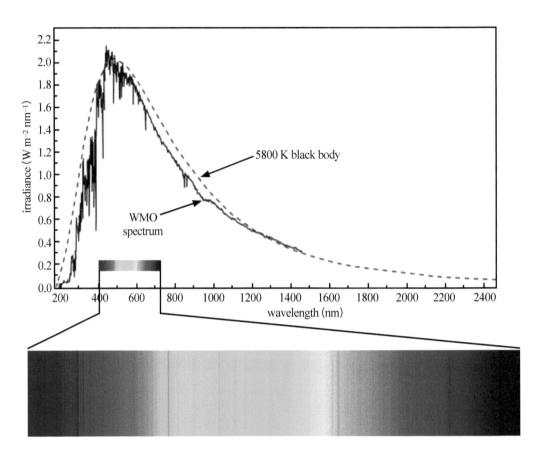

Fig. 1.6.3 The solar spectrum

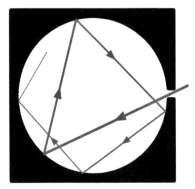

Fig. 1.6.4 Thermal radiation

Key terms

Black body: A body (or surface) which absorbs all the electromagnetic radiation that falls upon it. It also emits more radiation at any wavelength than a non-black body in the continuous spectrum.

Fig. 1.6.5 A cavity absorbs almost all incident radiation

These spectra contain a huge quantity of information about the star, in particular the temperature of its outer layer and its chemical composition. This information is contained in both the continuous and line spectra. Using the shape of the continuous spectrum and the positions of the dark lines (the *Fraunhofer* lines), astronomers can compare stars and also obtain evidence about their motion and even the age of the universe. This section will explore how such information is obtained.

1.6.1 Black body radiation

When a farrier heats up a horseshoe in a forge (Fig. 1.6.4), it starts to give out visible radiation. Initially it glows faintly dull red; as the temperature increases, the brightness increases and the colour changes from red to orange to yellow. The precise details of the spectrum of the emitted radiation vary from material to material: in general, the more a material absorbs radiation, the more it will emit. Scientists use the idea of a perfect **black body** as a theoretical standard against which to compare other bodies. A black body is a body (or surface) which absorbs all the electromagnetic radiation that falls upon it.

As an excellent approximation to perfect black body radiation, scientists have made measurements of the radiation coming from a small hole in the side of a furnace. This radiation is also known as **cavity radiation**. Why does a cavity behanve as a black body? Fig. 1.6.5 shows radiation entering such a cavity. It undergoes multiple reflections. If the material lining the cavity is very dark, it will absorb most of the radiation at each reflection so hardly any of the incident radiation will re-emerge from the aperture.

Because the furnace is hot, it also emits radiation, some of which escapes through the hole. The spectra of this radiation at various temperatures are in Fig. 1.6.6. These results fit in with the observations of the glowing objects such as the horseshoe in Fig. 1.6.4:

- Below about 1000°C no visible radiation is seen.

- At 1400°C a small amount of red light is emitted.

- At 1800°C much more visible radiation is emitted, mainly at the long wavelength (red) end but with some shorter wavelengths.

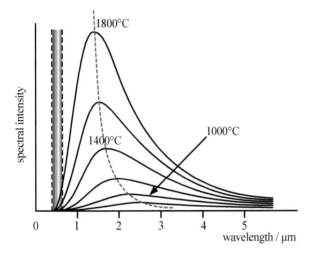

Fig. 1.6.6 Black body spectra

Studies of these spectra in the 19th century produced two empirical laws which were later explained theoretically by the German physicist, Max Planck.

The Wien displacement law

The wavelength λ_{max}, of peak emission from a black body is inversely proportional to the **absolute temperature**, T, of the body:

$$\lambda_{max} = \frac{W}{T}, \text{ where } W \text{ is the Wien constant, } 2.90 \times 10^{-3} \text{ m K.}$$

Stefan's law (or the Stefan–Boltzmann law)

The total power, P, of the electromagnetic radiant energy emitted by a black body of surface area A and temperature T is given by:

$P = A\sigma T^4$, where σ is the Stefan constant = 5.67×10^{-8} W m^{-2} K^{-4}.

Note that the temperature scale used in these laws is the **absolute** (or kelvin) **temperature** scale.

The temperature, T (in kelvin) is related to the temperature, θ (in °C) by $T / \text{K} = \theta/°\text{C} + 273.15$.

Note that, in fact, this equation is a definition of the celsius rather than the **kelvin temperature scale**.

>> Key terms

Absolute temperature, T, is expressed in kelvin (K).

Celsius temperature, θ, is defined by: $\theta /°\text{C} = T / \text{K} - 273.15$

On the **kelvin scale**, ice melts at 273.15 K, water boils at 373.15 K and absolute zero is 0 K.

>> Study point

Notice that the unit of the Wien constant is m K [metre kelvin] not mK [millikelvin]. The space is important!

>> **Study point**

Not all **radiation** is black body, e.g. the pulses emitted by pulsars arise from electrons spiralling around magnetic field lines (so-called *synchrotron radiation*).

 **Knowledge check**

The brightest star in Taurus, α Tau (see Fig. 1.6.1) is clearly reddish.

Explain how its temperature must compare with α Cen.

>> **Key terms**

Intensity, I: The power per unit area crossing a surface at right angles to the radiation. The unit is W m^{-2}.

Luminosity, L, of a star is the total power emitted in the form of electromagnetic radiation. The unit is W.

 **Knowledge check**

A candle gives out light with a power of 0.1 W. What will be the intensity of its radiation at a distance of:

(a) 1 m, (b) 10 m and (c) 1 km?

>> **Study point**

The luminosity and brightness of a star are not the same. A nearby low-luminosity star can appear as bright as a distant high-luminosity giant!

Example

The star alpha Centauri (α Cen) has a temperature of 5260 K.

(a) Calculate the wavelength of the peak of its spectrum.

(b) Explain why α Cen's colour is approximately white.

Answer

(a) Applying Wien's law: $\lambda_{max} = \dfrac{W}{T} = \dfrac{2.90 \times 10^{-3}\text{ m K}}{5260\text{ K}} = 5.51 \times 10^{-7}\text{ m}$

(b) The peak wavelength (~550 nm) is approximately in the middle of the visible spectrum (400–700 nm), so all colours are present at roughly the same strength.

The importance of these results for astronomers is that many astronomical objects emit thermal radiation, which approximates to black body radiation. Examples are: stars, with surface temperatures up to tens of thousands of kelvin; the accretion discs around black holes (up to 10^6 K); the cosmic microwave background radiation (2.713 K).

1.6.2 Luminosity, intensity and distance

The inverse square law

We first need to define a couple of terms, **intensity** and **luminosity**.

If we look back at the stars in Taurus in Fig. 1.6.1, we notice that they have a range of brightness (and colour). Are they really different or are they just at different distances – with the fainter-seeming ones just being further away? The diagram in Fig. 1.6.7 shows how radiation from a small source, such as a star, spreads out as it moves out. The further from the star, the bigger the area the same quantity of radiation has to cover, so the lower the **intensity** of the radiation.

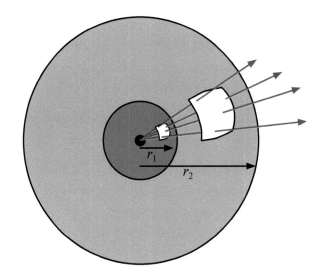

Fig. 1.6.7 Inverse square law

We can derive the inverse-square law, by considering the geometry of the situation, as follows:

Let L be the luminosity of the star.

As it crosses the first sphere, of radius r_1, this radiation is spread out over the surface of the sphere, which has an area of r_1, so the **intensity**, I, of the radiation is given by:

$$I = \frac{L}{4\pi r_1^2}.$$

At r_2 the intensity is

$$I = \frac{L}{4\pi r_2^2}.$$

So the intensity of the radiation decreases as the inverse square of the distance:

a star $10 \times$ as far away as an identical one will appear only $\frac{1}{100}$ as bright.

▶▶ Study point

The symbol ⊙ is commonly used for the Sun. So $L_\odot$ is the solar luminosity (3.85×10^{26} W). $M_\odot$ and $T_\odot$ are the mass and (surface) temperature of the Sun.

◀ Stretch & challenge

Measuring the distance to a star

Apart from the Sun, stars are vast distances away. How far? For distances up to ~1000 ly (light-years) astronomers can use the fact that nearby stars appear to shift position as the Earth moves around in its orbit. Fig. 1.6.8 shows this – but is rather exaggerated!

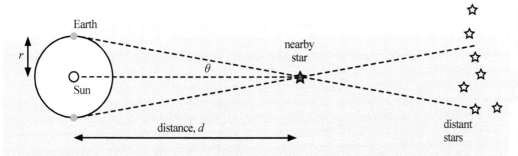

Fig. 1.6.8 Stellar parallax

Over a period of half a year, the nearby star appears to shift when seen against the background of much more distant stars because the Earth moves around in its orbit. If we know r, the radius of the orbit and we can measure the angle, θ, we can calculate the distance d to the star.

Using trigonometry: $\tan \theta = \frac{r}{d}$, so $d = \frac{r}{\tan \theta}$. As θ is very small (typically 10^{-8} rad) we can use the approximation $\tan \theta \approx \theta$ (with θ in radians) to calculate the distance using $d = \frac{r}{\theta/\text{rad}}$, using the Earth's orbital radius of 1.50×10^{11} km.

Using ground-based telescopes astronomers can measure θ to a precision of about 10^{-8} radians. The European Space Agency Gaia mission, the space telescope launched in December 2013, is measuring stellar parallaxes to ~5×10^{-11} rad enabling distances up to tens of thousands of light-years to be measured.

To measure larger distances, astronomers use objects of known brightness. If you know the true luminosity of an object and measure the intensity of radiation received, the inverse square law can be used to determine its distance. For distances to relatively nearby galaxies, astronomers can use Cepheid variable stars; for more distant galaxies they use Type 1a supernovae.

> The Gaia satellite measures the annual stellar parallax of a star to be 2.0×10^{-10} radian. Calculate the distance of the star (a) in m and (b) in ly.

▲ Knowledge check 1.6.3 ◀

Stars A and B appear to have the same brightness. Star B is twice as far away as star A. Compare the luminosity of the two stars.

▲ Knowledge check 1.6.4 ◀

The distance to star X is measured as 10 ly [1 ly = 9.5×10^{15} m]. The intensity of radiation from it is measured as 42.8 nW m^{-2}. Show that its luminosity is about 5×10^{27} W

◀ Maths tip

See Chapter 4 for angles in radians and for the small-angle approximation, $\theta = \tan \theta$

Fig. 1.6.9 An artist's impression of ESA's Gaia.

1.6.3 Black body radiation and stars

We are now in a position to interpret the graphical part of Fig. 1.6.3. The dotted line gives the shape of the spectrum for a perfect black body with a temperature of 5800 K. For wavelengths from 400–1000 nm, i.e. the visible and near infrared region, the overall shape of the observed spectrum matches the 5800 K spectrum reasonably closely. This suggests that the Sun and other stars emit radiation in a manner which is very similar to, but not exactly like, a black body. It might seem a little strange to regard the Sun as a *black* body. However, in radiation terms, that is exactly (or nearly) what it is! From the graph, the effective temperature of the Sun's photosphere (its outer layer) appears to be a little lower than 5800 K. Recent measurements give a best-fit estimate of the Sun's temperature of 5770 K. There is clearly more to be said about this. We'll consider the Fraunhofer lines in Section 1.6.4.

The assumption that we can treat stars as black bodies allows us to determine the temperature and diameter of a star from measurements of its spectrum, as long as we know its distance. The example does this for the Sun.

Example

Use the spectrum in Fig. 1.6.3 and the following data to estimate the Sun's:
(a) temperature, (b) power.

- The mean **intensity** of the solar radiation at the Earth = 1.36 kW m^{-2}.

- The mean radius of the Earth's orbit = 1.50×10^{11} m

Answer

(a) From Fig. 1.6.3 the peak wavelength λ_{max} = 500 nm

$$\therefore \text{ Using Wien's law, the temperature, } T = \frac{W}{\lambda_{max}} = \frac{2.90 \times 10^{-3} \text{ m K}}{500 \times 10^{-9} \text{ m}} = 5800 \text{ K}$$

(b) At the Earth's orbit, the Sun's radiation is spread out over the surface of a sphere of radius 1.50×10^{11} m. Area of a sphere = $4\pi r^2$.

$$\therefore \text{ Power of Sun} = 1.36 \times 10^3 \text{ W m}^{-2} \times 4\pi \times (1.50 \times 10^{11} \text{ m})^2 = 3.85 \times 10^{26} \text{ W}$$

1.6.4 Line spectra: Fraunhofer lines

(a) Emission spectra

Chemists identify ions using the flame test (see Table 1.6.1). The flame in Fig. 1.6.10 is 'brick red' revealing the presence of calcium ions. Similarly the glowing pink regions in the spiral arms of the Whirlpool galaxy, M51, in Fig. 1.6.11 pick out clouds of atomic hydrogen, that astronomers call HI clouds, which identify regions where active star formation is occurring.

The sources of the light could hardly be more different but, in fact, the physics by which the particles in the flame and the gas clouds emit light is the same in the two cases. The source of energy is different: the calcium ions in the flame derive their energy from impacts with other particles in the flame; the hydrogen atoms are energised by ultraviolet radiation from newly formed giant stars in the centre of the gas clouds.

1.6.5 Knowledge check

Use answers to (a) and (b) in the example together with the Stefan–Boltzmann law to estimate the diameter of the Sun.

1.6.6 Knowledge check

The mean apparent diameter of the Sun in the sky is 0.535°(9.34×10^{-3} rad). Show that this is consistent with the answer to Knowledge check 1.6.5.

1.6.7 Knowledge check

The peak wavelength of star X in Knowledge check 1.6.4, is 788 nm. Estimate the star's temperature and radius.

Fig. 1.6.10 Flame test showing calcium

Fig. 1.6.11 M51, the Whirlpool galaxy

These sources produce light which is very different from black body radiation, as the spectra of atomic hydrogen and calcium show: the spectra are called *line spectra* for obvious reasons. For comparison, Fig. 1.6.12 includes a black body spectrum corresponding approximately to the Sun's temperature – about 5800 K. The colour of the HI clouds is due to a combination of the red and blue lines; the 'brick-red' of the calcium flame from all the lines in its spectrum.

The reason that low density gases, whether in the Bunsen flame or in the galactic HI clouds, produce only discrete wavelengths rather than a continuous spectrum is explored in detail in Section 2.7.3. Importantly for astronomers (and chemists) different elements emit different combinations of wavelengths, so the lines act as a spectral fingerprint and we can use them to identify the gases present.

(b) Absorption spectra

In order to reach us, the Sun's radiation has to pass through the low pressure gas of its 'atmosphere' – the chromosphere and corona. These are normally only visible during a total solar eclipse because, although they emit light, the photosphere of the Sun is overwhelmingly brighter. Fig. 1.6.14 was taken during an eclipse in India in 1980 and Fig. 1.6.15 in France in 1999. Note that the pinkish colour of the *prominences* in the chromosphere is just like the HI regions in M51, because they are produced by the same process – they are glowing hydrogen gas.

Just as glowing hydrogen emits light of only a few characteristic wavelengths, the gas also absorbs light at just the same wavelengths. If visible radiation with a continuous spectrum passes through a gas, the gas absorbs just these wavelengths.

Fig. 1.6.13 shows the visible part of the **absorption spectrum** of hydrogen (it also emits and absorbs in the UV and IR). The diagram also shows the relationship between the image spectrum and the graphical representation.

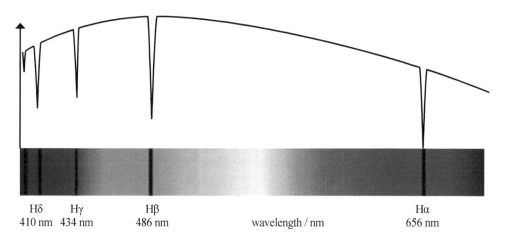

| Hδ | Hγ | Hβ | | Hα |
| 410 nm | 434 nm | 486 nm | wavelength / nm | 656 nm |

Fig. 1.6.13 Hydrogen absorption spectrum

The labels, Hα – Hδ, are the names that astronomers give to the absorption lines. The lines clearly form the same pattern as those of the emission spectrum in Fig. 1.6.12.

The solar spectrum in Fig. 1.6.3 shows a vast number of **Fraunhofer lines** because of the large number of elements present in the solar atmosphere. Fig. 1.6.16 shows a simplified solar spectrum with the most prominent Fraunhofer lines. Knowledge check 1.6.8 gives some wavelengths in the spectra of various elements, which you can use to identify elements present in the Sun.

Colour of the flame	Element
Colourless	Mg, Be
Red	Li
Crimson	Sr
Brick red	Ca
Red-purple	Rb
Purple	K
Yellow	Na
Apple green	Ba
Dark green	Cu
Blue	Cs

Table 1.6.1 Standard flame test colours

black body

hydrogen

calcium

Fig. 1.6.12 Atomic emission spectra

>> **Key term**

Absorption spectrum: The variation in intensity of radiation with wavelength due to absorption by a material.

Fig. 1.6.14 Sun's corona

Fig. 1.6.15 Sun's chromosphere

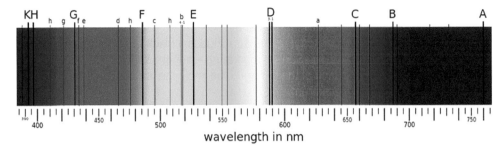

Fig. 1.6.16 Simplified solar spectrum

(c) 'Missing lines' and temperature

From Knowledge check 1.6.8, you'll notice that some of the lines which exist in the emission spectrum of an element (or ion) do not apparently occur in the solar spectrum, although other lines of the same element do. For example, the 470 nm line for atomic magnesium is missing but the 518 nm is identifiable. This is because, in order to absorb a particular photon the atom must be in the lower of two energy states which have a difference in energy equal to the energy of the photon. This is dealt with in detail in Section 2.7. If the temperature is too high, there will be very few atoms in the lower energy state (violent collisions will put them into a more excited state) so they won't be available for absorbing the photon; if the temperature is too low, even the lower energy state might be too high to have any significant population.

Observing which lines are present, and their prominence, gives astronomers information of the temperature of the gas which is responsible for the absorption spectrum.

1.6.5 Multiwavelength astronomy

The various regions of the electromagnetic spectrum give us information about different processes in the universe. We have seen that most of the power of the Sun is emitted in the form of near infrared, visible and near ultraviolet radiation. This is because the temperature of the photosphere is approximately 5800 K. The higher the temperature of an object, the shorter the wavelengths of the continuous spectrum the object emits.

The lower levels of the Sun's chromosphere have temperatures similar to that of the photosphere but the temperature rises with distance from the Sun's surface and the solar corona reaches over 10^6 K. In a solar flare, the temperature can reach tens of millions of K.

Some non-thermal processes result in the emission of radiation: 21 cm HI and synchrotron radiation. These can give us additional information about hydrogen clouds and about magnetic fields. So a study of radiation across the e-m spectrum provides us with much more information than observations in one spectral region alone.

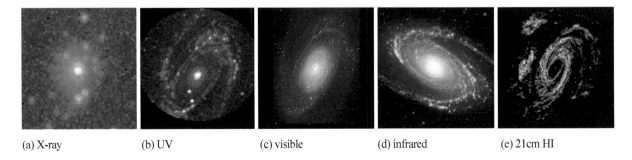

(a) X-ray   (b) UV   (c) visible   (d) infrared   (e) 21cm HI

Fig. 1.6.17 Images of M81 in various wavebands

Consider the images in Fig. 1.6.17 of the spiral galaxy, M81. The different regions of the electromagnetic spectrum reveal different processes.

Fig. 1.6.17(c) in visible light is the familiar astronomical image of a spiral galaxy. It is also high definition. The spiral arms are well shown, as is the central bulge. The stars in the centre are predominantly old, low mass stars and appear yellowish. Lanes of dust are also visible. The ultraviolet image, on the other hand, picks out hotter regions and the image shows knots of young giant stars forming well away from the centre. The infrared image shows regions where stars are heating up dust especially in the spiral arms.

Fig. 1.6.17(a) (the X-ray image) only displays very high temperature regions. The bright knot in the centre is heated by matter spiralling into the giant black hole at the heart of the galaxy. The other two bright blobs below are actually not part of the galaxy at all. These are vastly more distant *quasars* which happen to be behind M81. They are not visible in any of the other images. Fig. 1.6.17(e) shows neutral hydrogen by its $21 \, \text{cm}$ signature emission. It is clearly missing from the centre of the galaxy.

Some indication of the ability of $21 \, \text{cm}$ radio astronomy to reveal processes which are not detectable at other wavelengths is in Fig. 1.6.18. This shows M81 again but this time with some smaller neighbouring galaxies. Encounters between the galaxies have resulted in long filaments of hydrogen being pulled out into intergalactic space. Only the sensitivity of $21 \, \text{cm}$ radio telescopes allows this to be imaged and the dynamics of galactic tidal interactions to be studied.

>> **Study point**

Astronomers now increasingly refer to **multi-messenger astronomy**. This includes the information obtained from cosmic neutrinos and gravitational waves.

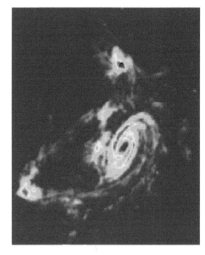

Fig. 1.6.18 Tidal effects in the M81 group

Test yourself 1.6

1. A white dwarf star has a temperature of $24 \, 000 \, \text{K}$ and a diameter of $14 \, 000 \, \text{km}$. Calculate:

   (a) its luminosity; and
   (b) the peak wavelength of its spectrum.

2. Without using a calculator, compare the luminosity and peak wavelength of the white dwarf in Q1 with those of the Sun. Take the temperature and diameter of the Sun to be $6000 \, \text{K}$ and $1.4 \, \text{million km}$ respectively.

3. By calculation suggest the regions of the e-m spectrum which are appropriate for studying processes which take place at:

   (a) $10 \, \text{K}$, (b) $10^3 \, \text{K}$, (c) $10^5 \, \text{K}$ (d) $10^7 \, \text{K}$.

4. Describe qualitatively how the scales of Fig. 1.6.6 would need to be changed to include a black body spectrum of $6000 \, \text{K}$ (the approximate temperature of the Sun).

5. A red giant star has a diameter 1000 times that of a red dwarf star of the same surface temperature. Compare their distances from the Earth given that the red giant appears 100 times as bright. Show your working.

6. The central bulges of spiral galaxies consist mainly of old stars. Very little star formation is taking place. How does this tie in with the absence of $21 \, \text{cm}$ emission from the centre of M81?

7. Interstellar dust particles are typically $0.1 - 1 \, \mu\text{m}$ in size. Stars are formed from cold molecular clouds which contain dust particles. Explain why stellar formation is most easily observed using infrared radiation.

8. The discs of dust and gas around young stars and from which planetary systems are thought to develop are heated up (to several $100 \, \text{K}$) by their parent star. Suggest how this is detected in the spectrum of the star.

9. The radiation from young hot stars heats up nearby clouds of atomic hydrogen (HI). Explain the appearance of the HI regions in M51 (Fig. 1.6.11) in terms of the emission spectrum of hydrogen (Fig. 1.6.12).

10. An X-ray pulsar is a neutron star (the remnant of a supernova) which pulls gas from the surface of its red giant companion star. This gas spirals in to the neutron star in an *accretion disc*, which is in the plane of equator of the star. The point of impact with the surface of the neutron star is heated to ~10^7 K, the position of which rotates with the star, which has a rotation period of less than 1 s. Describe how this would appear to a distant astronomer who observes from an angle well away from the axis of rotation.

11. Stars emit radiation as *black bodies*.

    (a) State what is meant by a *black body*.
    (b) State two laws of radiation obeyed by black bodies.

12. Light from the photosphere (surface) of a white star passes through a cloud of atomic hydrogen gas on its way to an observer. Explain the appearance of the spectrum of light observed.

13. Two identical spheres have temperatures of 3000 K and 6000 K. Use the laws of Wien and Stefan to compare the radiation emitted from the two spheres (assumed to behave as black bodies).

14. The peak wavelength of the black body spectrum of a star, X, is determined to be 200 nm.

    (a) State the region of the EM spectrum which includes 200 nm.
    (b) State the wavelength range of the visible spectrum.
    (c) Describe and account for the appearance of X.
    (d) Taking the Sun's surface temperature to be 6000 K and the peak wavelength its spectrum to be 500 nm, estimate the surface temperature of X.

15. A white dwarf star has a diameter of 2.0×10^7 m and a surface temperature of 50 000 K. Its distance from the Earth is 25 light-years. Calculate:

    (a) its luminosity;
    (b) the wavelength of the peak emission; and
    (c) the intensity of its radiation observed from the Earth (1 ly = 9.46×10^{15} m).

16. Two planets, A and B, orbit a star. Planet B is twice as far away from the star as A. Why is the mean surface temperature of planet B less than that of planet A? Can you estimate the ratio of these temperatures? State any assumptions you make.

17. The Sun has a surface temperature of about 6000 K. Its peak emission is about 500 nm.

    (a) State the region of the electromagnetic spectrum in which 500 nm radiation lies.
    (b) Use the data for the Sun to answer the following questions:

       (i) A blue giant star has a temperature of 12 000 K. What is its peak wavelength? Identify the electromagnetic spectrum region of this peak.
       (ii) A red giant star has a temperature of 3000 K. What is its peak wavelength? Identify the electromagnetic spectrum region of this peak.

    (c) The blue giant star has 10 times the diameter of the Sun; the red giant is 100 times as large as the Sun. Compare the luminosities of both these stars with the Sun.

Until the late 19th century, the atom was regarded as an elementary particle. The periodic table of the elements, first published by the Russian chemist, Dmitri Ivanovich Mendeleev in the 1860s, strongly suggested an underlying structure to atoms and by the end of the century, the negatively charged electron had been identified as a universal component of atoms. The positively charged atomic nucleus, which contained virtually all the mass of the atom, was discovered from the work of Rutherford, Geiger and Marsden between 1908 and 1913 and both its main constituents – protons and neutrons – were identified by the early 1930s and initially thought to be **elementary**.

The development of understanding of atomic structure since that time is shown schematically for a deuterium (heavy-hydrogen) atom in Fig. 1.7.1. With the development of quantum theory in the 1920s it was realised that electrons in an atom occupied a region a fraction of a nanometre across which is typically $100\,000 \times$ the diameter of the nucleus. The nucleus was shown in the 1930s to consist of protons (**p**) and neutrons (**n**), which are collectively known as nucleons. The results of collision experiments known as *deep inelastic scattering* in the 1960s and 1970s showed that nucleons are composed of 3 particles called quarks bound together by the so-called *strong interaction*. These quarks are thought to be elementary particles.

The middle decades of the 20th century produced a veritable smorgasbord of particles, called variously hadrons, fermions, bosons, mesons, nucleons, baryons, neutrinos…. The standard model of particle physics goes a long way towards simplifying this picture: this will be described briefly here.

>> **Key terms**

Elementary (fundamental) **particle**: a particle that is not a combination of other particles.

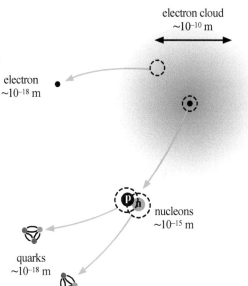

Fig. 1.7.1 The structure of a deuterium atom

1.7.1 The standard model – three generations of leptons and quarks

Almost all the normal matter in the universe (that is, ignoring the mysterious *dark matter* which is dealt with later) is composed of heavy protons and neutrons and the much lighter electrons. We infer the existence of other almost massless particles called neutrinos (see Section 1.7.4) and notice that, if we bash protons and neutrons together, we produce showers of other intermediate mass particles, which we call **mesons**. When cosmic ray particles collide with atoms in the upper atmosphere they produce showers of particles, called muons, which form part of the background radiation which we can detect using a Geiger-Müller tube. We now know that the electrons, muons and neutrinos (and some others) are elementary particles, which we call **leptons**. The other particles, the heavy ones, are called **hadrons**, and are not elementary – they are composed of combinations of **quarks**.

Table 1.7.1 contains the elementary particles in the **standard model**.

>> **Key terms**

Meson: A hadron consisting of a quark–antiquark pair.

Lepton: Low mass, elementary particle, e.g. electron, neutrino.

Quark: Elementary particle, not found in isolation, which combines with other quarks to form hadrons and baryons, e.g. up, down.

Hadron: High mass particle consisting of *quarks* and/or *anti-quarks*.

Table 1.7.1 Standard model particles

Generation	Leptons		Quarks	
1st	**electron** Symbol: e^- charge: $-e$	**electron neutrino** Symbol: ν_e charge: 0	**up** Symbol: u charge: $\frac{2}{3}e$	**down** Symbol: d charge: $-\frac{1}{3}e$
2nd	**muon** Symbol: μ^- charge: $-e$	**muon neutrino** Symbol: ν_μ charge: 0	**charm** Symbol: c charge: $\frac{2}{3}e$	**strange** Symbol: s charge: $-\frac{1}{3}e$
3rd	**tauon** Symbol: τ^- charge: $-e$	**tauon neutrino** Symbol: ν_τ charge: 0	**top** Symbol: t charge: $\frac{2}{3}e$	**bottom** Symbol: b charge: $-\frac{1}{3}e$

< **Top tip**

You should know that there are three generations of particles but you only need to be able to answer questions on the first generation.

1.7.2 Mass and energy in particle physics

(a) Mass–energy

To have even a basic understanding of subatomic particles we need to know about the relationship between mass and energy. The WJEC AS Physics course doesn't cover this, but the A level course does; questions in the A2 units involving the idea of mass–energy can be set in the context of particle physics.

Mass and energy are related through the Einstein mass–energy relationship:

$$E = mc^2$$

How are we to understand this equation? It really means that mass and energy are essentially the same – it's just that we are used to expressing them in different units, kg and J respectively. We'll use this idea several times in this topic.

Example

Calculate the mass–energy of an electron. The mass is 9.11×10^{-31} kg.

Answer

$$E = mc^2 = 9.11 \times 10^{-31} \text{ kg} \times (3.00 \times 10^8 \text{ m s}^{-1})^2 = 8.20 \times 10^{-14} \text{ J}$$

In atomic terms this is an enormous quantity of energy. It is nearly $40\,000$ times the energy needed to ionise a hydrogen atom, which is 2.18×10^{-18} J. So, when subatomic particles are annihilated, e.g. in interactions with their antiparticles, a very significant quantity of energy is released.

(b) Units of energy

Particle physicists invariably express energy of particles in terms of the **electron volt** (eV) or, more usually, its multiples: keV, MeV, GeV or TeV. This useful, non-SI, unit is defined as the energy transferred when an electron moves through a potential difference of 1 V (1.00 eV $= 1.60 \times 10^{-19}$ J).

1.7.3 Antiparticles

Antimatter is not only the stuff of science fiction. For each of the particles in Table 1.7.1, there is corresponding antiparticle with an identical mass; if the particle has a charge, the antiparticle has an equal and opposite charge. The symbol for most antiparticles that you will meet is formed by putting a bar over the symbol for the particle, e.g. $\bar{u}$, $\bar{v}_e$, $\bar{p}$ for the anti-up quark, the electron antineutrino [or 'anti-electron neutrino'] and the antiproton respectively. The exceptions are the antiparticles of the electron, muon and tauon which are written e^+, μ^+ and τ^+ respectively. The anti-electron has its own name: the *positron*.

When a particle and its antiparticle interact they annihilate each other; that is they disappear and their mass–energy manifests itself as two photons of electromagnetic radiation. These photons are given the symbol γ because they are at the very high energy end of the e-m spectrum. The total energy of the photons is equal to the sum of the mass–energy and kinetic energy of the annihilating particles.

1.7.1 Knowledge check

Show that the mass–energy of an electron is approximately 510 keV. (Use the answer to the Example.)

Key term

Electron volt: (eV) The energy transferred when an electron moves through a potential difference of 1 V. 1.00 eV $= 1.60 \times 10^{-19}$ J

Study point

If an electron and a positron annihilate, they produce two γ photons. These are emitted in opposite directions – otherwise momentum would not be conserved.

Example

An electron and a positron collide head on and annihilate. Each particle has a kinetic energy of 100 keV. Calculate the energy of each of the photons produced.

Answer

Using Knowledge check 1.7.1:

Electron mass–energy = 510 keV / c^2, so the mass–energy is 510 keV.

∴ Total energy = total mass–energy + total kinetic energy

$\quad\quad\quad\quad\quad$ = 2 × 510 keV + 2 × 100 keV

$\quad\quad\quad\quad\quad$ = 1220 keV

∴ Each photon has energy $\frac{1}{2}$ × 1220 keV = 610 keV

The opposite process can also happen: if it possesses enough energy, a high energy photon can create an electron–positron pair. It also needs to interact with another particle (usually an atomic nucleus) to enable energy and momentum to be simultaneously conserved. In Fig. 1.7.2 a high energy photon enters from the top and interacts with a hydrogen atom at A (it is in a bubble chamber, which consists of a tank of liquid hydrogen), ejecting a high energy electron, creating a low energy electron–positron pair and a second photon which continues to B where it creates a second (higher energy) e⁻e⁺ pair. A magnetic field at right angles to the page makes the charged particles travel in curves: the opposite charges of electrons and positrons results in the typical 'ram's horn' effect at A. (Note that the tracks of the photons do not show up in the bubble chamber image.)

1.7.4 The evidence for neutrinos

Neutrinos are neutral particles of very low mass which only interact via the weak force (see Section 1.7.6). This means that they need to come within ~10^{-18} m to interact, so interactions hardly ever happen: for example, a typical solar neutrino could expect to penetrate 1–2 light-years thickness of lead!

The first evidence for their existence came from studies in the 1930s of the energy spectrum of beta particles. Phosphorus-32, $^{32}_{15}$P, decays by β⁻ emission. Before neutrinos were known the complete reaction was expected to be:

$$^{32}_{15}\text{P} \longrightarrow\ ^{32}_{16}\text{S} + ^{0}_{-1}\text{e}$$

The energy release in the decay is 1.5 MeV. If we apply the principle of conservation of momentum, we can calculate that the beta particles should take nearly all the energy (>99.9%), with the much heavier sulfur nucleus taking a tiny fraction. Compare this with the actual energy spectrum in Fig.1.7.3: 1.5 MeV is indeed the maximum beta particle energy but there is a continuous spectrum of energies with the peak being less that 0.5 MeV. This energy spectrum is only possible if a third particle is also produced, which can share the energy with the beta particle. This particle is called the electron anti-neutrino (sometimes, loosely, a neutrino) and the complete reaction is:

$$^{32}_{15}\text{P} \longrightarrow\ ^{32}_{16}\text{S} + ^{0}_{-1}\text{e} + \bar{\nu}_e$$

The photograph in Fig. 1.7.4 is of the β-decay of a nucleus of **He-6**. As with Fig.1.7.2 this is in a bubble chamber. The curved section of track is the β-particle; the short fat track is the recoil of the resulting Li-7 nucleus. In order to conserve momentum a particle (a neutrino) must be emitted downwards in the photograph. The neutrino doesn't interact and so leaves no track.

Knowledge check 1.7.2

Calculate the frequency, wavelength and momentum of the photons in the example.

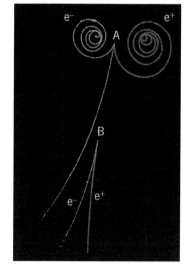

Fig. 1.7.2 e⁻e⁺ pair production

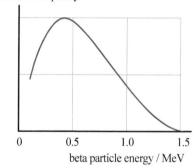

Fig. 1.7.3 P-32 β-spectrum

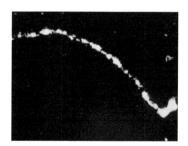

Fig. 1.7.4 He-6 decay in a bubble chamber

1.7.5 Building heavy particles

Electrons, being leptons, are elementary particles; that is they are not composed of other particles. **Hadrons**, e.g. protons and neutrons (which are together also called nucleons), on the other hand, are composed of quarks, bound together with the **strong force** (see Section 1.7.6). Evidence for the existence of quarks is indirect. Single quarks are never detected. They are always seen in combination.

Key terms

Baryon: Hadron composed of 3 quarks, e.g. proton, neutron.

Antibaryon: Hadron composed of 3 antiquarks, e.g. antiproton.

There are three different types of hadron:

- **Baryons**, such as protons and neutrons, are composed of three quarks. First generation baryons are composed entirely of a mixture of up (u) and down (d) quarks.
- **Antibaryons** such as antiprotons are composed of three antiquarks.
- Mesons, are composed of a quark and an antiquark.

The quark structure of protons and neutrons is shown schematically in Fig. 1.7.5 and can be summarised by

$$p = uud, \quad n = udd$$

Study point

Because mesons are composed of a quark and an antiquark, there is no need to define a separate category of 'antimeson'.

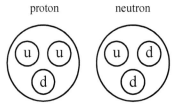

Fig. 1.7.5 Quark structure of nucleons

Note that, in particle physics, the proton has the symbol p rather than ^1_1H, which is usual in nuclear physics; the neutron is n rather than ^1_0n. Also the order of writing the quarks is arbitrary: $p = udu$ and $n = ddu$, etc., are perfectly good ways of writing the structure. Protons are the only stable baryons: there are theories which suggest they could be unstable with a half-life of around 10^{32} years!

Key terms

Pions: Have the following names and structure:

π^+ (pi plus) = $u\bar{d}$

π^- (pi minus) = $d\bar{u}$

π^0 (pi zero) = $u\bar{u}$ or $d\bar{d}$. A beam of π^0 mesons is composed of a mixture of the two!

Mesons are created in copious numbers when baryons are collided at moderate to high energies (more than a few hundred MeV). The first generation mesons are called **pions** (or pi mesons).

A typical meson-generating reaction is

$$p + p \longrightarrow p + n + \pi^+$$

which, at the quark level, can be written:

$$uud + uud \longrightarrow uud + udd + u\bar{d}$$

There are only six first-generation baryons. They are summarised in Table 1.7.2.

Study point

Notice that the charge on the two sides of the reaction is the same, as is the number of u-quarks (4); there are 2 d-quarks on the left and 3 d + 1 anti-d on the right. This is explored further in Section 1.7.6.

Family	baryons
Nucleons	p (uud); n (udd)
Δ particles	Δ^{++} (uuu); Δ^+ (uud); Δ^0 (udd); Δ^- (ddd)

Table 1.7.2 1st generation baryons

Study point

The antiproton and antineutron have the following structures:

$\bar{p} = \overline{uud}$; $\bar{n} = \overline{udd}$.

The symbol, Δ, is the Greek letter capital delta, so members of the Δ family are called delta double plus, delta plus, etc. Note that the quark structures of Δ^+ and Δ^0 are the same as those of p and n respectively but the mass–energy of the Δ^+ is 1232 MeV against 938 MeV for the proton. The Δ^+ can be regarded as an excited state of the proton: similarly for Δ^0 and the neutron.

1.7.6 Interactions (forces) between particles

Macroscopic objects are subject to two types of force: gravitational and electromagnetic. Subatomic particles are also affected by two other forces: the strong and the weak interactions. These are not experienced at all on the everyday scale because their range is so small. The four forces are summarised in Table 1.7.3 in order of increasing strength.

Interaction	Affects	Range	Comments
gravitational	all matter	infinite	negligible for subatomic particles
weak	all particles	~10^{-18} m	only significant when e-m and strong interactions not involved
electromagnetic (e-m)	all charged particles	infinite	also affects neutral hadrons because quarks have charges
strong	all quarks	~10^{-15} m	also affects interactions between hadrons (e.g. nuclear binding)

Table 1.7.3 Interactions summarised

The three non-gravitational forces have important roles in the stability of atoms:

- Electrons are bound to the nucleus by the electromagnetic force.
- Protons and neutrons are held together in the nucleus by the strong force which opposes the e-m repulsion of the protons.
- The weak force is responsible for the decay of neutrons in neutron-rich nuclei, giving rise to β^- decay.

Generally the interaction responsible for any particular interaction is the strongest one which is felt by all the particles on both sides of the equation. For example:

Consider the reaction in Section 1.7.5:

$$p + p \longrightarrow p + n + \pi^+$$

All the particles are composed of quarks and there is no change of quark flavour (see Section 1.7.7). Hence this reaction is controlled by the strong interaction meaning that, as long as there is enough energy to create the $\pi+$, the reaction is very likely to occur.

Neutrinos only feel the weak force, so any interaction with neutrinos in it (e.g. β-decay) must be weak. This results in the high ability of neutrinos to penetrate matter.

The different strengths of the interactions are illustrated by the different decay times of particles and the force responsible. The following are examples:

Strong	Δ^- (ddd) $\longrightarrow$ n + π^-	lifetime ~ 10^{-24} s
Electromagnetic	π^0 (u$\bar{\text{u}}$) $\longrightarrow$ γ + γ	lifetime ~ 10^{-12} s
Weak	n (udd) $\longrightarrow$ p + e$^-$ + $\bar{\nu}_e$	lifetime ~ 15 min

Note that neutron decay has an unusually long life-time. Most weak decays are of the order of $10^{-8} - 10^{-10}$ s.

Knowledge check 1.7.3

Heavier nuclei need a greater fraction of neutrons to overcome the increased e-m repulsion of the protons. Illustrate this from the fraction of neutrons in the stable nuclei, $^{12}_{6}$C, $^{56}_{26}$Fe and, $^{197}_{79}$Au.

Study point

The word *interaction* is often preferred to *force* because it has a wider implication than just attraction or repulsion. It includes the control of the creation of particles or their decay.

Study point

In considering which force is responsible for a reaction we also have to have an eye on the conservation laws (Section 1.7.7).

Top tip

In a decay reaction, the stronger the force, the shorter is the decay time.

In a collision reaction, the stronger the force, the more likely the reaction is to occur.

1.7.4 Knowledge check

In terms of energy conservation, why can a neutron (939.6 MeV) decay into a proton (938.3 MeV) an electron (511 MeV) and a neutrino, but an isolated proton cannot decay into a neutron, a positron and a neutrino?

1.7.7 Conservation laws in particle physics

The familiar conservation laws of mass, energy and momentum apply in particle physics, though they have to take account of the close-to-light-speed of the particles. Hence the formula for momentum is modified and instead of mass and energy separately, it is the mass–energy which is conserved. (This is not examined in A level Physics.)

(a) Conservation of charge, Q

Having observed trillions of reactions, physicists have never observed an example of the violation of the conservation of charge.

So the following reaction is observed:

$$\text{p} + \text{p} \longrightarrow \text{p} + \text{n} + \pi^+$$

Q values:

$$1 + 1 \longrightarrow 1 + 0 + 1$$

The total Q value on each side $= 2$. Hence charge is conserved.

But the following reaction is **never** observed:

$$\text{p} + \text{p} \longrightarrow \text{p} + \text{n} + \pi^-$$

because the total Q value would change from 2 to 0, i.e. it would not be conserved.

(b) Conservation of lepton number, L

The lepton number, L, is defined for the first generation of leptons as shown in Table 1.7.4. Other first generation particles, photons, quarks (and hence baryons and mesons) are assigned a lepton number of 0. Experimentally, we find that lepton number is always conserved, i.e. in any observed reaction the total lepton number stays the same.

So the following reaction has **never** been observed:

$$\text{p} + \text{e}^- \longrightarrow \text{n} + \pi^0$$

L values:

$$0 + 1 \neq 0 + 0$$

But free neutrons decay by the reaction

$$\text{n} \longrightarrow \text{p} + \text{e}^- + \bar{\nu}_e$$

L values:

$$0 = 0 + 1 + (-1)$$

and the following reverse reaction **is** observed

$$\text{p} + \text{e}^- \longrightarrow \text{n} + \nu_e$$

>> **Study point**

Note that the values of charge are given in terms of e. So:

$Q_{\text{proton}} = +1$; $Q_{\text{neutron}} = 0$;

$Q_{\text{electron}} = -1$, etc.

Particle	L
e^-	1
e^+	-1
ν_e	1
$\bar{\nu}_e$	-1

Table 1.7.4

1.7.5 Knowledge check

Show that the reaction

$\text{p} + \text{e}^- \longrightarrow \text{n} + \nu_e$

conserves both charge and lepton number.

(c) Conservation of baryon number, B

Similarly to lepton number we define a baryon number, B. Each baryon, e.g. proton, has $B = 1$; antibaryons have $B = -1$; leptons and mesons have $B = 0$. Again, baryon number is always conserved.

The following reaction could not happen:

$$\text{p} + \pi^- \longrightarrow \text{n} + \text{n}$$

even if the **p** and π^- had enough kinetic energy and even though it conserves charge and lepton number. Why not? Because the total baryon number on the left is 1 and on the right it is 2!

The conservation of baryon number is really a special case of the conservation of quark number, q. Looking again at our 'impossible reaction' and assigning a quark number of -1 to antiquarks, we can tally the quarks as follows:

Left-hand side: $q = 3 + (1 - 1) = 3$ Right-hand side: $q = 3 + 3 = 6$

So this 'impossible reaction' would violate the conservation of quark number. Looking again at a reaction which **does** exist:

$$p + p \longrightarrow p + n + \pi^+$$

the quark totals are the same (6) for the reactants and the products. We can analyse the same reaction in terms of the **up quark number (U)** and the **down quark number (D)**. Writing this reaction at the quark level:

$$uud + uud \longrightarrow uud + udd + u\bar{d}$$

U	2	2	2	1	1	Total 4
D	1	1	1	2	-1	Total 2

So, both the up and down quark numbers are conserved. This is true for all changes controlled by the strong and electromagnetic interactions.

However, individual quark numbers **can be changed by ± 1 in weak interactions**.

Looking again at neutron decay:

$$n \, (udd) \longrightarrow p + e^- + \bar{v}_e$$

Writing this in terms of quarks: $udd \longrightarrow uud + e^- + \bar{v}_e,$

$$U = 1 \quad 2 \quad 0 \quad 0$$
$$D = 2 \quad 1 \quad 0 \quad 0$$

We see that U changes from 1 to 2 and D from 2 to 1. The total quark number q is 3 on both sides but one of the quarks has changed its flavour from down to up.

Knowledge check 1.7.6

For the decay: $\pi^+ \longrightarrow e^+ + v_e,$

(a) Explain what interaction is responsible.

(b) Show what conservation laws are demonstrated.

≫ Study point

The different types of quark, up, down, etc., are said to possess different **flavours**. This strange (!) use of the word is possibly related to the German sour milk product (quark) which is often flavoured with fruit.

Top tip

Indications of a weak force are:

1. Neutrinos are involved: they don't feel the e-m force or the strong force.

2. If quarks are involved, a change in quark flavour occurs.

3. If it is a decay, the lifetime is greater than $\sim 10^{-10}$ s.

Test yourself 1.7

1. Some particles can be described as *elementary* particles; others are *composite* particles. Distinguish between these two types of particle using the proton and the electron as examples.

2. State the difference between baryons and mesons in terms of quark make up.

3. Compare protons and electrons in terms of their structure and interactions.

4. Momentum and mass–energy are always conserved in particle interactions. State to what extent charge, baryon number, lepton number and quark number are conserved.

5. A particle decays by the weak interaction. What conservation laws apply?

6. A baryon with a charge of $+1$ interacts strongly (i.e. with the strong interaction) with a meson of charge -1. What predictions can you make about the products of the interaction? Explain your answer.

7. A neutron and an anti-proton collide. The interaction is strong. What predictions can you make about the products of the interaction? Explain your answer.

8. A first generation meson has a charge of $+1$. State its quark structure.

9. A first generation baryon has a charge of −1. State its quark structure.

10. An electron neutrino collides with a neutron and two different first generation particles are produced.

   (a) Identify the family of particles (leptons, baryons, mesons) to which each of the colliding particles belongs.
   (b) Use conservation laws, which you should identify, to identify the products of the reaction as far as you are able.
   (c) Identify the interaction (strong, e-m, weak) responsible for this interaction. Give reasons.

11. An electron and a positron annihilate each other and produce two photons.

   (a) Explain which interaction (strong, e-m, weak) is involved.
   (b) Why cannot such an annihilation event produce just one photon? (You may assume that the electron and positron had negligible momentum before their interaction.)

12. A proton, and a helium nucleus are each accelerated through a potential difference of 500 V. State the increase in kinetic energy of each:

   (a) in eV and
   (b) in J.

13. The radioactive isotope $^{13}_{7}$N decays by the emission of a positron, when one of the protons in its nucleus transforms into a neutron. The equation is:

$$p \longrightarrow n + e^+ + X$$

   where X is an unidentified particle. The half-life of the decay is 10.1 minutes.

   (a) Identify X and justify your choice in terms of relevant conservation laws.
   (b) State which of the interactions controls this decay. Justify your choice.

14. (a) Give the decay equation for neutron decay.
   (b) Explain what is meant by *the conservation of lepton number*. Illustrate your answer using the decay of a neutron.
   (c) Explain what other conservation laws are illustrated by the neutron decay equation.
   (d)

   ◀ **Stretch & challenge**

   The mass-energies of protons, neutrons and electrons are 938.3 MeV, 939.6 MeV and 0.5 MeV respectively. The mass–energy of a neutrino is negligible. Use this information to explain why, although an isolated neutron can decay into a proton, the decay of a free proton has never been observed.

15. The delta particles usually decay into either a proton or a neutron plus a charged pion (π^+ or π^-) in about 10^{-24} s. See the end of Section 1.7.6 for the decay of Δ^-. This decay can be written in two ways:

$$\Delta^- \longrightarrow n + \pi^- \quad \text{and} \quad ddd \longrightarrow udd + \bar{u}d$$

Write equations for the decays of the other Δ particles in the same way. What indications are there that these decays take place by the strong interaction?

Unit 1 equations

The WJEC Data Booklet contains the equations you may need to use in the examination. The symbols in the equations are not identified in the Data Booklet: they are standard symbols and you are expected to recognise them. The equations below are the ones needed for Unit 1.

Equation	Description
$\rho = \dfrac{m}{V}$	The density equation: ρ = density, m = mass, V = volume
$v = u + at$ $x = \dfrac{1}{2}\left(u + v\right)t$ $x = ut + \dfrac{1}{2}at^2$ $v^2 = u^2 + 2ax$	The kinematic equations for constant acceleration. The symbols are: t = time x = displacement (x = 0 when t = 0) u = initial velocity v = final velocity
$\Sigma F = ma$	Equation from Newton's 2nd law of motion. ΣF = resultant force m = mass, a = acceleration
$p = mv$	The definition of momentum, p, with m = mass and v = velocity
$W = Fx\cos\theta$	The definition of the work, W, done by a force: F = force, x = displacement, θ = angle between F and x.
$\Delta E = mg\Delta h$	Increase in gravitational potential energy, ΔE
$E_\mathrm{p} = \dfrac{1}{2}kx^2$	Elastic potential energy, E_p. k = spring constant, x = extension
$E_\mathrm{k} = \dfrac{1}{2}mv^2$	Kinetic energy, E_k
$Fx = \dfrac{1}{2}mv^2 - \dfrac{1}{2}mu^2$	Work done = change in kinetic energy
$P = \dfrac{W}{t} = \dfrac{\Delta E}{t}$	Power $= \dfrac{\text{work}}{\text{time}} = \dfrac{\text{energy transfer}}{\text{time}}$
efficiency $= \dfrac{\text{useful energy transfer}}{\text{total energy input}} \times 100\%$	
$F = kx$	Hooke's law. F = force, k = spring constant, x = extension
$\sigma = \dfrac{F}{A}$	Definition of stress, σ. F = force, A = cross-sectional area
$\varepsilon = \dfrac{\Delta \ell}{\ell}$	Definition of strain, ε. $\Delta \ell$ = increase in length, ℓ
$E = \dfrac{\sigma}{\varepsilon}$	Definition of the Young modulus
$W = \dfrac{1}{2}Fx$	Work, W, done in stretching a Hookean material. F = final (maximum) force, x = extension
$\lambda_{\max} = \dfrac{W}{T}$	Wien's law. $\lambda_{\max}$ = wavelength of maximum emission, W = Wien's constant, T = kelvin temperature
$P = A\sigma T^4$	Stefan-Boltzmann law. P = emitted power, A = surface area, σ = Stefan's constant, T = kelvin temperature

Unit 1

1 (a) State, in words, the equation used to calculate the moment of a force about a point. **[1]**

(b) The picture and diagram show a window hinged at the upper surface. The window is opened by pushing on the horizontal metal bar attached to its lower surface. Holes are drilled into the metal bar so that the window can be supported at various opening positions, one of which is shown below and labelled as Position 1. The hinge provides no resistance to the movement of the window.

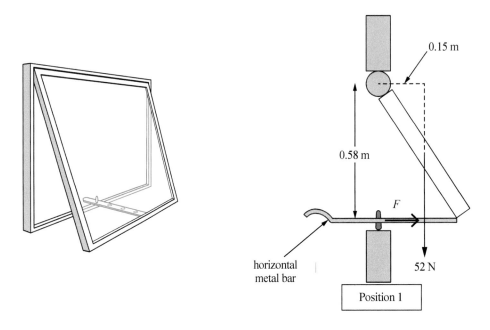

(i) Show that the clockwise moment produced by the weight of the window is approximately **8 N m**. **[1]**

(ii) Hence, calculate the force, F, the metal bar exerts on the window. **[2]**

(c) Tom and Bethan discuss how the force in the metal bar changes with changing positions. Tom thinks that the force in the bar is greater when the window is in Position 2, whereas Bethan believes that the force is greater when the window is in Position 1. Discuss who is correct, giving a detailed explanation in terms of moments. Assume the metal bar is horizontal in both positions. **[4]**

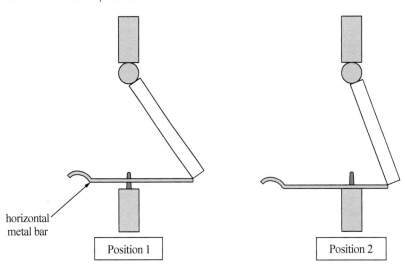

(Total 8 marks)

[WJEC AS Physics Unit 1 2018 Q1]

2 **(a)** Describe a method to investigate the force–extension properties of rubber in the form of an elastic band as it is loaded. You should describe how the extension of the rubber is accurately measured. **[3]**

(b) The results from such an experiment for a rubber band of unstretched length **8.0 cm** are plotted in a graph.

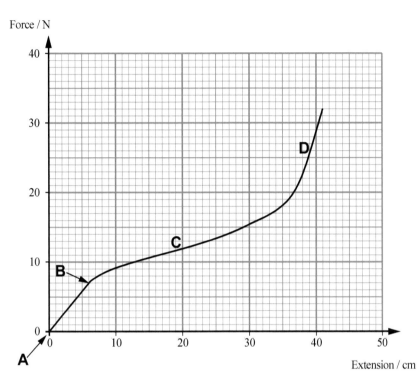

(i) Calculate the strain in the rubber at point **B**. **[1]**

(ii) Determine the Young modulus of the rubber in the region **AB**. Assume the band has a total cross-sectional area of 0.050 cm^2. **[3]**

(c) By referring to the molecular structure of rubber, explain why the gradient at **C** is less than the gradient at **D**. **[3]**

(Total 10 marks)

[*WJEC AS Physics Unit 1 2018 Q3*]

3 **(a)** The table shows information about some sub-atomic particles.

Particle	Symbol	Quark combination	Charge/e	Baryon number
proton	p	uud	+1	1
delta particle	Δ^{++}	uuu	....	....
electron	e^-	no quarks present	....	....
pion	π^-	....	−1	....

(i) Complete the table. **[3]**

(ii) Identify the lepton in the table. **[1]**

(b) JJ Thomson, when studying the properties of cathode rays in 1897, discovered the electron. In the early 20th century, Ernest Rutherford, carrying out a series of experiments on radioactive substances, discovered the proton. The following interaction between protons and electrons has been observed by using high energy particle accelerators.

$$e^- + p \longrightarrow e^- + \Delta^{++} + \pi^-$$

Show how charge and lepton number are conserved in the above interaction. **[2]**

(c) The Δ^{++} decays in about 6×10^{-24} s as shown below.

$$\Delta^{++} \longrightarrow p + \pi^+$$

(i) Show clearly that both up-quark number and down-quark number are conserved in this decay. **[2]**

(ii) Give **two** reasons for believing that this decay is a strong force interaction. **[2]**

(d) During a press conference, the spokesman for a nuclear research centre was asked the question:

'You have discovered many new particles, none of which have had any discernible impact on society. How do you justify the huge expense of continuing with these experiments?'

In response, the spokesman referred to the work of JJ Thomson and Ernest Rutherford.

Suggest why the spokesman responded in this way. **[2]**

(Total 12 marks)
[*WJEC AS Physics Unit 1 2018 Q4*]

4 The diagram shows part of a rollercoaster ride at a theme park.

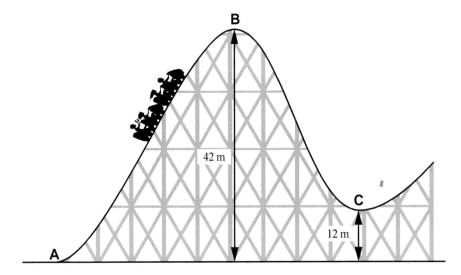

(a) A motor with a power output of $65\ \text{kW}$ and a chain mechanism pulls the carriages of mass $2600\ \text{kg}$ from **A** to **B** in a time of 32 s.

(i) Show that the work done by the motor in 32 seconds is approximately 2 MJ. **[1]**

(ii) Hence calculate the efficiency of the mechanism, assuming the carriages are momentarily at rest at **B**. **[3]**

(b) At **B**, the carriages become disconnected from the motor and the carriages move under the influence of gravity for the rest of the ride. In moving from **B** to **C**, a distance along the track of 36 m, the carriages experience a mean resistive force of 2.8 kN. Calculate the speed of the carriages at **C**. **[5]**

(Total 9 marks)

[WJEC AS Physics Unit 1 2018 Q7]

5 (a) A black body graph of spectral intensity against wavelength for a star is shown. A magnified section, showing the finer detail of the spectrum is also given. An associated line spectrum is also shown.

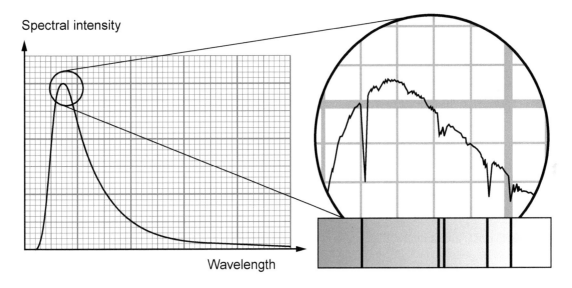

Explain how the graph and the spectra can be used to provide information about the star and the elements from which it is made. **[6 QER]**

(b) (i) Altair is the brightest star in the Aquila constellation. It is 1.58×10^{17} m away, and the intensity of its electromagnetic radiation reaching the Earth is 1.32×10^{-8} W m^{-2}.

Show that its luminosity is approximately 4×10^{27} W. **[3]**

(ii) Calculate Altair's diameter given that its surface temperature is 7700 K. **[3]**

(Total 12 marks)

[WJEC AS Physics Unit 1 2018 Q6]

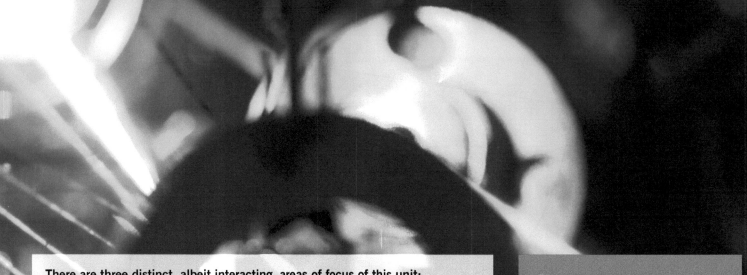

Unit 2

Electricity and light

There are three distinct, albeit interacting, areas of focus of this unit:

- Electricity

  The nature of an electric current is explored, together with the way in which different materials and devices respond to currents. The properties of circuits and power supplies are examined, allowing students to predict their behaviours.

- Waves

  Like electricity, wave motion forms a cornerstone of modern physics. Waves are classified and their properties examined in mathematical detail. The wave model of light is used to explain the phenomena of refraction, diffraction and interference.

- Photons

  We now understand that, as well as having wave properties, electromagnetic radiation behaves like a stream of particles, called photons. Evidence for this is presented and this model is used, together with knowledge of atoms, to account for the atomic absorption spectra, which were introduced in Unit 1. The basics of laser operation are explored.

Content

2.1 Conduction of electricity
2.2 Resistance
2.3 Direct current circuits
2.4 The nature of waves
2.5 Wave properties
2.6 Refraction of light
2.7 Photons
2.8 Lasers

Practical work

Unit 2 provides a wealth of opportunities, especially in the electricity and wave sections, for students to continue developing their practical skills.

2.1 Conduction of electricity

This short section gives the elementary facts about electric charge (forces between charges, why we say that electrons have a *negative* charge, conservation of charge and so on). We're then able to discuss moving charges and to relate quantitatively the velocity of charges moving in a wire to the current in the wire.

2.1.1 Electric charge

This section is mainly about *electric charge* (simply charge from now on) flowing in electrical conductors. An electrical *conductor*, as opposed to an electrical *insulator* such as air, is indeed defined as a material, or a piece of material, through which a charge can flow.

Charge was first investigated as *static* (i.e. stationary) – residing on the surfaces of materials that had been rubbed with other materials. (If the rubbed materials are conductors, they have to be held with insulators, so that their charges aren't conducted away, by hands, for example.)

Two sorts of charge were proposed. These were enough to account for the attractive and repulsive forces observed between *any* rubbed materials.

- Glass, after rubbing with silk, was said to have a *positive* charge.
- Amber, after rubbing with fur, was said to have a *negative* charge.

The attractive and repulsive forces then fitted the rule:

Like charges (e.g. two positives) repel, unlike charges attract (see Fig. 2.1.2).

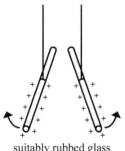
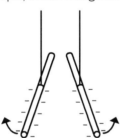
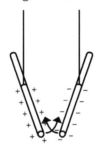

suitably rubbed glass (or perspex...) rods repel

suitably rubbed amber (or polythene...) rods repel

suitably rubbed glass (or perspex) rod and amber (or polythene...) rod attract

Fig. 2.1.2 Attraction and repulsion of charged objects

Positive and *negative* are apt names. For one thing, the different charges can cancel, or neutralise, as when oppositely charged metals touch.

Over a century after these discoveries had been made, the particles inside the atom were discovered. Protons have a positive charge (according to the glass-rubbed-with-silk definition) and electrons a negative charge. So we now picture the charging of the glass as some electrons being rubbed off some of the surface atoms of the glass on to the silk.

Charge can be quantified, and an amount of charge is a scalar quantity usually denoted by Q or q. The charge on a proton, that is the amount of charge that a proton has, is denoted by e. An electron's charge is $-e$. For practical purposes, we use a unit far larger than e.

The SI unit of charge, Q, is the coulomb (C). The coulomb is now defined as

$$1\ \mathrm{C} = 6.241\,509\,074 \times 10^{18} \times e.\ \text{(See the Study point.)}$$

Fig. 2.1.1 Charles-Augustin de Coulomb

Knowledge check 2.1.1

The base of a typical thunder-cloud has a negative charge of 150 C. Calculate the number of excess electrons that it carries.

Knowledge check 2.1.2

A polythene rod acquires a negative charge of 3.2 nC when rubbed with fur. Explain what happens in terms of electrons, calculating how many are involved.

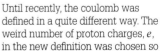

So the charge on a proton is $e = \dfrac{1\ C}{6.2415 \times 10^{18}} = 1.602 \times 10^{-19}\ C$

Although positive and negative charges can 'neutralise' each other, we have:

The law of conservation of charge
The net charge in a system remains constant (provided that charges can't enter or leave).

This law has no exceptions as far as we know. It applies even when particles are created or destroyed, for example when a neutron (no charge) 'decays' into a proton and an electron (with equal and opposite charges) and an antineutrino (no charge). A more mundane case of charge conservation would be the touching together of two metal spheres, one with a positive charge, the other with a negative. Net charge remains the same; the 'neutralisation' – complete or partial – is simply due to redistribution of free electrons.

2.1.2 Electric current

Charge flows through wires in an electric circuit. We say that there is an electric current in the circuit. The flow is invisible, so how do we know that it takes place at all? One indicative experiment uses the apparatus shown in Fig. 2.1.3.

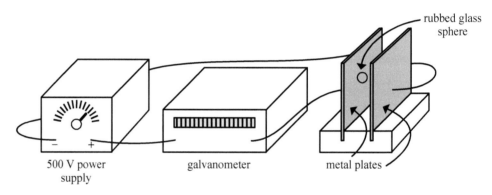

Fig. 2.1.3 Charge flow and current

Study point

The ampere is named after André-Marie Ampère, who discovered (in the 1820s) several magnetic effects of electric currents.

When the power supply is turned on, the galvanometer (a sensitive detector of electric current) deflects briefly. If a small glass ball (hollow like a Christmas tree bauble for low mass) is now rubbed with silk and dangled on a long insulating thread in the gap, it accelerates away from one plate (which must therefore be charged positively) and towards the other (charged negatively).

The galvanometer deflection coincides with the plates acquiring charges – which must be via the connecting wires. The power supply urges electrons in one direction through the wires and its own internal conducting pathway so that some electrons are taken from one plate, and extra electrons are deposited on the other.

(a) Conventional current

Before the discovery of protons and electrons, it wasn't known whether it was positive or negative charge that flowed in conductors. Scientists made a convention (an agreement) to assume it was positive. *In circuit diagrams, arrows denoting currents still show the direction in which positive charge would flow.*

We now know that in metals it is electrons that flow – in the opposite direction to the conventional current! The positive nucleus of each atom is surrounded by most of the atom's electrons, making a positive ion. The ions vibrate randomly but about fixed positions in a regular crystal *lattice* (Fig. 2.1.4). In most metals only a small proportion of the electrons are free to flow. For example, in copper, each atom contributes one electron to the 'pool' of free electrons.

From now on we shall be dealing with circuits having complete conductive paths. The simplest is shown in Fig. 2.1.5. Charge flows continuously. Note that symbols used in circuit diagrams will be labelled *the first time* they are used. You need to learn any that you don't know.

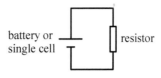

Fig. 2.1.5 Simple circuit

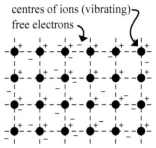

Fig. 2.1.4 Highly stylised metal structure

(b) Current as a measurable quantity

Electric current is defined as follows:

The **electric current**, I, through a conductor is the rate of flow of charge (the charge passing per unit time through a cross-section of the conductor).

In symbols: $I = \dfrac{\Delta Q}{\Delta t}$

Unit: ampere = A = C s⁻¹

Example: using the definition of current

By using a variable power supply, the current through an electric lamp is made to vary as shown in Fig. 2.1.6. Calculate the charge that passes through the lamp:

(a) between 40 s and 100 s

(b) between 0 and 40 s.

Fig. 2.1.6 Current–time graph

Answer

(a) $I = \dfrac{\Delta Q}{\Delta t}$ so $\Delta Q = I\,\Delta t = 2.0\text{ A} \times 60\text{ s} = 120\text{ C}$

Notice that this is the 'area' under the graph line between 40 s and 100 s, calculated not literally in area units but in units read from the scales!

(b) We can't straightforwardly use $\Delta Q = I\,\Delta t$ in which $\Delta t = 40$ s, because I keeps changing over that time. But we can use the 'area' under graph line method.

Thus $\Delta Q = \frac{1}{2} \times 2.0\text{ A} \times 40\text{ s} = 40\text{ C}$

Ammeters

An ammeter (Ⓐ) is an instrument for measuring current. The current to be measured must be routed through it. Analogue (pointer-and-scale) and digital types work on quite different principles, but both have low resistance conductive pathways for the current, as we don't want their inclusion in a circuit to reduce the current significantly.

Current is the same all around a series circuit

This means, for example, that in Fig. 2.1.7, $I_1 = I_2 = I_3$. This follows from the law of conservation of charge. Take $I_1 = I_2$; it means that the rate of flow of charge into the resistor is the same as the rate of flow of charge out of it; charge can't disappear or be created. But couldn't free electrons pile up in the resistor? No: excess free electrons, all having the same sign of charge (negative) would repel one another.

> **Study point**
>
> We all fall into the trap sometimes, but try to avoid writing 'Current flows…' If it meant anything it would mean 'rate of flow of charge flows'.

> **Knowledge check** `2.1.3`
>
> A rechargeable battery is labelled '2400 mAh'. This means that, after charging, it can sustain a current of 2400 mA for 1 hour, 1200 mA for 2 hours and so on. Calculate:
>
> (a) the charge, in coulomb, that passes through the battery before it needs recharging
>
> (b) the time for which a current of 500 mA could be sustained.

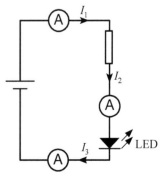

Fig. 2.1.7 Current in a series circuit

Experimentally, you would expect to find that all three ammeters read the same. If they didn't, what do you do before claiming that you've disproved $I_1 = I_2 = I_3$? You could swap round the positions of the ammeters and see whether the readings followed the meters. If so, what would you deduce? Or you could just put a single ammeter in different places in turn!

2.1.3 How current depends on drift velocity of free electrons in a metal

(a) Thermal motion and drift velocity

Free electrons share in the random thermal energy of the metal. They continually collide with the vibrating ions in the metal, undergoing changes in their directions of motion as well as in their speeds. (A typical thermal speed for an electron at room temperature is 100 km s^{-1}.) This random motion doesn't produce a flow of charge along the wire.

Charge *does* flow along the wire if we connect a battery across its ends, as electrons acquire a so-called *drift* velocity along the wire. Although this drift velocity is superimposed on their random thermal velocity, in what follows we can forget about the random motion and treat all electrons as moving with the mean drift velocity.

(b) Derivation of the equation $I = nAve$

Suppose that the mean drift velocity of free electrons in the wire is v.

Then in a time Δt the free electrons in a length $v\Delta t$ of wire, that is in a volume $Av\Delta t$, will pass through the shaded cross-section, of area A. (See Fig. 2.1.8.)

Fig. 2.1.8 To help show $I = nAve$

Now let n denote the *free electron concentration* in the metal, that is the number of free electrons per unit volume.

Then, in volume $Av\Delta t$ of wire there are $nAv\Delta t$ electrons.

So, in time Δt, the charge, ΔQ, passing through the cross-section is $\Delta Q = nAve\Delta t$

But current, $I = \dfrac{\Delta Q}{\Delta t}$ So $I = nAve$

Note: An electron's charge is $-e$, so we could have written the last equation as $I = -nAve$. However, the minus is usually omitted and we simply *remember* that the conventional current (see section 2.1.2) is in the opposite direction to the electron drift velocity.

Example: using $I = nAve$

An insulated copper wire in a car's headlamp circuit carries a current of 8.0 A. The diameter of the copper conductor is 2.5 mm. Calculate the drift velocity of the free electrons in the wire. The free electron concentration in copper is $8.47 \times 10^{28} \text{ m}^{-3}$.

Answer

$I = nAve$ so $v = \dfrac{I}{nAe} = \dfrac{I}{n\pi\left(\frac{d}{2}\right)^2 e}$ in which d is the diameter of the wire.

So, substituting the numbers,

$$v = \dfrac{8.0 \text{ A}}{8.47 \times 10^{28} \text{ m}^{-3}\pi\left(\dfrac{2.5 \times 10^{-3} \text{ m}}{2}\right)^2 1.60 \times 10^{-19} \text{ C}} = 1.2 \times 10^{-4} \text{ m s}^{-1}$$

> **2.1.4 Knowledge check**
>
> Suppose the copper wire in the example of using $I = nAve$ had only half the diameter. Calculate v for the same current. Try to do this by noting the single factor that needs inserting, rather than by doing the same calculation all over again with 1.25×10^{-3} m instead of 2.5×10^{-3} m!

Comments on example

- Check that the units work out.

- Note how low the drift velocity is, even though **8.0 A** is a fair-sized current. It would take a free electron over 2 hours to travel 1 metre through the wire!

- However, the free electrons start drifting through the wire within nanoseconds of closing the switch that turns on the headlights. The so-called 'electric field' that sets them drifting travels along the wire at almost the speed of light.

Test yourself 2.1

1. Calculate the charge on a $^{235}_{92}\text{U}$ nucleus.

2. When a Van de Graaff generator is ready to produce sparks, the negative charge per m^2 on the surface of its dome is approximately $10\ \mu\text{C m}^{-2}$.

(a) Calculate:
   (i) the number of excess electrons per m^2 on the surface of its dome
   (ii) the number of excess electrons on the surface of its dome, approximating the dome as a sphere of radius **13 cm**.
(b) The *total* number of electrons in the metal of the dome is approximately 1×10^{13} times larger than the answer to (a) (ii). Explain how it is ever possible for the dome to be electrically neutral.

3. The artificial radioactive material $^{99}_{43}\text{Tc}$ is β⁻ radioactive. A sample of 1.0 mg of $^{99}_{43}\text{Tc}$ has an activity of **640 MBq**, i.e. it gives out **640 million** β⁻ particles every second.

Calculate the electric current between the $^{99}_{43}\text{Tc}$ sample and its support for the sample to remain electrically neutral. State the direction of the current.

4. The current supplied by a **Ni-Cd** rechargeable battery is monitored over its discharge and this graph is produced.

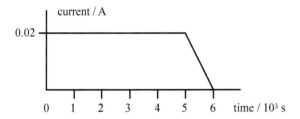

(a) Calculate the total charge passing through the battery.
(b) The battery is labelled as having a capacity of **100 mAh** (milliampere hour). Comment on this.

◄Stretch & challenge

1. 69.1% of natural copper atoms are $^{63}_{29}\text{Cu}$, each of which has a mass of 1.046×10^{-25} kg, and 30.9% $^{65}_{29}\text{Cu}$, with a mass of 1.079×10^{-25} kg.

The density of copper is 8930 kg m⁻³. Calculate the total charge carried by all the electrons in a **1 cm** cube of copper.

2. This question involves some concepts from Section 2.2.

A capacitor, C, (see question 5) is initially uncharged. It is connected into the circuit shown:

When the switch is closed, the pd across the resistor varies with time as follows:

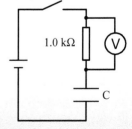

time / s	2.0	5.0	7.0	10.0	12.0	15.0	20.0
pd / V	6.53	3.54	2.02	1.19	0.78	0.41	0.14

(a) By drawing a suitable graph, estimate the total charge transferred around the circuit as the capacitor charges up.
(b) Determine the half-life of the *I–t* relationship.
(c) What fraction of the total charge is transferred in the half-life you determined in part (b)?

5. A capacitor is a device for storing separated charge. A student connects a resistor across a charged capacitor, which discharges. The charge, Q, on one of its plates is plotted against time (see graph).

Use the graph to calculate:

(a) The mean current between $t = 0$ and 15 s.
(b) The initial current, i.e. the current at $t = 0$ s.

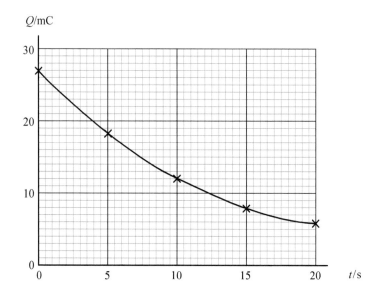

6. The student in Q5 notices that the graph looked similar to a radioactive decay graph. This means that the time taken for Q to halve should always be the same. By taking readings from the graph, show that this appears to be true and find the value of this half-life.

7. The current in a metal wire of cross-sectional area A is given by $I = nAve$.

(a) State the meanings of n and v.
(b) Two copper wires, P and Q, are connected in series. The diameter of P is 3.0 times the diameter of Q.

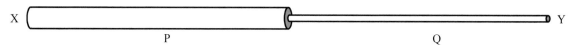

(i) List the quantities in the equation that have the same value in P and Q.
(ii) Deduce how the value of v in Q compares with the value of v in P, explaining your reasoning.

8. Aluminium has an ionic radius of **63 pm**. Metallic aluminium has 3 free electrons per aluminium ion.

(a) By considering the volume of an aluminium ion, estimate the number of free electrons per m^3 of metallic aluminium.
(b) Calculate the drift velocity of free electrons in a **1 cm** diameter aluminium conductor carrying a current of **1 kA**.

9. Germanium is a semiconductor. It has many fewer mobile charge carriers than do metallic conductors. In germanium at room temperature there are approximately 18 mobile charge carriers for every million atoms of germanium. A 1 mm diameter germanium wire carries a current of **10 mA**.

Estimate the drift velocity of the charge carriers from the following data:

Mass of a germanium atom = 1.20×10^{-25} kg;

Density of germanium = 5.3×10^3 kg m^{-3}

10. A 1 mm diameter copper wire carries an alternating current which varies as shown in the graph on the right. The free electron concentration is 8.5×10^{28} m^{-3}.

(a) Sketch a graph of how the electron drift velocity varies with time.
(b) Use your graph to estimate the distance drifted by a free electron in half a cycle. Comment on your answer.

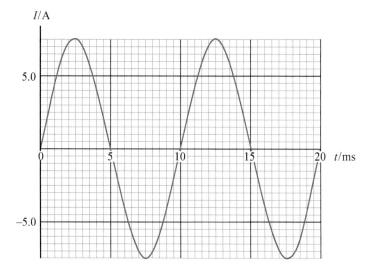

2.2 Resistance

Having studied current as rate of flow of charge, we now turn our attention to the accompanying *energy* transfers. The key idea is potential difference (pd), and we get to grips with this first of all. It enables us to define the very useful concept of *resistance*, and we discuss in some detail what gives a metal wire its resistance. Finally, we look briefly at the extraordinary phenomenon of superconductivity.

2.2.1 Potential difference

A voltmeter (ⓥ) reads the *potential difference* (pd) between two points, X and Y, in a circuit – as long as one of its leads (wires) is connected to X and the other to Y! So, in Fig. 2.2.1, the voltmeter will tell us the pd across the resistor, *not* across the filament lamp. If we want the pd across a particular component, the voltmeter must be connected across that component.

X is said to be at a higher potential than Y if the voltmeter gives a positive reading when the meter lead that is red or labelled '+' is connected to X.

Suppose the voltmeter reads **6.0 V**. This tells us that *for every coulomb passing between X and Y, 6.0 J of work is done, resulting in 6.0 J of energy changing category: from electrical potential energy to another form.*

Voltage is a term often used to mean potential difference. A formal definition of *potential difference* is given in the Key term box above.

If there is a resistor between **X** and **Y**, as in Fig. 2.2.1, charge does work as it travels through the resistor. The free electrons continually collide with ions (more about this in Section 2.2.4) and these collisions amount to a resistive force against which the work is done. As it moves through the resistor, charge loses electrical potential energy and the resistor acquires random thermal energy (the ions vibrate more vigorously!). This process is often called *energy dissipation*. The random thermal energy soon escapes to the surroundings as heat.

Examples

1. If the pd across a resistor is **6.0 V**, and the current through it is **1.5 A**, calculate how much energy is dissipated in it every minute.

Answer

Energy dissipation = Energy transferred per unit charge × charge passing

$$= \text{pd} \times (\text{current} \times \text{time})$$

$$= 6.0 \text{ V} \times 1.5 \text{ A} \times 60 \text{ s}$$

$$= 540 \text{ J}$$

2. Referring to Fig. 2.1.3, the glass ball loses electrical potential energy as it goes across the gap from the positive plate to the negative. It gains kinetic energy. (This is the principle of the particle accelerator.)

Suppose the ball has a mass of **10 g** and a charge of **6.0 nC**, and that it starts from rest and reaches a speed of **0.060 m s⁻¹**. What would be the pd between the plates?

Answer

$$\text{pd} = \frac{\text{Electrical PE lost}}{\text{charge passing}} = \frac{\text{KE gained}}{\text{charge passing}} = \frac{\frac{1}{2} \times 0.010 \times 0.060^2 \text{ J}}{6.0 \times 10^{-9} \text{ C}} = 3000 \text{ V}.$$

Key term

Potential difference, *V*: The work done between two points, X and Y, that is the loss of electrical potential energy, per unit charge passing between X and Y,

Unit: volt (V) = J C⁻¹.

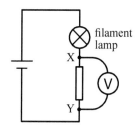

filament lamp

Fig. 2.2.1 The voltmeter: knowing its place

Study point

The symbol for pd is V (or ΔV) and, like all symbols for physical quantities, the V is in italics. The symbol for its unit is V, an upright or 'Roman' letter, like all symbols for units.

Study point

To help avoid mistakes, connect the voltmeter last when wiring a circuit.

Knowledge check 2.2.1

A typical flash of lightning has been estimated to carry a mean current of 30 kA for a time of 0.5 ms and to dissipate 450 MJ. Calculate the pd.

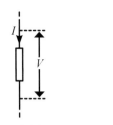

Fig. 2.2.2 Electrical power

2.2.2 Electrical power

Here's a reminder of what *power* means in Physics:

Power is the rate of doing work or rate of transfer of energy.

So Power $= \dfrac{\text{work done}}{\text{time taken}} = \dfrac{\text{energy transferred}}{\text{time taken}}$

Unit: watt $= \mathbf{W} = \mathbf{J\ s^{-1}}$

Now, generalising the reasoning in Example 1 in Section 2.2.1, when there is a current I for a time Δt through a conductor across which the pd is V (see Fig. 2.2.2), then

Work done = Work done per unit charge × charge passing

So   Work done $= V \times I\Delta t$

So   Power $= \dfrac{\text{work done}}{\text{time taken}} = \dfrac{V \times I\Delta t}{\Delta t}$ that is $P = VI$

Example

In a simple particle accelerator, a beam of protons passes from a metal plate X to a metal plate Y. The pd between these plates is $150\ \text{kV}$, with plate X at the higher potential. The number of protons leaving X (and arriving at Y) per second is $7.0 \times 10^{16}\ \text{s}^{-1}$. Calculate the electrical power.

Answer

Current $=$ charge flowing per unit time $= 7.0 \times 10^{16}\ \text{s}^{-1} \times e$

$\qquad = 7.0 \times 10^{16}\ \text{s}^{-1} \times 1.60 \times 10^{-19}\ \text{C} = 11.2\ \text{mA}$

$\therefore$ Power, $P = VI = 150\ \text{kV} \times 11.2\ \text{mA} = 1.7\ \text{kW}$.

2.2.3 I against V (I–V) graphs for conductors

The current through a conductor and the pd across it are related. When one is zero, so is the other. When one increases, so (almost always) does the other. We're using 'conductor' to mean anything that conducts, for example a piece of wire, a resistor, a filament lamp, a diode connected in the 'forward' direction, such as the light-emitting diode shown in Fig. 2.1.7.

Water flow analogy

The electric current through a conductor is rather like the rate of flow of water from one tank to another through a pipe (Fig. 2.2.3). The *height difference* between water levels in the tanks is analogous to *potential difference* in an electric circuit. When the height difference is zero the flow rate is zero. The greater the height difference the greater the flow rate. A pump and pipe to take water from the bottom tank back the top to keep the water levels constant could be included to act like a battery in an electric circuit. You should now study Fig. 2.2.4.

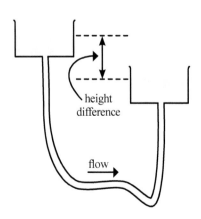

Fig. 2.2.3 Water flow analogy

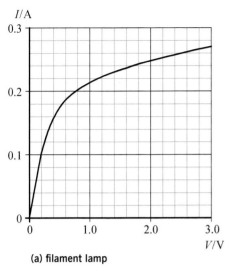

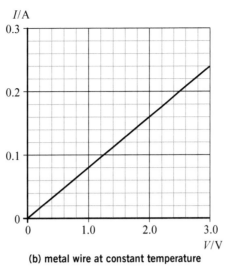

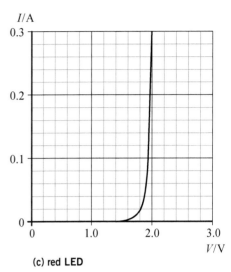

(a) filament lamp

(b) metal wire at constant temperature

(c) red LED

Fig. 2.2.4 *I–V* graphs for various conductors

Fig. 2.2.4(b) for the metal wire at constant temperature stands out because it is a straight line through the origin. The current is proportional to the pd applied. If the pd is doubled, the current doubles, and so on. The conductor is said to be *ohmic*, because it obeys Ohm's law.

> **Ohm's law**
> The current through a conductor is proportional to the pd across it.

The law is obeyed by most single-substance conductors at constant temperature, for example metal wires and carbon resistors.

As you can see from the other two graphs, some quite well-known conducting devices are *non*-ohmic. You should try to remember the shape of the graph for the filament lamp (Fig. 2.2.4 (a)). Some explanation for its shape is given in Section 2.2.6.

2.2.4 Resistance

The **resistance**, R, of a conductor is defined as

$$R = \frac{\text{pd across conductor}}{\text{current through conductor}} \qquad \text{or, in symbols,} \qquad R = \frac{V}{I}$$

Unit: $V\,A^{-1} = \text{ohm} = \Omega$

A conductor obeying Ohm's law has a *constant* resistance whatever pd we put across it – as long as it *continues* to obey Ohm's law!

To convince yourself that this claim is true, calculate the resistance at (say) three different voltages for the metal wire whose *I–V* graph is given in Fig. 2.2.4 (b). If this is too easy, go to the Stretch & challenge.

A conductor that doesn't obey Ohm's law has a resistance that *does* depend on the applied pd. By considering the ratio V/I you should be able to deduce from the graph (Fig. 2.2.4 (c)) that the LED's resistance falls (dramatically) with increasing pd.

For example, at 1.8 V, $R \cong \dfrac{1.8\ \text{V}}{0.016\ \text{A}} \cong 100\ \Omega$ (1 significant figure),

whereas, *as you should check*, at 2.0 V, $R = 6.7\ \Omega$.

◀Stretch & challenge

Use *similar triangles* on an *I–V* graph that is a straight line through the origin to show that R must be constant for an ohmic conductor.

Knowledge check 2.2.4 ◀

For the metal wire in Fig. 2.2.4(b), calculate the pd needed for a current of 0.40 A.

Knowledge check 2.2.5 ◀

There is a current of $2.5\ \mu$A through a resistor when a pd of 9.0 V is placed across it. Calculate the resistance of the resistor, giving your answer in MΩ.

Knowledge check 2.2.6 ◀

Calculate the current through a 4.7 kΩ resistor when a pd of 12 V is placed across it. Give your answer in mA.

2.2.7 Knowledge check

Calculate the resistance of the filament lamp from the graph in Fig. 2.2.4 (a), when the applied pd is: (a) 0.20 V, (b) 3.0 V.

The filament lamp's resistance (Fig. 2.2.4 (a)), on the other hand, *rises* at larger pds, though up to about 0.2 V it hardly changes. At these very small pds, the lamp obeys Ohm's law because the temperature of the filament (a very thin metal wire) is almost constant.

Note that resistance is simply V/I; it is *not* defined in terms of graph *gradients*. It's true that for an ohmic conductor, the resistance *does* equal the reciprocal of the gradient of the I–V graph, but this is not so for non-ohmic conductors except at very low pds.

(a) Resistance in a metal: free electron picture

We start by recalling (see Sections 2.1.2 and 2.1.3) that if charge flows in a metal, it is by the drift of free electrons. For a drift to occur through a metal wire, each free electron needs a force on it, urging it along the wire. Putting it another way, there must be a pd across the wire. [A pd implies *work* done on the charges, which implies forces acting on them in the direction of their motion.]

There seems to be a flaw in the argument just presented: a constant force on a free electron will give it not a constant velocity but a constant acceleration. Yes – until the electron collides with one of the vibrating ions. [The high thermal speeds of the free electrons will make such collisions very frequent.] The collision wipes out the electron's acquired drift velocity: on average it has to accelerate again from rest. And so on. The result (see Fig. 2.2.5 for crude representation) is a particular *mean* drift velocity, v, and, since $I = nAve$, a particular current, for a given pd placed across the wire. In other words the wire will have a finite *resistance* – due, essentially to collisions between free electrons and vibrating ions.

drift velocity

v

0

time

Fig. 2.2.5 Mean drift velocity, v (stylised)

2.2.8 Knowledge check

Explain why applying a constant pd to a metal wire (at constant temperature) results in a constant drift velocity of free electrons.

(b) Energy dissipation in a metal: free electron picture

The vibrating ions in the metal impede the drift of free electrons, but the drift of free electrons has an effect on the ions! The drift velocity gives the free electrons slightly more kinetic energy than their usual random thermal energy, so they hit the ions harder (on average). This gives the ions (and indeed the free electrons) more *random* vibrational energy, so the wire gets hotter. This electrical heating is sometimes called Joule heating, after James Prescott Joule who helped establish the idea of conservation of energy. The energy transfers in electrical heating are described more broadly in Section 2.2.1.

Joule heating, you might not be surprised to read, is the principle of the electric heater, hair-dryer, toaster and 'conventional' oven. Tungsten filament lamps, in which the filament becomes white hot (or *incandescent*), were common in homes up until a few years ago. They are now rightly condemned as energy wasters, as they give out much more infrared radiation than they do light.

Fig. 2.2.6 James Prescott Joule

(c) Useful equations for power dissipation

Suppose we know that the current through a 5.0 Ω resistor is 0.30 A. How much power is dissipated in it? We could do this is two stages:

pd across resistor is $V = IR = 0.30 \text{ A} \times 5.0 \text{ Ω} = 1.5 \text{ V}$,

so $P = VI = 1.5 \text{ V} \times 0.30 \text{ A} = 0.45 \text{ W}$.

We don't really need, though, to *work out* the pd, because, sticking a little longer with symbols,

$P = VI = (IR)I$, that is $P = I^2R$

Similarly, if we know the resistance, R, of a conductor and the pd, V, across it, we have

$P = VI = V \times \dfrac{V}{R} = \dfrac{V^2}{R}$.

So, summarising: $\qquad P = IV = I^2R = \dfrac{V^2}{R}$

Knowledge check 2.2.9

Calculate the current needed for a $2.0\ \Omega$ heating coil to produce $50\ \text{W}$ of power.

Example

A resistor of resistance $47\ \Omega$ has a maximum power rating of $5.0\ \text{W}$. Calculate the maximum pd that can safely be applied across it.

Answer

So, $P = \dfrac{V^2}{R}$

So, $V = \sqrt{PR} = \sqrt{5.0\ \text{W} \times 47\ \Omega} = 15\ \text{V}$ (2 s.f.)

2.2.5 Resistivity

What are the factors that determine the resistance of a piece of wire?

- Suppose there is a current I in a wire of length ℓ when a pd V is placed across it. Energy is changing category (from electrical potential energy to thermal energy) uniformly along its length. The pd across half of it (Fig. 2.2.7) must therefore be $V/2$.

  So the resistance of length $l/2$ must be:

  $\dfrac{V/2}{I} = \dfrac{1}{2}\dfrac{V}{I} = \dfrac{1}{2} \times$ resistance of length ℓ.

  Generalising, a wire's resistance, R, is proportional to its length, ℓ.

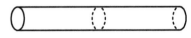

Fig. 2.2.7 Halves of a wire

- But we can also regard the piece of wire as made up of two 'length-ways halves' each with half the cross-sectional area of the original (Fig. 2.2.8). Each will carry current (half that of the whole), so

  Resistance of wire area $\dfrac{A}{2}$ must be $\dfrac{V}{I/2} = 2\dfrac{V}{I} = 2 \times$ resistance of wire area A.

  Generalising, a wire's resistance is inversely proportional to its cross-sectional area, A. (The shape of cross-section doesn't matter, as the distribution of current is determined by free electron concentration, which is the same throughout the metal whatever its shape.)

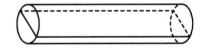

Fig. 2.2.8 Wire split lengthways

We can incorporate the dependencies on ℓ and A in a single equation as follows:

The resistance, R, of a wire of length ℓ and cross-sectional area A is given by

$$R = \dfrac{\rho\ell}{A} \qquad \text{equivalently} \qquad \rho = \dfrac{RA}{\ell}$$

ρ is a constant for the material of the wire at a given temperature, called its **resistivity**. Unit of resistivity: $\Omega\ \text{m}$

Knowledge check 2.2.10

An aquarium heater consists of a coil of wire inside an insulating, waterproof tube. The heater supplies $50\ \text{W}$ of heat when connected to the $230\ \text{V}$ 'mains'. The wire is $0.122\ \text{mm}$ in diameter and made of the alloy 'nichrome', with a resistivity of $1.25\ \mu\Omega\ \text{m}$ at its working temperature. Calculate the resistance, and hence the length, of the wire.

2.2.11 Knowledge check

Inside the pvc sheath of a flexible connecting wire there are 19 strands of copper wire, each of diameter 0.30 mm and length 0.50 m. Calculate the total cross-sectional area and hence the resistance of the wire.

Stretch & challenge

A metal wire is stretched so that its length increases by 1.0%. If its volume and resistivity are unchanged, by what percentage does its resistance increase?

2.2.12 Knowledge check

A tungsten filament lamp is rated at 240 V, 60 W. At 0 °C its resistance is measured to be 67 Ω. Calculate how many times its resistance increases between 0 °C and its working temperature. Hence estimate the working temperature, assuming that for every °C rise in temperature the resistance increases by 0.0045 of its resistance at 0 °C.

Either equation serves as the *definition* of resistivity, ρ, provided that the meanings of the other letters are stated!

The lower the resistivity, the better the material conducts electricity. The best conventional conductor is silver, which has a resistivity of $1.59 \times 10^{-8}\ \Omega\,\text{m}$, at 20°C, followed by copper ($1.68 \times 10^{-8}\ \Omega\,\text{m}$), gold ($2.44 \times 10^{-8}\ \Omega\,\text{m}$) and aluminium ($2.82 \times 10^{-8}\ \Omega\,\text{m}$). Iron ($9.72 \times 10^{-8}\ \Omega\,\text{m}$) is unimpressive. Contrast metals with insulators: the resistivity of sulfur is about $1 \times 10^{15}\ \Omega\,\text{m}$.

Measuring the resistivity of a metal in the form of a wire is in Section 2.2.8.

Example

A metal wire, 2.00 m long and 0.400 mm in diameter is found to have a resistance of 1.50 Ω at 18 °C. Determine the resistivity of the metal at this temperature.

Answer

$$\rho = \frac{RA}{\ell} = \frac{1.50\ \Omega \times \pi\ (0.200 \times 10^{-3}\ \text{m})^2}{2.00\ \text{m}} = 9.42 \times 10^{-8} \times \frac{\Omega\,\text{m}^2}{\text{m}} = 9.42 \times 10^{-8}\ \Omega\,\text{m}$$

Note how the units of resistance, area and length have been inserted along with the numbers, and how they have been multiplied and divided as algebraic quantities to yield Ω m as the unit of resistivity.

2.2.6 How the resistance of a metal depends on temperature

A metal wire's resistance increases with temperature. This can be investigated using simple apparatus over the temperature range 0°C to 100°C as described in Section 2.2.8. The effect is almost entirely due to the metal's change in resistivity, ρ, since thermal expansion of the wire makes only very small fractional changes to ℓ and A.

The sketch-graph in Fig. 2.2.9 shows a typical relationship, as found by experiment over a wider range of temperatures than obtainable in a non-specialist laboratory.

- Note that we have used the celsius temperature, θ.
- On the vertical axis is the resistivity, ρ, divided by a constant, ρ_0, the resistivity at 0°C. (It's not essential to divide ρ by ρ_0; the benefit of so doing is that the graph will fit most pure metals at least roughly.)

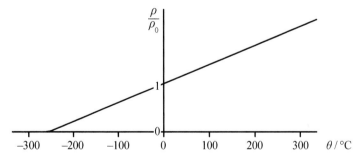

Fig. 2.2.9 Resistivity of a pure metal wire against temperature

The graph is nearly straight over quite a range of temperatures (at least from −100 °C to +200 °C). Here the gradient is different for different metals, but is somewhere around 0.004 °C^{-1} for most pure metals. For alloys it is lower.

The resistivity of metals at very low temperatures is considered later.

Resistance and temperature for a metal: free electron explanation

For a given pd applied across a wire, the mean drift velocity of the free electrons is limited by their collisions with the vibrating ions. The higher the temperature, the greater the vibration amplitude and the shorter the mean time between collisions. This reduces the mean drift velocity, v, and, because $I = nAve$, the current (for a given pd). So, because $R = V/I$, the resistance increases!

Note that in a metal, the free electron concentration, n, doesn't depend on temperature.

Example: the shape of the I–V graph for a filament lamp (Fig. 2.2.4 (a))

Applying a large enough pd causes Joule heating (by harder collisions between free electrons and ions). The temperature rise increases the filament's resistance (because of the increased vibration amplitude of the ions)! So when we double the pd, the current goes up – but less than double. Note that the resistance of a filament at its operating temperature of perhaps 2500 °C may be more than 10 times its resistance at room temperature; see Knowledge check 2.2.12.

2.2.7 Superconductivity

In 1911, the Dutch physicist Heike Kamerlingh Onnes cooled a wire made of (frozen) mercury to lower and lower temperatures, and made the astonishing discovery that, at –269.0 °C, its resistance suddenly dropped to zero, or at least became too low to measure. It had become, as we now say, *superconducting*. Mercury was the first **superconductor** to be found.

Physicists have now observed superconductivity in many metals. The transition temperatures are all within a few degrees of absolute zero (–273.15 °C). Note the sketch-graph of resistance against temperature in Fig. 2.2.10. Among metals which *haven't* been made to superconduct, despite cooling to within a minute fraction of a degree above absolute zero, are copper, silver and gold – the best conductors at ordinary temperatures! See Fig. 2.2.11.

In 1986, it was discovered that certain ceramic materials could be made to superconduct, with transition temperatures much higher than those of metals, and mostly somewhat above –196 °C, the boiling point of liquid nitrogen. This is significant because liquid nitrogen can therefore be used as a (relatively cheap) coolant, to keep these so-called *high temperature superconductors* superconducting.

Is the resistance of a superconductor really zero? A current once started in a ring of superconducting metal has been found not to diminish noticeably over periods of years, even with no applied potential difference!

Uses of superconductors

Superconducting wires will carry currents without dissipating any energy at all. As well as the energy *saving*, there is no unwanted heat to get rid of.

There is a limit to the current that a superconducting wire can carry. This is not because it gets hot (there's no Joule heating!) but because it gives rise to a magnetic field, and too great a magnetic field makes a superconductor 'go normal' *even at temperatures below* θ_c.

Several prototype electrical power transmission cables have been set up using 'high temperature' (ceramic) superconductors. Keeping the whole length of the cable very cold is seriously expensive, but the energy savings could make such systems economic.

⟨ Link ⟩

See Resistance in a metal: free electron picture in Section 2.2.4.

Knowledge check 2.2.13

Explain, in terms of free electrons, why the resistance of a lamp filament increases when the pd across it is increased.

Knowledge check 2.2.14

A student writes: 'When the pd across a filament lamp is increased the filament's resistance increases so the current decreases.' Rewrite this statement so that it is correct.

≫ Key term

Superconductor: Material that loses all its electrical resistance below a certain temperature, the *superconducting transition temperature* (or superconducting critical temperature), θ_c.

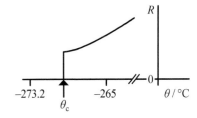

Fig. 2.2.10 Superconducting transition

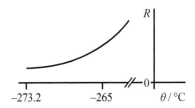

Fig. 2.2.11 Non-superconducting metal

Fig. 2.2.12 MRI magnet coil

Electromagnets producing large magnetic fields over quite large volumes of space are needed in MRI (magnetic resonance imaging) machines for medical diagnosis, for most types of particle accelerator, and for magnetically levitating vehicles. Superconducting wires are routinely used for the coils of these electromagnets. Whereas the coils of conventional electromagnets need iron cores, superconducting coils don't, because much higher currents are possible. In an MRI machine, this leaves room for the patient – quite an advantage. (To prevent the superconductor going normal because of the magnetic field, it has to be cooled *well below* its transition temperature.)

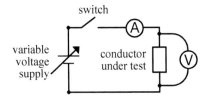

Fig. 2.2.13 Circuit for obtaining V and I readings

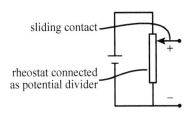

Fig. 2.2.14 A variable voltage supply

2.2.8 Specified practical work

(a) Investigation of the *I–V* characteristics of the filament of a lamp and a metal wire at constant temperature

We place the conductor under test in the circuit shown in Fig. 2.2.13. Note the ammeter in series with it and the voltmeter across it.

- The variable voltage supply could be a purpose-made mains-powered unit or the 'potential divider' circuit shown in Fig. 2.2.14. How this works will be explained in Section 2.3.3.
- *Starting from zero*, we increase the pd in steps from zero up to the maximum allowed for the conductor under test (e.g. a filament lamp might be labelled '3 V, 0.45 A'), taking readings of I and V each time.
- It's usual to present the results on an *I–V* graph: I (vertical) against V. We take extra readings where the local shape of the graph is not clear, for example where it bends. For a curved graph we'd expect to take about ten pairs of readings, for a straight graph, perhaps fewer.

(b) Determining the resistivity of a metal

It's easiest to investigate an uninsulated (bare) wire made of an alloy (e.g. constantan) with a relatively high resistivity. Essentially, we need to measure the wire's resistance, R, length, ℓ and diameter, d, since:

$$R = \frac{\rho\ell}{A} = \frac{\rho\ell}{\pi\left(\frac{d}{2}\right)^2} \text{ that is } R = \frac{4\rho\ell}{\pi d^2} \text{ so } \rho = \frac{\pi d^2 R}{4\ell}.$$

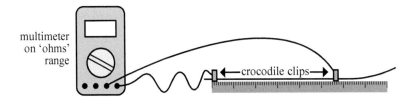

Fig. 2.2.15 Measuring resistance and length of thin wire

R We can use a digital meter on its *ohms* range. Its zero error (what it reads when the crocodile clips at the ends of its leads are held together) must be subtracted from any resistance reading. Alternatively we could use a battery, ammeter and voltmeter, connected as suggested in Fig. 2.2.16.

ℓ The length of wire between the crocodile clips is measured with a metre rule. We need to minimise uncertainties due to parallax, and to the wire not being straight.

d The wire is likely to have a diameter of less than **0.3 mm**, so an absolute uncertainty of **0.01 mm** would constitute an uncertainty of more than **3%** in d and of more than **6%** in d^2, since it is d^2 that we see in the equation for ρ. We certainly need an instrument with a resolution no coarser than **0.01 mm**: either electronic calipers or micrometer screw gauge will serve. We take the mean of five or six measurements spaced along the length of the wire, and across different diameters and at right angles (Fig. 2.2.17).

We could determine ρ by putting one set of mean measurements into the equation. However, it's instructive to measure the resistance of progressively longer lengths of the same wire, and to plot a graph of R against ℓ as in Fig. 2.2.18.

Since $R = \dfrac{4\rho\ell}{\pi d^2}$, we expect the graph to be a straight line through the origin, with a gradient of $\dfrac{4\rho}{\pi d^2}$. So $\rho = \dfrac{\pi d^2}{4} \times$ gradient. See Study point.

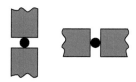

Fig. 2.2.16 Alternative circuit for measuring resistance

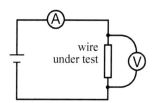

Fig. 2.2.17 Measurement at right angles

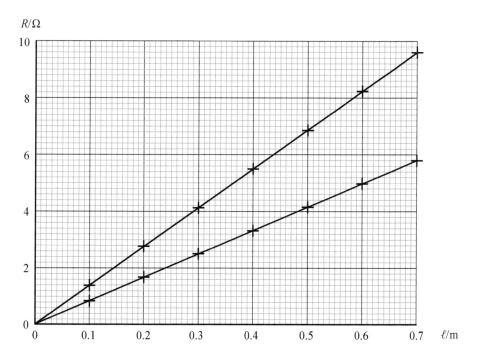

Fig. 2.2.18 Resistance against length for constantan wires

>> **Study point**

The mean value of d, the diameter of the wire for which the **top** R–ℓ graph in Fig. 2.2.18 was plotted, was 0.215 mm ($\pm$0.01 mm). The gradient of the graph is

$$\frac{(9.60 - 0.00)\ \Omega}{(0.700 - 0.000)\ \text{m}} = 13.7\ \Omega\ \text{m}^{-1}$$

This gives a resistivity of

$$\rho = \frac{\pi d^2}{4} \times \text{gradient}$$

$$= \frac{\pi (0.000215\ \text{m})^2}{4} \times 13.7\ \Omega\ \text{m}^{-1}$$

$$= 5.0 \times 10^{-7}\ \Omega\ \text{m}$$

Knowledge check 2.2.15

The **lower** line in Fig. 2.2.18 is plotted from R and ℓ measurements taken from a thicker piece of constantan wire. Determine the gradient of the graph, and hence the diameter of the wire, taking the resistivity of constantan to be $5.0 \times 10^{-7}\ \Omega\ \text{m}$.

◀ Stretch & challenge

To achieve the same rises in temperature as the water gets hotter, longer and longer heating times will be needed. Suggest why.

(c) Investigation of the variation of resistance with temperature for a metal wire

The apparatus is shown in Fig. 2.2.19. We could use either a low-power electric kettle or a beaker of water heated by a Bunsen burner. Instead of measuring the coil's resistance with a multimeter on its ohms range, we could use the circuit in Fig. 2.2.15 and calculate the resistance from $R = V/I$.

- Start with no heat applied (Bunsen off or kettle turned off at the mains), and a mixture of crushed ice and water surrounding the coil of wire.

- After stirring, read the temperature and the resistance. Don't forget to subtract from the multimeter reading the zero error, which is the reading when its probes are touched together.

- Apply heat (switch on the kettle?) for long enough for the ice to melt.

- With the heating off, stir the water gently with a long stick until the temperature stabilises. Then take another pair of readings.

- Keep repeating the process, aiming for temperature rises of between 10° and 15° each time, until 100 °C is reached.

- Plot a graph of resistance against temperature. The graph is expected to be a straight line of positive gradient. It is worth calculating the temperature at which the resistance would be zero if the straight line relationship continued down to low temperatures.

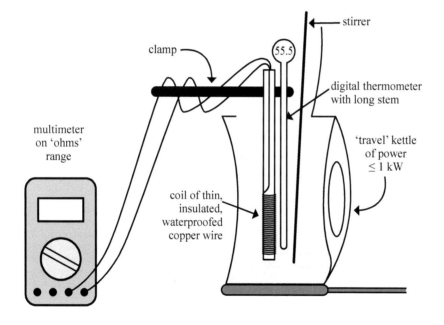

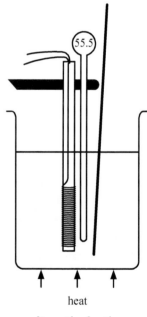

alternative heating

Fig. 2.2.19 Dependence of resistance on temperature

Test yourself 2.2

1. (a) An LED (a light-emitting diode) has a pd of 1.6 V applied across it. Explain this statement in terms of *energy*.

   (b) For this pd, there is a current of 12 mA through the LED. Calculate the energy transferred in the LED in half an hour.

2. An electric kettle is labelled '230 V, 2.5 kW'.

   (a) Explain, in terms of *energy*, what 2.5 kW means.

   (b) Calculate the current taken by the kettle at 230 V.

   (c) Calculate the resistance of the heating element (a coil of electrically insulated wire in a waterproof sheath).

   (d) In the UK, a typical 'mains' pd is 240 V. In Germany, a typical mains pd is 220 V. Calculate the actual power of the kettle when used (i) in the UK, (ii) in Germany.

   (e) Explain what difference in the performance of the kettle a user might notice in the two countries.

3. (a) State Ohm's law.

   (b) The graph shows how the current through the metal filament of a lamp depends on the potential difference applied across it.

   Calculate the resistance of the filament for pds of (i) 0.60 V, (ii) 6.0 V.

   (c) Discuss to what extent, if at all, Ohm's law applies to the filament.

   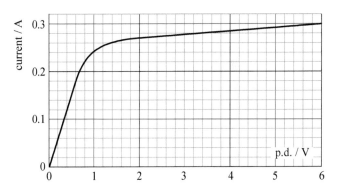

4. The diagram shows an electron gun, a device operating in a vacuum. It consists of a heated metal coil, the cathode, **K**, which gives off electrons, and a thimble-shaped anode, **A**. The electrons are accelerated by a pd between **A** and **K**. Most of the electrons hit the anode but a small fraction emerges in a narrow beam through a hole in the anode.

   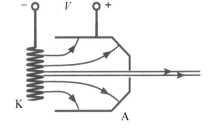

   The electron gun is set up with a pd between its terminals of 2500 V. The cathode gives off 5.0×10^{15} electrons per second, 95% of which hit the anode. Calculate:

   (a) the current in the wire between **A** and the + terminal; (b) the power supplied by the voltage source; (c) the energy transfer when an electron travels from **K** to **A**; (d) the speed attained by an electron reaching **A** ($m_e = 9.11 \times 10^{-31}$ kg); (e) the momentum of an electron reaching **A**.

5. An iron wire of square cross-section is shown. The resistivity of iron at room temperature is 9.7×10^{-8} Ω m.

   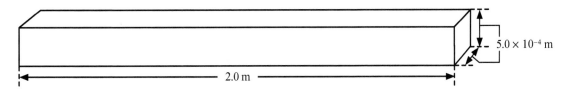

   (a) Calculate the resistance of the wire at room temperature.

   (b) Liam calculates the resistance of another wire, *each* of whose linear dimensions is twice that of the wire shown in the diagram, and finds it to be *half* that of the wire in the diagram. Explain why Liam should have expected this result.

6. (a) A flexible wire is made of 7 identical, parallel strands of copper wire, **each** of length 15.0 m and **diameter** 0.12×10^{-3} m. The resistivity of copper is 1.8×10^{-8} Ω m.

    (i) Calculate the resistance of a single strand.

    (ii) Using your answer to (a)(i), find the combined resistance of the 7-strand wire, treating the strands as resistances in parallel.

    (iii) Explain how the resistance of the 7-strand wire could be calculated from the data at the start of the question, without first finding the resistance of a single strand.

(b) (i) Show, in clear steps, that P/ℓ, the power dissipated per unit length of wire, is given by:

$$\frac{P}{\ell} = \frac{\rho I^2}{A}$$

    in which ρ, I and A have their usual meanings.

    (ii) **X** and **Y** are wires made of the same material. **X** has diameter d and carries current I. **Y** has diameter $2d$ and carries current $2I$. Use the formula in (b)(i) to show which, if either, of **X** and **Y** has the greater power dissipation per unit length.

7. Two students investigate the current–voltage (I–V) relationship for a filament car headlamp bulb which is labelled 12 V, 24 W. They set up a suitable circuit to explore both the low and high voltage variation. They expected the filament to obey Ohm's law for low voltages but not for high voltages.

(a) Explain briefly why they expected two different behaviours.

(b) Their power supply only gave outputs in 2.0 V steps. Draw a circuit they could use to investigate the current for lower voltages and explain how it works.

(c) Their results are given in the table. Plot a graph of I against V and estimate the value of V at which the behaviour changes.

(d) Calculate the resistance of the filament at low voltages and at the operating voltage.

(e) The students read that, in the high voltage region, the relationship between I and V is $I = kV^n$ and that the value of n is approximately 0.6. Plot a graph of I against $V^{0.6}$ to investigate this and discuss the extent to which the line agrees with $I = kV^{0.6}$ for voltages above the transition. Refine your answer to the transition voltage between the ohmic and non-ohmic behaviours.

V/V	I/A	V/V	I/A
0.00	0.000	2.00	0.741
0.25	0.117	4.00	1.078
0.50	0.234	6.00	1.342
0.75	0.352	8.00	1.568
1.00	0.469	10.00	1.768
1.50	0.634	12.00	1.951

(f) [For A level candidates] Plot a suitable log-log graph to determine a more accurate value for n. Use the results to obtain a value for k.

2.3 Direct current circuits

We shall be dealing with circuits that can be resolved into series and/or parallel combinations of conducting elements (for example, resistors, lamps) and a power supply. We show how to calculate currents and pds in these circuits, and finish by considering two important cases, the potential divider and the power supply with internal resistance.

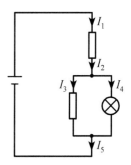

Fig. 2.3.1 Branches in a circuit

2.3.1 Currents and pds in series and parallel circuits

(a) Currents

We've already seen (in Section 2.1.2) how the law of conservation of charge shows that current must be the same all round a simple *series* circuit (that is all components in a ring with no branches). But suppose there *are* branches, so we have components *in parallel* as in the lower resistor and the filament lamp in Fig. 2.3.1. Because charge is conserved, the currents (rates of flow of charge) through these two components must add up to the current in the circuit 'before' and 'after' it branched. In other words:

$$I_1 = I_2 = (I_3 + I_4) = I_5$$

This illustrates the simple rule (officially known as *Kirchhoff's first law*): *The sum of currents coming into a point is equal to the sum of the currents going out from it.*

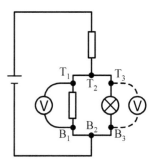

Fig. 2.3.2 Voltmeter choice

(b) Potential differences

We start by looking at the parallel combination in Fig. 2.3.2. Whether the voltmeter is in the left-hand position or the right-hand position it must read the same, because T_1 and T_3 are joined by a wire (of negligible resistance), and so are effectively the same point. Similarly with B_1 and B_3.

In other words, when components are in parallel, the pd is the same across both of them: there is only *one* pd! For the parallel combination in Fig. 2.3.3, this pd has been labelled V_2.

More fundamentally: pds arise from forces on free electrons due to distributions of charge brought about by the battery. The forces do work on free electrons going from one point to another (e.g. from B_2 to T_2). The amount of work is independent of the route taken between the points, just like the work done on us by the pull of the Earth when we change levels by using a staircase instead of a sloping ramp.

Finally, consider pds across components in series, as shown in Fig. 2.3.3. For an electron going from Y to X, the work done on it as it goes through either the lower resistor or the filament lamp is eV_2 and as it goes through the top resistor is eV_1. So the total amount of work as it goes from Y to X is $(eV_1 + eV_2)$, but it is also eV_3. So $eV_1 + eV_2 = eV_3$. Dividing through by e:

$$V_1 + V_2 = V_3.$$

Thus pds in series add together.

The rules for pds in circuits with series and parallel components are, then, a consequence of the conservation of *energy*.

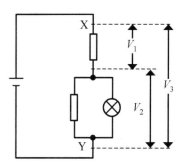

Fig. 2.3.3 Pds in series

Knowledge check 2.3.1

For the circuit in Fig. 2.3.4, determine:

(a) the currents through R_2 and R_3

(b) the pd across R_1 and the pd across R_2

(c) the resistances R_1, R_2 and R_3.

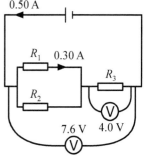

Fig. 2.3.4

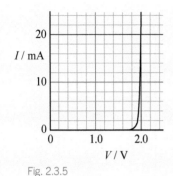

Fig. 2.3.5

Example

The red LED whose *I–V* graph is given in Fig. 2.3.5 has the desired brightness with a current of **20 mA**. Calculate the resistance of the resistor that must be placed in series with the LED to run it at this current from a **6.0 V** supply.

Answer

We first put the information on a circuit diagram (Fig. 2.3.6). But Fig. 2.3.5 shows that for a current of **20 mA**, a pd of **2.0 V** (to 2 s.f.) is needed across the LED. Since pds in series add, the pd across the resistor needs to be (6.0 V – 2.0 V) so its value, *R*, must be:

$$R = \frac{4.0\ \text{V}}{0.020\ \text{A}} = 200\ \Omega$$

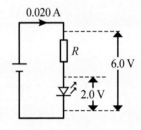

Fig. 2.3.6 Data to find R

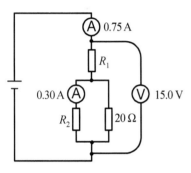

Fig. 2.3.7 Meters and their readings

(c) A note on the resistance of ammeters and voltmeters

The quantities alongside the meters in Fig. 2.3.7 are their readings. It wouldn't be silly to ask: are these the pd and currents *before the meters were connected*? The answer is yes, but only if:

- The voltmeter's resistance is so high that the current through it is negligible.
- Each ammeter's resistance is so low (compared with those of the resistors) that inserting it in series doesn't lower the current.

With modern meters, these are usually safe assumptions. For example, a typical auto-ranging digital multimeter on its 'dc volts' range has a resistance of **10 MΩ** or more. On its 10 A range its resistance is usually a small fraction of an ohm. See tip.

Multi-step example

Determine the resistances R_1 and R_2 in Fig. 2.3.7.

[This example, unlike the last, is rather contrived, but it does demonstrate the use of just about all the basic circuit rules.]

Answer

We know the currents through R_1 and R_2, but we don't know the pds across either of them – yet. We start by calculating one of the things we can calculate immediately. This gives us another, and so on.

- For the 20 Ω resistor, current = 0.75 A – 0.30 A = 0.45 A.
- So the pd across the 20 Ω resistor is $V = IR = 0.45\ \text{A} \times 20\ \Omega = 9.0\ \text{V}$.
- This is also the pd across R_2.
- So $R_2 = \dfrac{9.0\ \text{V}}{0.30\ \text{A}} = 30\ \Omega$.
- But R_1 is in series with the parallel combination, across which the pd is **9.0 V**. So pd across R_1 is $V = 15.0\ \text{V} - 9.0\ \text{V} = 6.0\ \text{V}$.
- So $R_1 = \dfrac{6.0\ \text{V}}{0.75\ \text{A}} = 8.0\ \Omega$.

This is straightforward as long as you are very careful to associate each pd with the circuit component(s) to which it belongs, i.e. make your working clear.

 **2.3.2 Knowledge check**

Calculate *R* in Fig. 2.3.8.

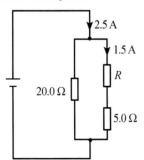

Fig. 2.3.8 Practice circuit

2.3.2 Formulae for resistances in series and in parallel

(a) Resistances in series

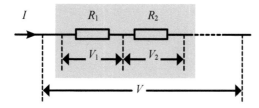

Fig. 2.3.9 Resistances in series

Look at Fig. 2.3.9. If meters were placed to measure the pd V and the current I shown in Fig. 2.3.9 then V / I would give us the combined or equivalent resistance, R, of everything in the grey box, that is of the resistors in series.

But we have shown that pds in series add up, that is:

$$V = V_1 + V_2$$

The current, I is the same all through, so the equation can be rewritten:

$$IR = IR_1 + IR_2$$

Dividing through by I: $R = R_1 + R_2$

The right-hand side can be extended to include any number of resistances in series. The equation may seem too obvious to need a derivation, but note that resistances don't *always* add (see below).

Knowledge check 2.3.3

(a) Write down the resistance of three $12\ \Omega$ resistors in parallel.

(b) How would you use four $12\ \Omega$ resistors to make a combination with a resistance of $16\ \Omega$?

(b) Resistances in parallel

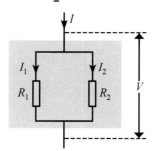

Fig. 2.3.10 Resistances in parallel

Knowledge check 2.3.4

Calculate currents x and y.

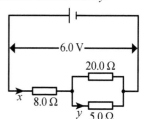

The resistors in Fig. 2.3.10 are in parallel. This time it is the currents through the individual resistors which add to give the current, I, entering and leaving the combination.

So $I = I_1 + I_2 + ...$

There is only one pd, V, so $\dfrac{V}{R} = \dfrac{V}{R_1} + \dfrac{V}{R_2} + ...$

Dividing through by V gives $\dfrac{1}{R} = \dfrac{1}{R_1} + \dfrac{1}{R_2} + ...$

Here, R is the equivalent (or combined) resistance of the parallel combination, as would be found by measuring V and I and evaluating V/I. (Just for interest: for resistances in parallel it is the conductances, defined by I/V, that add.)

Stretch & challenge

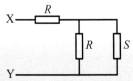

Let g be the ratio $\frac{S}{R}$. Determine the value of g for which the resistance measured between X and Y is simply S. You should find yourself solving a quadratic equation in g. [Then, look up *golden ratio* for interest.]

Example

Determine the resistance of resistors of $3.0\ \Omega$ and $4.0\ \Omega$ in parallel.

Answer

Leaving out units, $\frac{1}{R} = \frac{1}{3} + \frac{1}{4}$

We could find R very quickly with a calculator, and that's good if the answer is all we want, but for our purposes it's more instructive to add the fractions by putting each over the *common denominator* of 3×4.

Doing so, $\frac{1}{R} = \frac{4}{3\times4} + \frac{3}{4\times3} = \frac{3+4}{3\times4} = \frac{7}{12}$

Thus $R = \frac{12}{7} = 1.71\ \Omega$.

Notice that the combined resistance is less than either of the individual resistances. That is as it should be because, with two resistors connected between the same two points, the current will be greater (for the same pd) than if there were only one.

Product divided by sum for two resistances in parallel

It's useful to re-do the above example in algebra for two resistances, R_1 and R_2.

$$\frac{1}{R} = \frac{1}{R_1} + \frac{1}{R_2} = \frac{R_2 + R_1}{R_1 R_2} \qquad \text{so} \qquad R = \frac{R_2 R_1}{R_1 + R_2} = \frac{\text{product of } R_1 \text{ and } R_2}{\text{sum of } R_1 \text{ and } R_2}.$$

This result (which is not in the WJEC Data Booklet) is easy to remember (the *units* tell you that the product must be on the top), *and* easy to use *and* without a final reciprocal to forget to take! However, it only works for two resistances at a time.

n equal resistances in parallel

You should show for yourself that the equivalent resistance is given by

$$R = \frac{1}{n} \times \text{one individual resistance.}$$

Resistor networks can't always be resolved into combinations of resistors in series and parallel but all the ones you will meet at AS can be.

2.3.3 The potential divider

This name is given to resistances connected in series in order to 'divide up' the pd, V_{total}, placed across the combination. In Fig. 2.3.11 we have

$$V_1 = IR_1,\ V_2 = IR_2 \ldots \qquad V_{\text{total}} = I(R_1 + R_2 + \ldots.) \qquad \text{that is} \qquad V_{\text{total}} = IR_{\text{total}}.$$

By division, $\dfrac{V_1}{V_2} = \dfrac{R_1}{R_2}$ and so on, $\qquad \dfrac{V_1}{V_{\text{total}}} = \dfrac{R_1}{R_{\text{total}}}, \qquad \dfrac{V_2}{V_{\text{total}}} = \dfrac{R_2}{R_{\text{total}}} \qquad$ and so on.

The ratio of pds is simply equal to the ratio of the resistances across which the pds would be measured!

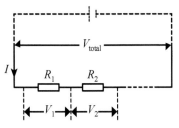

Fig. 2.3.11 Potential divider (general)

(a) Using a potential divider to give a desired output pd

If we place an *input* pd, V_{in}, across two resistors, as in Fig. 2.3.12, we can obtain an *output* pd across either of the resistors (we've picked R_2). By choosing the resistors correctly we can have any V_{out} we like (as long as $V_{out} \leq V_{in}$) since

$$\frac{V_{out}}{V_{in}} = \frac{R_2}{R_{total}} \qquad \text{that is} \qquad V_{out} = \frac{R_2}{R_{total}} \times V_{in}.$$

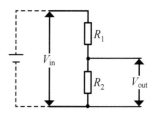

Fig. 2.3.12 Potential divider

As an example, we'll select R_1 and R_2 in order that $V_{out} = 3.0$ V when $V_{in} = 9.0$ V.

$$\frac{R_2}{R_{total}} = \frac{3.0 \text{ V}}{9.0 \text{ V}} = \frac{1}{3}.$$

So it looks as though we could have $R_1 = 2.0 \ \Omega$, $R_2 = 1.0 \ \Omega$, or $R_1 = 30 \ \Omega$, $R_2 = 15 \ \Omega$, or $R_1 = 2000 \ \Omega$, $R_2 = 1000 \ \Omega$ and so on.

In practice we'd avoid very *low* resistances, so as not to tax the power supply and overheat R_1 and R_2. Consider the combination $R_1 = 2.0 \ \Omega$, $R_2 = 1.0 \ \Omega$. In this case the total power dissipation in the resistors would be $\frac{V^2}{R} = \frac{(9.0 \text{ V})^2}{3.0 \ \Omega} = 27$ **W**: equal to the power of a small soldering iron!

R_1 and R_2 and could also be too *high* as explained in the section below.

Knowledge check 2.3.5

If $R_1 = 30 \ \Omega$, $R_2 = 15 \ \Omega$, and $V_{in} = 9.0$ V in the potential divider of Fig. 2.3.12, calculate the percentage by which V_{out} falls when R_L, a load of $1000 \ \Omega$, is connected across R_2.

The loaded potential divider

Our last example showed how to use a potential divider to produce a V_{out} of 3.0 V from a 9.0 V supply (V_{in}). But why would we want to do this? Probably because we need to supply 3.0 V to some device. The device would be connected as a *load*, R_L, across the output terminals, that is across R_2, as in Fig.. 2.3.13.

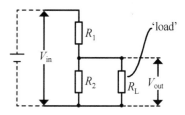

Fig. 2.3.13 Loaded potential divider

Now we hit a snag. R_2 has in effect been replaced by a lower resistance, that of R_2 and R_L in parallel. This makes the output voltage lower than planned.

Suppose $V_{in} = 9.0$ V, $R_1 = 2000 \ \Omega$, $R_2 = 1000 \ \Omega$, $R_L = 1000 \ \Omega$.

The resistance of R_2 and R_L in parallel is $500 \ \Omega$, so this replaces R_2, giving

$$V_{out} = \frac{500 \ \Omega}{R_{total}} \times V_{in} = \frac{500 \ \Omega}{2000 \ \Omega + 500 \ \Omega} \times 9.0 \text{ V} = 1.8 \text{ V}$$

This is much less than the 3.0 V planned. However, if R_1 and R_2 are chosen to be much lower than the load resistance, R_L, the reduction in output voltage is very small. See for yourself by doing Knowledge check 2.3.5.

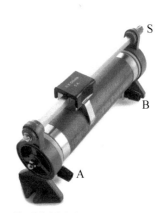

Fig. 2.3.14 A rheostat

(b) A variable potential divider

The idea is simple: we make the ratio of R_1 to R_2 variable. This can be done using an ordinary laboratory rheostat (Fig. 2.3.14). Note that it has three terminals. The lower two – call them 'A' and 'B' – connect to the ends of a single-layer coil of bare wire of high resistivity. So there's a fixed resistance (often about $15 \ \Omega$) between A and B. The top terminal, 'S', connects to a *sliding contact* which can press against the coil anywhere along its length, so 'dividing' it into two portions, AS and SB, gives our old friends R_2 and R_1, but whose ratio we can vary.

Fig. 2.3.15 incorporates the circuit symbol for a variable potential divider. Note carefully how the connections to it correspond to the terminals on the rheostat, and how V_{in} is connected across AB. V_{out} will be ⅓ of V_{in} if the sliding contact is ⅓ of the way between A and B, and so on.

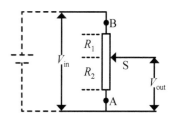

Fig. 2.3.15 Variable potential divider

Study point

ntc thermistors are made of a semiconductor material, usually a metal oxide with deliberately added 'impurity' atoms. When the temperature is increased the number of mobile charge carriers increases.

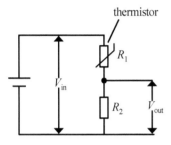

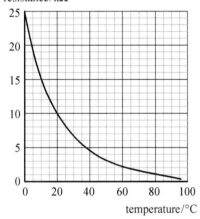

Fig. 2.3.16 Thermistor as part of potential divider

resistance/kΩ

Fig. 2.3.17 Resistance against temperature for a thermistor

▲ 2.3.6 Knowledge check

Following on from the example, if R_2 were changed to 2.4 kΩ, determine the temperature at which the alarm would sound.

Variable potential dividers have many uses in electronics. The usual form consists of a carbon 'track' forming a major arc of a circle, and a sliding contact which can be moved by turning a spindle.

(c) Potential dividers incorporating resistive sensors

Sensors are devices that 'respond' to changes in their surroundings. We shall consider two that respond by changing their resistance. *Thermistors* do so when the temperature changes, and *light-dependent resistors* (LDRs) do so when the light level changes. By making a thermistor or an LDR one of the resistances in a potential divider, the change will result in a change in output pd – useful for 'triggering' digital systems, alarms and so on.

Thermistor circuit

We shall consider only 'ntc' (negative temperature coefficient) thermistors, whose resistance goes down as the temperature goes up.

Fig. 2.3.16 shows a thermistor used as one of the resistances in a potential divider circuit – note the (not very intuitive) thermistor symbol. When the temperature rises, its resistance, R_1, will decrease, but R_2 will stay (almost) the same, so V_{out} will increase.

Example

In Fig. 2.3.16, V_{in} is 9.0 V, and V_{out} is to be connected to the input of an alarm that is triggered when V_{out} reaches 2.5 V. The resistance–temperature relationship for the thermistor is shown in Fig. 2.3.17. Calculate the value of the fixed resistor, R_2, needed in order for the alarm to be triggered when the temperature reaches 40 °C.

Answer

We first note from the graph that at 40 °C, the thermistor resistance, R_1, is 4.5 kΩ.

Since $\dfrac{R_2}{R_1} = \dfrac{\text{pd across } R_2}{\text{pd across } R_1}$ we need $\dfrac{R_2}{4.5 \text{ kΩ}} = \dfrac{2.5 \text{ V}}{9.0 \text{ V} - 2.5 \text{ V}}$

so $R_2 = 4.5 \text{ kΩ} \times \dfrac{2.5}{6.5} = 1.7 \text{ kΩ}$

Note that:

1. Below 40 °C, $R_1 > 4.5$ kΩ, so with $R_2 = 1.7$ kΩ, we have $V_{out} < 2.5$ V.

2. We could have used $\dfrac{V_{out}}{V_{in}} = \dfrac{R_2}{R_{total}}$ with $R_{total} = R_2 + 4.5$ kΩ, but the maths is messier.

3. We have assumed that the alarm has a very high input resistance, so it doesn't 'load' the potential divider significantly.

LDR circuit

Certain semiconductors, such as cadmium sulfide, have very high resistivities in the dark, but conduct better and better as the light level increases. An LDR (*light-dependent resistor* or *photoresistor*) is made by depositing a layer of such a semiconductor as a zigzag 'track' on an insulating substrate enclosed in a transparent case (Fig. 2.3.18).

Fig. 2.3.19 shows an LDR used as one of the resistors in a potential divider circuit. The LDR simply replaces the thermistor in Fig. 2.3.16, so the brighter the light falling on the LDR, the higher V_{out}. The circle is often omitted from the LDR symbol.

2.3.4 Power supplies

Here we consider the role of a battery, or other power supply, in a circuit. Strictly, a *battery* is a series combination of cells, but a single cell is often, loosely, called a battery. A cell consists of two *electrodes* made of different conducting materials separated by a conducting liquid or paste *electrolyte* in what we hope is a leak-proof case. Fig. 2.3.20 is a simplified diagram of the popular *alkaline* cell.

Fig. 2.3.21 shows a cell (the dotted box) with a *load* connected across its terminals. The load could be a filament lamp, a resistor, an LED in series with a protection resistor. We shall assume the load to have a resistance R.

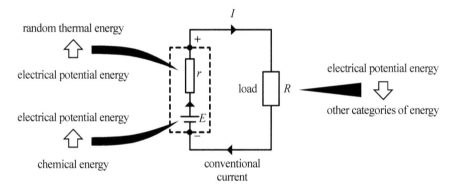

Fig. 2.3.21 Energy transfers in a circuit with a cell

The cell pumps charge around the circuit. Outside the cell the conventional current is from the cell's positive terminal (marked '+') to its negative. From there the current continues inside the cell from the negative terminal back to the positive (by means of the movement of charged 'ions').

(a) The emf of a cell

As the charge flows, energy transfers take place. At the interfaces between the electrodes and the electrolyte the charge picks up electrical potential energy, transferred from chemical energy. In Fig. 2.3.21 we've denoted this process by the conventional two-stroke cell or battery symbol. A specific amount of energy is transferred per unit charge (i.e. coulomb) passing, and we make the following definition:

The **emf**, E, of a source is the energy transferred from some other form (e.g. chemical) to electrical potential energy per coulomb of charge flowing through the source.
Unit: $\text{J C}^{-1} = \text{volt} = \text{V}$

Fig. 2.3.18 An LDR

Knowledge check 2.3.7

State how the V_{out} would behave if the LDR and the fixed resistor were swapped round in Fig. 2.3.19.

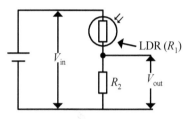

Fig. 2.3.19 LDR as part of potential divider

►►► Study point

For interest only: In an LDR, the current increases with increasing light intensity because photons can supply enough energy to knock a small proportion of the electrons out of the bonds between atoms, creating free electrons (and so-called positive 'holes').

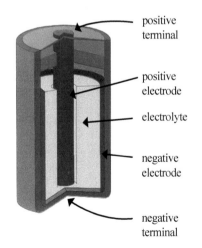

Fig. 2.3.20 A cell (simplified)

►►► Study point

Emf stands for *electromotive force*, a silly name for a quantity with the unit **V**, so it is better abbreviated than used in full. *Electromotance* is a sensible proposed alternative.

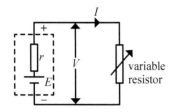

Fig. 2.3.22 Cell with a load of variable resistance

(b) The internal resistance of a cell

The *purpose* of the cell is to enable a *useful* transfer of energy in the load, as shown in Fig. 2.3.21. However, this is accompanied by an unavoidable 'dissipation' of energy in the cell itself (as ions bump their way through the electrolyte). In fact the cell behaves as if it had an approximately constant *internal resistance*, r. It often helps to include r in circuit diagrams, as we've done in Fig. 2.3.22, but note that r can't be measured directly (e.g. using a multimeter on its 'ohms' range). This is because the internal resistance is inseparable from the charge-pumping role of the cell, as we've indicated by putting both E and r inside the dotted box.

(c) $V = E - Ir$

The internal resistance makes its presence known when we vary the resistance of the load that we connect across the cell terminals (Fig. 2.3.22). If we start with a very high resistance and gradually lower it, not only does the current increase (as you should expect), but the pd, V, across the cell terminals (that is across the load) decreases. [In an ideal world this wouldn't happen: if you buy a cell labelled 1.6 V, that's the pd you want the cell to give you, whatever the load!] In fact we find that:

$$V = E - Ir$$

We can deduce this equation from the *principle of conservation of energy*, as each term in the equation is an amount of energy per unit charge. This is how it works:

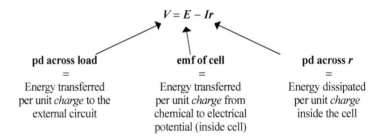

pd across load	**emf of cell**	**pd across r**
=	=	=
Energy transferred per unit *charge* to the external circuit	Energy transferred per unit *charge* from chemical to electrical potential (inside cell)	Energy dissipated per unit *charge* inside the cell

(d) Open circuit terminal pd

The equation $V = E - Ir$ shows that as we decrease the current, I (by increasing the load resistance) the pd, Ir, across r decreases, allowing a greater pd, V, across the load, that is across the cell terminals. For an infinite load resistance, I is zero, so $V = E$. But how do we make the load resistance infinite? Simply by leaving the cell 'open circuit', that is not connected to anything (except a voltmeter, with a typical resistance of 10 MΩ). We say that the *open circuit* terminal pd of the cell is equal to the emf.

(e) Power

Multiplying through the 'voltage' equation ($V = E - Ir$) by I, we get a useful equation for *power*:

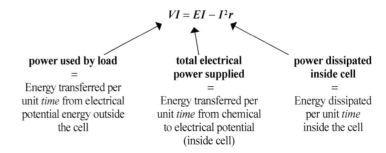

power used by load	**total electrical power supplied**	**power dissipated inside cell**
=	=	=
Energy transferred per unit *time* from electrical potential energy outside the cell	Energy transferred per unit *time* from chemical to electrical potential (inside cell)	Energy dissipated per unit *time* inside the cell

▶ **2.3.8 Knowledge check**

Calculate (a) the current and (b) the total power generated by the cell, when a 5.0 Ω resistor is connected across a cell of emf 1.50 V and internal resistance 0.50 Ω.

Hint: Consider total resistance in the circuit.

Example 1

A cell has an emf of 1.62 V. When a $1.50\ \Omega$ resistor is connected across its terminals, the pd falls to 1.39 V. Calculate: (a) the cell's internal resistance, (b) the fraction of the total power dissipated in the internal resistance.

Answer

(a) We first put the data on a diagram, as shown. Considering the $1.50\ \Omega$ load:

$$I = \frac{1.39\ \text{V}}{1.50\ \Omega} = 0.927\ \text{A}.$$

So (see diagram)

$$r = \frac{1.62\ \text{V} - 1.39\ \text{V}}{0.927\ \text{A}} = 0.25\ \Omega.$$

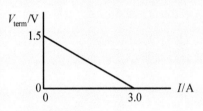

(b)

$$\frac{\text{power in } r}{\text{power generated}} = \frac{rI^2}{EI} = \frac{rI}{E} = \frac{1.62\ \text{V} - 1.39\ \text{V}}{1.62\ \text{V}} = 0.14.$$

Example 2

Sketch a graph of V against I for a cell of emf 1.50 V and internal resistance $0.50\ \Omega$.

Answer

We have $V = 1.50\ \text{V} - 0.50\ \Omega \times I$.
V falls linearly with I. When $I = 0$, $V = 1.50$ V; when $I = 1.00$ A, $V = 1.00$ V and so on, hence the graph sketched.

Note that $I_{\max} = \dfrac{E}{r}$, corresponding to zero external resistance; we say that the cell is *short-circuited*. Cells get hot and run out of energy quickly when short-circuited. They don't like it.

(f) Batteries

Cells are often connected in series as a 'battery' to produce a larger emf. In practice there is little point in using anything but identical cells.

> Emf of battery = sum of emfs of cells in series

The positive of one cell must be connected to the negative of the next, and so on. The emf of any cell connected the wrong way round counts as a negative emf. See example in the Study point.

Note also that:

> Internal resistance of battery = sum of internal resistances of cells

Here there are no minus signs even when a cell is the wrong way round.

2.3.5 Specified practical work: Experimental determination of the internal resistance of a cell

Two methods are given here:

Method (a) uses $V = E - Ir$ directly.

Method (b), the one in the lab book, uses an equation easily derived from $V = E - Ir$.

It may be more convenient to work with a battery of cells rather than a single cell. In that case we can divide the values of E and r that we find for the battery by the number of cells, to give the mean E and r of each cell.

(a) By measuring V and I

We set up the circuit in Fig. 2.3.23. The variable resistor could be a laboratory rheostat. It could also be a selection of different resistors, which could be used singly or in combinations. Ideally it should be possible to go down to an ohm or lower. We don't make use of the actual values of resistance.

The idea is to take several pairs of readings of V and I (at least 7 is good), and to plot a graph of V against I. Because the relationship between V and I is

$$V = E - Ir$$

a linear graph is expected with a negative gradient, which is equal to $-r$, and an intercept of E on the V-axis.

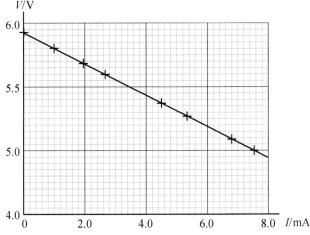

Fig. 2.3.24 Typical V–I graph for a power supply of high internal resistance

The first reading should be with the switch open, so that $I = 0$. If an approximate value of the internal resistance is not known, trial readings need to be taken to establish a value of current which significantly reduces V and then a set of readings taken with roughly equally spaced currents up to several times this value.

The graph in Fig. 2.3.24 shows a typical set of results for a power supply with a high internal resistance.

(b) By measuring I and the external resistance

This time we use different resistors, or combinations of resistors *whose values we know or can measure*, as the 'load' resistance R. The payoff is that we don't need to measure both I and V. That's why Fig. 2.3.25 shows no voltmeter! For several known values of R we measure and record the current, I. We plot a graph of R against $1/I$.

 **Knowledge check**

(a) Show that the power dissipated in a resistance R connected across the terminals of a cell of emf E and internal resistance r is given by

$$P = \frac{E^2 R}{(R+r)^2}$$

(b) For a cell of emf 1.50 V and internal resistance 0.50 Ω sketch a graph of P against R. Recommended values of R to consider are 0, 0.25 Ω, 0.50 Ω, 1.0 Ω, 2.0 Ω.

(c) For maximum power in R, how would it seem that the value of R compares with r?

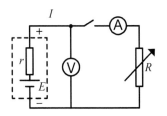

Fig. 2.3.23 Method (a) for determining r of a cell

Practical check

The switch should be open when not taking readings or adjusting the variable resistor; real cells suffer from a downward drift of emf under load – a nuisance we try to minimise.

Practical check

In this experiment, there is little value in taking repeat readings. It is better to take more pairs of values.

Knowledge check

For the power supply with the V–I graph in Fig. 2.3.24

(a) determine the emf and the internal resistance and

(b) hence calculate the current and terminal pd for an external (load) resistance of 200 Ω.

Here's why we do so:

For the load resistance, $V = IR$.

Substituting for V in $V = E - Ir$,

we obtain $IR = E - Ir$.

Dividing through by I, $R = \dfrac{E}{I} - r$.

Comparing $R = \dfrac{E}{I} - r$ with $y = mx + c$, we see that

the graph should be a straight line

with intercept $-r$ and gradient E.

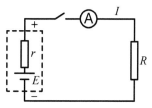

Fig. 2.3.25 Method (b) for determining the internal resistance of a cell

Test yourself 2.3

1. (a) For the circuit shown, calculate:
   (i) the pd between **B** and **C**,
   (ii) the current, x,
   (iii) the current, y.

   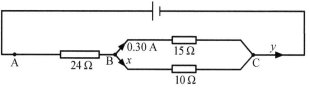

   (b) Hence show clearly that the pd between **A** and **C** is 22.5 V.

   (c) (i) Use resistance formulae to calculate the resistance of the combination of three resistors in the diagram.
   (ii) Check your answer to (c) (i) using a current and a pd chosen from parts (a) and (b), explaining your reasoning.

2. In the circuit shown, the power supply has negligible internal resistance. Determine the unknown resistance, R to 2 s.f.

   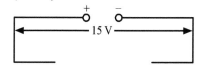

3. (a) A teacher gives a student a sealed box with **two** terminals, containing a combination of two resistors. One has a known value of $10\ \Omega$, and the other has an unknown resistance, X. The student finds that when a pd of $15\ \text{V}$ is placed across the terminals there is a current of $2.5\ \text{A}$ through the combination.
   (i) Calculate the resistance of the combination. (ii) Explain why the resistors cannot possibly be in series.
   (b) (i) Add the resistor combination to complete the diagram, and show clearly the currents in each resistor. (ii) **Hence** calculate the resistance, X. (iii) Use the appropriate resistance combination formula to check whether your answers to (a) (i) and (b) (ii) are consistent.

   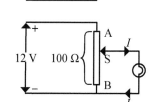

4. The pd across a typical light-emitting diode (LED) when emitting light is roughly $2\ \text{V}$ over a wide range of currents. Calculate a suitable value of resistor to put in series with an LED in order to run it from a $5.0\ \text{V}$ supply at its recommended current of $15\ \text{mA}$.

5. (a) The diagram shows a variable potential divider using a carbon track, **AB**, of resistance $100\ \Omega$. An 'input' of $12\ \text{V}$ is applied across **AB**. The sliding contact, **S**, is positioned so that the pd between **S** and **B** is $8.4\ \text{V}$. Calculate: (i) the resistance of the carbon track between **S** and **B**; (ii) the pd between **A** and **S**.

   (b) A filament lamp is now connected between **S** and **B**, with **S** in the same position as in part (a). There is a current, I, through the lamp as shown.
   State, **with reasons**, whether the following quantities increase, decrease or stay the same when the lamp is connected: (i) the current through the track between **A** and **S**; (ii) the pd between **A** and **S**; (iii) the pd between **S** and **B**.

6. The diagram shows a circuit used to investigate how the pd across the terminals of a cell depends on the current. Four pairs of readings were taken. They are plotted on the grid as A, B, C and D.

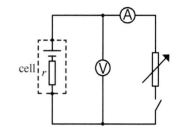

(a) (i) For which one of the points was the switch open?
   (ii) Define the emf of a cell.
   (iii) Write down the emf of the cell above.

(b) (i) For point D, calculate the pd across the internal resistance.
   (ii) **Hence** calculate the internal resistance of the cell.

(c) (i) Complete the graph and hence calculate the quantity:

$$\frac{\text{emf of cell}}{\text{maximum current cell can supply}}$$

   (ii) What is the physical significance of this quantity?

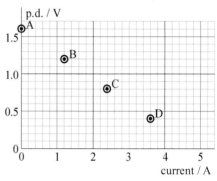

7. In an experiment to determine the emf, E, and internal resistance, r, of a battery, the battery, a switch, an ammeter and a 22 Ω resistor were connected in series (Fig. 2.3.25). With the switch closed the current, I, was recorded. This was repeated with decreasing values of resistor, down to 1.5 Ω.

Resistor value, R / Ω	22.0	15.0	10.0	4.7	1.5
Current reading, I / A	0.203	0.289	0.414	0.762	1.548

(a) Derive the expected relationship between current, I, and resistor value, R, in the form $R = \dfrac{E}{I} - r$; (b) Plot a graph of R against $1/I$. There is no need to plot error bars; (c) Discuss whether or not your graph supports the equation in (a); (d) Use the graph to determine values for E and r; (e) The switch was left open except when readings were being taken. Suggest why.

8. This question relates to circuits containing a 6.0 V, 60 mA indicator bulb, with the characteristic shown, and a 100 Ω resistor.

(a) Use the graph to determine the resistance of the bulb when:
   (i) it is operating at its rated voltage and
   (ii) when the pd across it is 2.0 V.

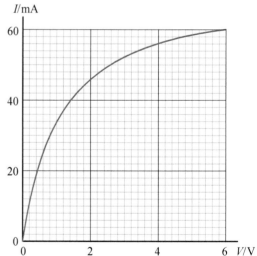

(b) Calculate:
   (i) the total pd across the bulb and resistor in series with a current of 45 mA, and
   (ii) the total current in a parallel combination of the two with a pd of 4.0 V across them.

(c) The bulb and resistor are connected in parallel and a total current of 80 mA is in the pair. Determine the current in each and the pd across the pair. [Hint: first draw a current–voltage graph for the resistor on the same axes as for the bulb.]

(d) A pd of 6.0 V is applied across the bulb and resistor in series. Determine the pd across each and the current.

2.4 The nature of waves

How can energy be transferred from one place to another? Here are some examples.

1   A bullet fired from a gun carries kinetic energy.
2   Oil or gas flowing in a pipeline carries some kinetic energy, but, more importantly, chemical energy.
3   Seismic waves travelling through the Earth's crust, and tsunamis travelling through the sea, transport terrifying amounts of energy spread over short time intervals.
4   Sound waves travelling through air carry energy which can activate neurons in our inner ears, or generate electric 'signals' in microphones.

In 1 and 2, matter ('stuff') travels from one place to another taking energy with it. 3 and 4 are quite different: a disturbance, carrying energy, travels (or *propagates*) through a *medium* (solid, liquid or gas) as a *progressive wave*. The particles of the medium are temporarily displaced from their usual positions as the wave passes, generally vibrating or oscillating about these positions. To sum up:

A progressive wave is a pattern of disturbances travelling through a medium and carrying energy with it, involving the particles of the medium oscillating about their equilibrium positions.

When a major tsunami reaches the shore the water particles may be displaced by many metres, but the wave itself may have travelled for tens or hundreds of kilometres.

2.4.1 How waves propagate

All mechanical waves (such as seismic waves, sound waves and water waves) propagate in the same general way. A wave *source* – for a sound wave this might be the cone of a loudspeaker – displaces the medium next to it, which exerts a force on the adjacent part of the medium, causing it to accelerate from rest and become displaced, exerting a force on the 'next' part, and so on.

Usually, the *speed* at which a wave propagates depends on the properties of the medium rather than the nature of the disturbance produced by the source. See Knowledge checks 2.4.2 and 2.4.3.

Fig. 2.4.1 A tsunami

Knowledge check 2.4.1

Give two more examples of energy transfer: one enabling life on Earth, and one developed by human beings.

Knowledge check 2.4.2

The speed of transverse waves on a stretched rope is given by

$$v = \left(\frac{T}{\mu}\right)^n$$

in which μ is the mass per unit length of rope, and T is the tension in it. By considering *units*, find the value of n.

Knowledge check 2.4.3

The speed of sound in air is given by

$$v = \sqrt{\frac{1.40p}{\rho}}$$

in which p is the air pressure. ρ is another property of the air. Suggest which, and confirm your suggestion by checking units.

Stretch & challenge

Although you won't be tested on this, it's worth looking at how a wave travels in a taut rope. This is arguably the simplest example of a wave. Fig. 2.4.2 (a) shows the rope after one end has been sharply displaced by a small amount in a direction at right angles to the rope itself. Fig. 2.4.2 (b) is a free-body diagram for a small portion **P** of the rope. It shows the resultant force on **P** to be (almost) at right angles or *transverse* to the line of the rope, and so **P** will be accelerating (from rest) in that direction, and will shortly be displaced, exerting a transverse force on the 'next' portion. In that way the transverse displacement is passed along the rope. See Fig. 2.4.2 (c).

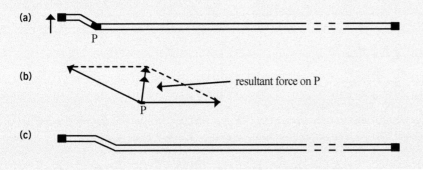

Fig. 2.4.2 Propagation of transverse waves on a rope

2.4.2 Transverse waves

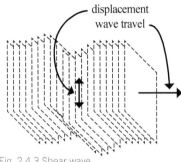

Fig. 2.4.3 Shear wave

We have already met one example: a **transverse wave** in a taut rope, string or wire (Fig. 2.4.2). Another case is the *secondary* (S) or *shear wave* (Fig. 2.4.3) that travels through the Earth's crust from an underground event, such as the slipping of one rock mass against another.

Light and other **electromagnetic (e-m) waves** are also transverse. They are special as they can travel through a vacuum, at a speed, denoted by c, of 2.998×10^8 m s^{-1}. Their speed in air is the same to four significant figures. Even though they don't need a medium, e-m waves behave just like other transverse waves in many ways. But, you may be thinking, in a vacuum there are no particles to oscillate!

Instead of *particles* oscillating as an e-m wave travels, it is vector quantities called *electric field strength* and *magnetic field strength* that oscillate at each point. Both vectors oscillate at right angles to the direction of travel of the wave (and to each other). Their oscillations *are* the wave. When an e-m wave interacts with matter (for example stimulating cells in our retinas) it is usually by means of its *electric* field (a quantity closely related to potential difference). Indeed, when we talk of *displacement* for an e-m wave, by convention we mean the *electric* field strength. The *magnetic* field strength vector is always present in an e-m wave, but we shall not refer to it again.

(a) Polarisation of transverse waves

Fig. 2.4.4 (a) is a snapshot of a rope that is being wiggled at one end in an oscillating motion. The transverse waves are **polarised** (strictly, *linearly polarised*).

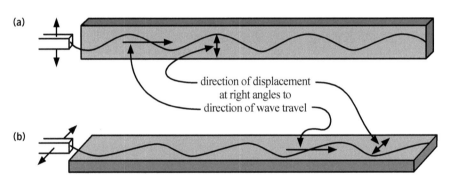

Fig. 2.4.4 Polarised transverse waves on a rope

Light from most 'ordinary' sources (such as the Sun, flames, filament lamps and LEDs) is *unpolarised*. It consists of a random succession of oscillation sequences with electric fields at different angles, though all are at right angles to the propagation direction. This is shown symbolically in Fig. 2.4.5 (a).

Most lasers produce *polarised* light: the electric field oscillations are confined to a single direction at right angles to the direction of travel, as in Fig. 2.4.5 (b) or (c). (The precise term is *linearly polarised*, but we shall simply use *polarised*.)

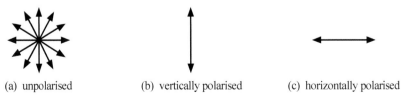

     (a) unpolarised   (b) vertically polarised   (c) horizontally polarised

Fig. 2.4.5 Light coming out of page towards viewer: directions of electrical field oscillations

We can find out whether light is polarised or unpolarised by putting a *polarising filter* (or *'polaroid'*) in the path of the light. The filter is a plane sheet of material containing special long straight molecules arranged parallel to, and close to, each other. Electric field components parallel to the molecules are blocked; components perpendicular to the molecules are transmitted (allowed through).

Look carefully at Fig. 2.4.6, which shows the effect on **vertically polarised light** of turning the filter through different angles.

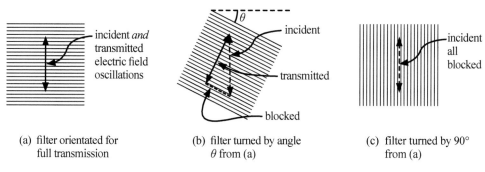

(a) filter orientated for full transmission

(b) filter turned by angle θ from (a)

(c) filter turned by 90° from (a)

Fig. 2.4.6 Effect of polarising filter on vertically polarised light

When **unpolarised light** is viewed through a polarising filter, its brightness is reduced (because electric field components parallel to the molecules are removed). The filter polarises the light – makes it into polarised light.

If the filter is rotated, the brightness of the transmitted light does not change. Electric field components parallel to the newly orientated molecules are removed, but this will take just the same amount of energy out of the light as before rotation. See Fig. 2.4.7.

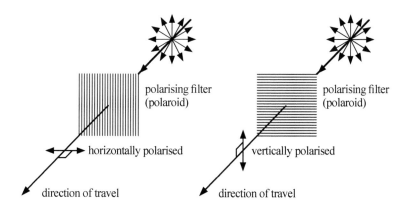

Fig. 2.4.7 Using a polarising filter to polarise unpolarised light

Table 2.4.1 summarises what is observed through a polarising filter when it is gradually turned through a whole revolution starting from the angle for maximum observed brightness.

	0°	90°	180°	270°	360°
Polarised light incident	brightest	dark	brightest	dark	brightest
Unpolarised light incident	constant brightness, but darker than without filter				

Stretch & challenge

The molecules in a polaroid are special because electrons can move along them. Electric field components parallel to the molecules exert forces on the electrons that make them oscillate back and forth along the molecules, colliding with atoms and transferring energy out of the wave into random thermal energy in the polaroid. Electric field components at right angles to the molecules are unaffected as the forces they exert on the electrons hardly cause them to move, because the electrons are confined to their own molecules.

Knowledge check 2.4.4

Unpolarised light passes through two polarising filters, one after the other. The molecules of the two filters are parallel at first. The first filter is fixed and the second one slowly rotated.

Describe how the brightness varies as the second polaroid is rotated through one revolution.

2.4.3 Longitudinal waves

Transverse (shear) waves can't travel through gases or liquids because these lack rigidity: successive layers (Fig. 2.4.3) can't exert transverse forces on each other without slippage and energy dissipation (transfer to random thermal energy).

By contrast, **longitudinal waves** can travel through a gas or liquid, as well as through a solid.

The *primary* (*P*) *waves* travelling through the crust, mantle and core of the Earth from an underground event are longitudinal.

Sound waves are longitudinal. They are generated by vibrating objects, such as the cones of loudspeakers. Even in a rock concert, except very close to the loudspeakers, the maximum displacement of air due to the sound waves is less than 0.1 mm.

A longitudinal wave travels along a stretched 'Slinky' spring, if one end is moved back and forth parallel to the spring (Fig. 2.4.8). On a Slinky we can actually see the *rarefactions* and *compressions* that are also present in a sound wave, but remember that the Slinky wave travels in one dimension, whereas sound propagation is three dimensional.

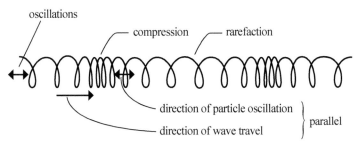

Fig. 2.4.8 Longitudinal wave on a Slinky

2.4.4 Waves from oscillating sources

We shall now assume that *all* our wave sources are *oscillating* (vibrating) *sinusoidally*, as shown in the graph of displacement against time (Fig. 2.4.9). *Sinusoidal* describes the graph shape. Note that it is *periodic* – repeating the same *cycle* over and over again. The displacements of the particles in the medium will also vary sinusoidally with time, but the graph will shift further to the right the further a particle is from the source, as each feature (such as a peak) will arrive later.

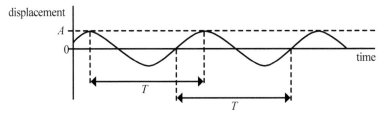

Fig. 2.4.9 Displacement–time graph for an oscillating particle

Referring to Fig. 2.4.9, here are some basic definitions that you should learn:

The **amplitude**, *A*, of an oscillation is the maximum value of the displacement from its mean position. Unit: m

The **frequency**, *f*, of a wave is the number of cycles of a wave that pass a given point per second [or the number of cycles per second performed by any particle in the medium through which the wave is passing]. Unit: Hz

The **period**, *T*, of a wave is the time for one cycle. Unit: s

Note that the definitions of **period** and **frequency** refer to a *cycle*. This can be defined as the smallest portion of an oscillation, starting at any point, that repeats itself exactly. We can demonstrate the relationship between them by an easy example.

Frequency and period are simply related.

Suppose $T = 0.10$ s, then clearly there will be 10 cycles per second, so the *frequency, f,* is 10 Hz.

Generalising, $\qquad f = \dfrac{1}{T} \qquad$ and $\qquad T = \dfrac{1}{f}$

Note that all particles of the medium oscillate with the frequency of the wave source; cycles of oscillation can't be lost or gained as the wave travels!

Knowledge check 2.4.7

In Fig. 2.4.10 the distance PR between particles P and R (assuming very small displacements) is $\frac{3}{4}\lambda$. Express in terms of λ, in similar fashion, (a) the distance RQ, (b) the distance PS.

Example

A student, looking over the railings on a pier, observes the sea below. She estimates that as each wave passes, the sea level rises by 1.1 m from its lowest point to its highest. She times 24 cycles of oscillation in 2.0 minutes. Calculate:

(a) the wave amplitude,   (b) the wave frequency,   (c) the time taken for the sea level to go from its lowest to its highest point.

Answer

(a) amplitude = maximum displacement from the mean = $\dfrac{1.1 \text{ m}}{2} = 0.55$ m

(b) frequency = $\dfrac{\text{number of cycles}}{\text{time taken}} = \dfrac{24}{120 \text{ s}} = 0.20$ Hz

(c) time required = $\dfrac{1}{2}T = \dfrac{1}{2} \times \dfrac{1}{f} = \dfrac{1}{2} \times \dfrac{1}{0.2 \text{ Hz}} = \dfrac{1}{2} \times 5.0 \text{ s} = 2.5$ s

2.4.5 Snapshot of a wave

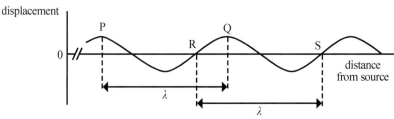

Fig. 2.4.10 Displacement against distance from source at some particular time

Fig. 2.4.10 looks like Fig. 2.4.9 but note that, now, *distance from the source* at one particular time is along the horizontal axis! We'll call this sort of graph a *snapshot*. For a wave in a stretched rope it could be a single flash photograph of the rope, with axes added.

For waves that spread out as they travel further from the source, the more distant particles will oscillate with smaller amplitudes, as the energy is more thinly spread. Our graphs won't usually show this.

Particle Q is oscillating **in phase** with particle P. Another example of in-phase particles is the pair R and S. The distance between P and Q or R and S is the **wavelength** (λ).

▶▶ **Key terms**

Oscillations of the same frequency are **in phase** if they are at the same point in their cycles at the same time.

The **wavelength**, λ, of a progressive wave is the minimum distance (measured along the direction of propagation) between two points on the wave oscillating in phase.
Unit: m

2.4.6 Speed of a wave; $v = f\lambda$

In a progressive wave the whole pattern of disturbances in the medium is moving away from the source. That's what we mean by the wave travelling. In Fig. 2.4.11 the direction of travel is to the right. The full line is a snapshot at time $t = 0$ and the dashed line is a snapshot at $t = \dfrac{T}{4}$, that is a quarter of a cycle later. The whole pattern has moved away from the source by a distance $\dfrac{\lambda}{4}$.

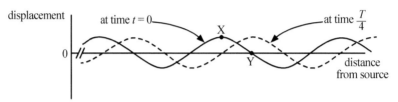

Fig. 2.4.11 Snapshots of wave at time 0 and time T/4

Suppose we let a whole period, T, elapse between one snapshot and the next. Each particle will oscillate through one extra cycle, bringing it back to where it started. So the snapshot will be unchanged. Yet the pattern will have moved – away from the source. It must, then, have moved on by a whole wavelength, λ.

So: wave *speed*, $v = \dfrac{\text{distance gone}}{\text{time taken}} = \dfrac{\lambda}{T} = \dfrac{1}{T}\lambda = f\lambda$

We have, then, the well-known equation $v = f\lambda$.

The equation $v = \dfrac{\lambda}{T}$ can be useful but is not given to you in the Data Booklet.

Remember that the frequency is that of the wave source and that the wave speed, v, is usually a constant for a given medium, independent of frequency.

Example

Sketch displacement–time graphs starting at $t = 0$ for particles X and Y in Fig. 2.4.11, and comment on the phase difference.

Answer

Extrapolating from the positions of X and Y at time $t = 0$ and time $t = \dfrac{T}{4}$ we have the graphs shown in Fig. 2.4.12.

Note that each peak (or any other feature) of Y's graph occurs time $T/4$ later than the corresponding feature on X's graph. We say Y has a phase lag of a quarter of a cycle behind X. This phase lag is just what we'd expect: the wave reaches Y *after* it reaches X!

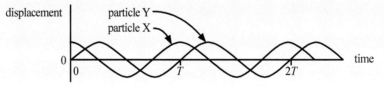

Fig. 2.4.12 Displacement–time for particles X and Y in Fig. 2.4.11

Key term

The **speed**, v, of wave is the distance that the wave profile moves per unit time. Unit: m s^{-1}

2.4.8 **Knowledge check**

Water waves from a source of frequency 3.0 Hz travel across a pond at a speed of 3.6 m s^{-1}. Calculate the shortest distance between water particles oscillating in antiphase (exactly half a cycle out of phase with each other).

2.4.9 **Knowledge check**

Mobile phones transmit and receive e-m waves of frequency in the order of 900 MHz, using a built-in aerial – a metal rod. Ideally this ought to be at least a quarter of a wavelength long. Investigate whether such an aerial could fit in a mobile phone.

2.4.7 Wavefront diagrams

The moving ridges and troughs that we see when water waves are travelling are examples of **wavefronts**.

The definition (see Key term) is geared to waves such as sound and light that travel outwards from a source, that is in three dimensions. Water waves, like waves on a drum-skin, travel in two dimensions only, so their wavefronts are really lines (often curved lines) rather than surfaces.

- We usually draw wavefronts at intervals of one wavelength, like the peaks (crests) of a water wave.
- The direction of travel of a wavefront at any point is at right angles to the wavefront through that point. (See Fig. 2.4.13).
- Wavefronts from a small source are spherical. A long way from the source they are therefore almost plane (flat) over any small region. For example, the wavefronts of light reaching the Earth from a star will be, to all intents and purposes, plane.

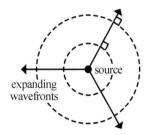

Fig. 2.4.13 Wavefronts from a point source

2.4.8 Specified practical work: measurement of the intensity variations for polarisation

A qualitative investigation of the effect of a polarising filter on a polarised beam of light requires nothing more than two filters and a light source, which could be one of the laboratory room lights or from the window (Fig. 2.4.14).

The light emerging from the first filter is polarised, i.e. the oscillations of the electric field are all in the same direction, which is at right angles to the direction of propagation.

It is useful to put a reference mark on one of the filters, so that you can keep track of the angle through which it is rotated.

As the second filter is rotated, the intensity of the transmitted light varies smoothly; two maxima and two minima are observed per rotation. These are equally spaced, i.e. there is 90° between each maximum and adjacent minima.

The effect can also be investigated quantitatively using the setup shown in Figs 2.4.15 and 2.4.16.

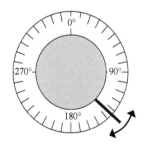

Fig. 2.4.14 Investigating polarisation

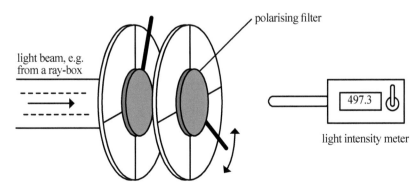

Fig. 2.4.15 Investigating polarisation (quantitative)

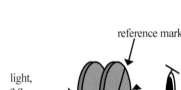

Fig. 2.4.16 Rotatable polarising filter

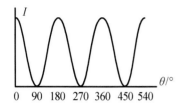

Fig. 2.4.17 Variation of intensity with angle

The light transmitted through a 100% polarising filter (see Study point) will be polarised with all its oscillations in the same direction, called the plane of polarisation or direction of polarisation. If a second such filter is placed in the path of the light, any emerging light will be 100% polarised in the direction defined by *this* filter (that is at right angles to its molecules); the component parallel to the molecules is absorbed. Fig. 2.4.6 (b) (p131) shows the action of the second filter on vertically polarised light (emerging from the first filter, which is not shown). Note how the light transmitted by the second filter, at angle θ to the first, has a smaller amplitude than the light incident on it.

Using simple trig, if the angle between the polarisation directions of the two filters is θ, the amplitude of the light which is transmitted through the second filter is $A \cos \theta$, where A is the amplitude of the light between the filters. As with all wave motions, the energy carried by the light is proportional to the square of the amplitude, so the intensity, I, of the transmitted light varies as: $I \propto \cos^2 \theta$ assuming a 100% effective filter. This relationship is shown in Fig. 2.4.17, which can be plotted experimentally using the set-up shown in Fig. 2.4.15 and Fig. 2.4.16.

Test yourself 2.4

1. A student, asked for the difference between transverse and longitudinal waves, wrote the following sentences.

   'A transverse wave moves at right angles to its direction of travel. A longitudinal wave moves in the same direction as its direction of travel.'

   Explain why this is not a good answer and give a better one.

2. The WJEC definition of the wavelength of a progressive wave is 'the *minimum* distance (*measured along the direction of propagation*) between two points on the wave oscillating in phase'.

   Explain briefly why the following word and phrase are important in the definition:

   (a) *minimum*
   (b) *measured along the direction of propagation*

3. Two students are given a snapshot of several cycles of a wave travelling along a rope. They are asked to describe, in words, how to measure the wavelength. They give the following answers:

   Answer 1: Measure the distance from one peak to the next one. This is the wavelength.
   Answer 2: Find two places where the displacements are the same and measure the distance between them. This is the wavelength.

   Explain why answer 1 is better than answer 2.

4. The diagram shows two snapshots of the same transverse wave, travelling in the positive x direction (with expanded scale for displacement, y), the fainter snapshot being 0.25 ms later than the bolder.

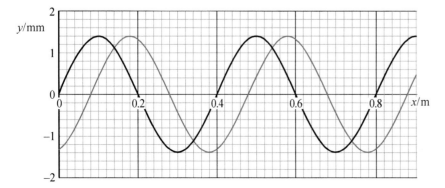

(a) (i) Give the amplitude of the wave.
   (ii) Give the wavelength.

(b) (i) Show that the lowest wave speed consistent with the graphs is 320 m s^{-1}.
   (ii) Calculate the next highest wave speed that is consistent with the graphs.

(c) Assuming that the wave speed is 320 m s^{-1}, calculate:
   (i) the period,
   (ii) the frequency.

5. Consider the wave in Q4 at the instant of the bolder line.

(a) State the values of x for which the vertical velocity of the particles in the medium is:
   (i) zero, (ii) maximum positive and (iii) maximum negative.
(b) [Tricky] Assuming a wave speed of 320 m s^{-1}, use the graph to estimate the maximum vertical velocity of the particles.

6. For the wave in Q4, give the x co-ordinates of all points in between $x = 0$ and $x = 2.0 \text{ m}$ which are oscillating (i) in phase with and (ii) in antiphase with the particle at $x = 0.15 \text{ m}$.

7. A progressive wave is travelling from left to right. Displacement–time graphs are given below for the same time interval for two points, A and B, in the path of the wave. B is 0.30 m to the right of A.

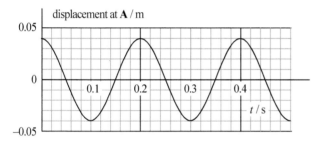

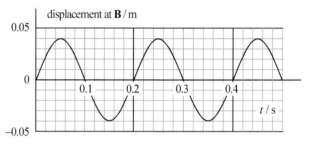

(a) (i) Write down the value of the wave's *amplitude*.
   (ii) Determine the *frequency*.

(b) (i) By comparing the two displacement–time graphs, determine the shortest time the wave could be taking to go from A to B.
   (ii) Hence show that 6.0 m s^{-1} is a possible speed for the wave.
   (iii) Explain why a speed of 1.2 m s^{-1} would also be consistent with the graphs.

(c) (i) Define the *wavelength* of a wave.
   (ii) Taking the wave speed as 6.0 m s^{-1}, calculate the wavelength of the above wave.

8. The diagram shows a snapshot of a wave at time $t = 0$. The wave is moving to the right with a speed of 5.00 m s^{-1}.

(a) Calculate the frequency, f, and period, T, of the wave.

(b) Sketch a graph to show the vertical velocity of the particle at $x = 2.0 \text{ m}$ between $t = 0$ and $t = 1.00 \text{ s}$.

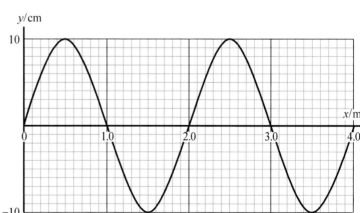

9. A beam consists of a mixture of 70% polarised and 30% unpolarised light. A perfect polarising filter (i.e. it transmits 100% of the light vibrating in one plane of polarisation and 0% of the light vibrating at right angles) is put into the beam and rotated about its axis. Sketch a graph to show how the intensity of the transmitted radiation varies with the angle θ of the filter. Assume intensity is minimum when $\theta = 0$.

10. The diagram shows a straight wavefront in a ripple tank approaching (from the left) a submerged lens-shaped platform, which has the effect of slowing the wave as it passes over the platform, and deforming its wavefront.

    Copy the diagram, leaving space to the right.

    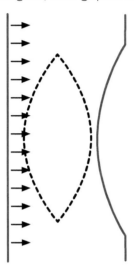

    (a) Explain briefly the shape of the right-hand wavefront,
    (b) Add arrows to this wavefront to show the direction(s) of propagation,
    (c) Hence show the positions of the wavefront at two more successive instants. The time intervals between the four wavefront positions should be approximately equal.
    (d) Label a point that could sensibly be called the focus.

11. Unpolarised light passes through two polarising filters, one after the other. The molecules of the two filters are parallel at first. The first filter is fixed and the second one slowly rotated.

    (a) The polarising filters are said to be crossed when their molecules are at right angles to each other. Explain why no light emerges from the second polaroid when the polaroids are crossed.
    (b) If another polaroid, with its molecules at 45° to those of the crossed polaroids, is placed between the crossed polaroids, some light is again transmitted through the combination. Explain why this happens.

    [Hint: Fig. 2.4.6 (b) may help.]

12. [For A level students]. The equation $y = 10 \cos 6.28x$ (with y in cm and x in m) represents a wave at time $t = 0$. The wave propagates with a speed of 800 m s^{-1}. A student suggests that the equation of motion of a particle at $x = 2.5$ m is given by $y = A \cos (\omega t + \phi)$. Find the values of A, ω and ϕ that make this statement correct. Give your reasoning.

2.5 Wave properties

2.5.1 Diffraction

Diffraction can sometimes be seen near the coast. When there is a rock in the path of waves at sea, after passing the rock to either side, the wavefronts spread back into the 'shadow' of the rock, as sketched in Fig. 2.5.1.

We shall now concentrate on diffraction around the edges of a slit. Fig. 2.5.2 shows straight wavefronts in a 'ripple tank' approaching a slit, and having gone through, spreading out beyond its edges.

When the slit's width is equal to, or less than, the wavelength, the diffracted wavefronts at some distance from the slit are more or less semicircular as in Fig. 2.5.3 (a), though the amplitude is greater in the middle than at the edges.

When the slit width is several times greater than the wavelength as in Fig. 2.5.3 (b), there is a main or central beam of diffracted waves that spreads through only a small angle either side of the 'straight-through' direction. There are also 'side' beams of much smaller amplitude than the main beam. These side beams can be seen faintly in the ripple tank image (Fig. 2.5.2). We'll see later how, in principle, this rather complicated behaviour arises.

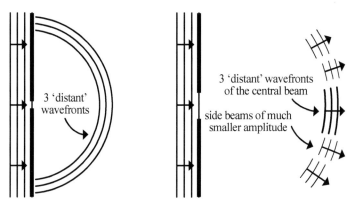

Fig. 2.5.3 (a) Diffraction: slit width ≤ wavelength

Fig. 2.5.3 (b) Diffraction: slit width >> wavelength

We don't usually observe *light* diffracting around obstacles such as the sides of holes or slits. Indeed a pinhole camera relies on light *not* spreading much, and so not changing its direction noticeably, as it passes through the pinhole. Close examination of the picture might reveal some blurring due to diffraction, but the set-up in Fig. 2.5.4 demonstrates diffraction of light unmistakeably.

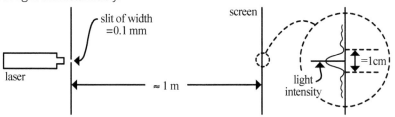

Fig. 2.5.4 Demonstrating diffraction of light at a slit

The very small angular spread of the main beam through this narrow slit, compared to that in Fig. 2.5.2, suggests that the wavelength of light is very small indeed. We shall soon describe how to confirm this.

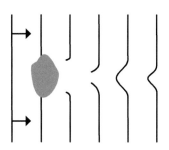

Fig. 2.5.1 Water waves diffracting round a rock

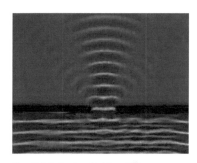

Fig. 2.5.2 Single slit diffraction

Knowledge check 2.5.1

In Fig. 2.5.2 the main beam spreads through an angle of roughly 20°. By making measurements with a ruler estimate how many times the slit width is greater than the wavelength.

Knowledge check 2.5.2

You can hear the news blaring out from a neighbour's radio when the windows are open, but you can't always make out the words. Explain this in terms of diffraction.

Hint: consonants contain higher frequencies of sound (> 3000 Hz).

Speed of sound = 330 m s^{-1}.

 **Key term**

Principle of superposition: the resultant displacement at each point is the vector sum of the displacements that each wave passing through the point would produce by itself.

2.5.3 Knowledge check

In the example, all traces of the pulses seem to have vanished at $t = 1.0$ s. How can they re-appear? Where has the energy gone?

Hint: it's the *displacement* that's zero at $t = 1.0$ s.

2.5.4 Knowledge check

Suppose two beams of light cross over: how, if at all, will they have been affected by each other *beyond* the region of overlap?

Hint: apply the principle of superposition to the region beyond the overlap.

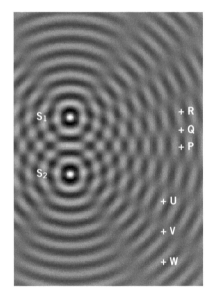

Fig. 2.5.6 Two-source interference

2.5.2 Interference

Interference is probably the most distinctive wave effect: if you can demonstrate it taking place with some unknown 'radiation', you know that waves are involved!

Interference is what happens when waves from more than one source, or waves travelling by different paths from the same source, *superpose* or 'overlap' in the same region. The behaviour is governed by the **principle of superposition**.

Example

Suppose two identical but inverted pulses are travelling along a string at 1.0 m s^{-1} in opposite directions, as shown at time $t = 0$ in Fig. 2.5.5 (a).

What does the string look like at times $t = 1.0$ s and $t = 2.0$ s?

Answer

Applying the principle of superposition, we get the situations shown in Fig. 2.5.5 (b)

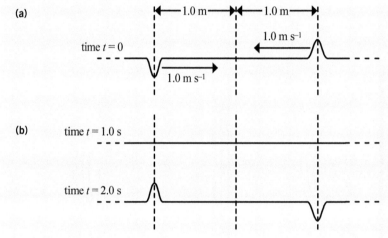

Fig. 2.5.5 Superposition in progress

(a) The two-source interference pattern

Fig. 2.5.6 is a snapshot of a ripple tank when two small rods, S_1 and S_2, vibrating *in phase*, move up and down in contact with the water surface. These are the wave sources. You can see beams of waves, separated by 'channels' of very low amplitude. Each beam has alternating peaks and troughs, seen as bright and dark elongated blobs.

The crosses marked P and R are examples of points where the wave amplitude is (locally) highest, because waves from S_1 and S_2 arrive **in phase** there, and interfere **constructively**. This is shown graphically in Fig. 2.5.7. (Note that the *displacement* isn't necessarily a maximum at P and R *at the instant of the snapshot*.)

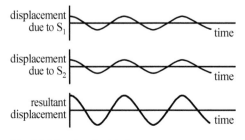

Fig. 2.5.7 Constructive interference

Where the *amplitude* is lowest in the two-source pattern, at Q, for example, the waves from S_1 and S_2 are arriving in antiphase (half a cycle out of phase) and are interfering *destructively*: Fig. 2.5.8.

Note that cancellation can never take place if the component waves are vibrating at right angles to each other. To produce an interference pattern, transverse waves cannot be polarised at right angles to each other.

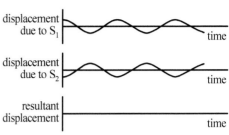

Fig. 2.5.8 Destructive interference

(b) Path difference

Even without seeing the pattern, we can predict where, in relation to the sources, S_1 and S_2, there will be constructive or destructive interference. For example, we know that there must be constructive interference at points such as P in Fig. 2.5.6, which are equal distances from S_1 and S_2. This is because S_1 and S_2 are oscillating in phase, so the waves from the two sources, having travelled by *paths*, S_1P and S_2P, will *arrive* in phase at P. The same applies for all points on the central axis (the central line through the point midway between S_1 and S_2 and perpendicular to S_1S_2).

For point R, the paths are S_1R and S_2R. The *path difference*, $S_2R - S_1R$ is 1 wavelength, so waves from S_2 arrive at R a whole cycle later than waves from S_1. This means that they arrive at R *in phase* with the waves from S_1, so there will be constructive interference at R.

For point Q, the path difference $S_1Q - S_2Q$ is half a wavelength, so waves arrive at Q in antiphase and will destructively interfere.

Knowledge check 2.5.5

In Fig. 2.5.6 determine λ by measuring the distance along a beam between centres of dark blobs separated by a number of wavelengths, and dividing by that number. Now measure S_1R and S_2R and comment on whether your results agree with the path difference rules.

The general rules for waves from in-phase sources are:

For constructive interference at a point X,

$$\text{path difference, } |S_1X - S_2X| = 0, \lambda, 2\lambda, 3\lambda \ldots$$

That is: path difference, $|S_1X - S_2X| = n\lambda$, where $n = 0, 1, 2, 3 \ldots$

For destructive interference at a point X,

$$\text{path difference, } |S_1X - S_2X| = \frac{\lambda}{2}, \frac{3\lambda}{2}, \frac{5\lambda}{2} \ldots$$

That is: path difference, $|S_1X - S_2X| = (n + \frac{1}{2})\lambda$, where $n = 0, 1, 2, 3 \ldots$

Knowledge check 2.5.6

In Fig. 2.5.6 determine the path differences for points U and V in terms of wavelength.

Example

Referring to point W in Fig. 2.5.6, determine the path difference, $|S_1W - S_2W|$ in terms of λ.

Answer

W is on the mid-line of a beam of waves, so there is constructive interference here. So $|S_1W - S_2W| = 0$, λ, 2λ or 3λ .... The point P is on the perpendicular bisector of S_1S_2 so P is equidistant from the two points, i.e. the path difference at P is zero. So counting the centre beam as 'beam zero', we find W on the third beam 'out', so $|S_1W - S_2W| = 3\lambda$.

◀ **Stretch & challenge**

(a) Referring to Fig. 2.5.9, two transmitters, S_1 and S_2, are sending out radio waves of wavelength 24 m in phase and both vertically polarised. Determine what type of interference there will be at X if $a = 120$ m, MX = 600 m and $\theta = 30°$.

(b) There is a simple approximate formula for the path difference in a two-source set-up. Referring to Fig. 2.5.9:

$$S_2X - S_1X \approx a \sin \theta$$

It applies accurately only if MX $\gg a$.

Test this formula by comparing the value of $S_2X - S_1X$ that it gives with the value found in (a).

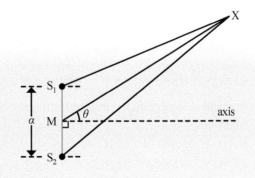

Fig. 2.5.9 Calculating path difference

2.5.3. Young's fringes experiment

In the early 1800s, Thomas Young investigated light passing through two parallel slits (or sometimes two pinholes) close together. He observed a pattern of light and dark *fringes* (stripes) on a screen placed some distance from the slits. He recognised this pattern as a section through an interference pattern, and deduced that light was wave-like. He ran various checks, such as moving the screen nearer the slits and finding that the fringes became closer together. (Compare vertical sections through Fig. 2.5.6, at different distances from the sources.) Young was able to determine the wavelengths of different colours of light.

Here is a modern version of Young's experiment.

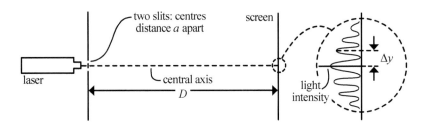

Fig. 2.5.10 Modern Young's fringes experiment

With $a = 0.5$ mm and $D = 1.5$ m, and using red light, the fringe separation, Δy (between the centres of adjacent bright fringes or adjacent dark fringes) is found to be about 2 mm.

How does it work? The slits act as sources. Light passing through them spreads out slightly by *diffraction* so, if the slits are about 0.1 mm wide, there will be a few millimetres on the screen 'above' and 'below' the axis, where light from both slits overlaps and interferes. (This would still happen without the screen!)

The wavelength of light can be found from the equation:

$$\lambda = \frac{a\Delta y}{D}$$

This is an approximation based on the *path difference* rules. As long as $a \ll D$ and $\Delta y \ll D$, as in the set-up described, the equation is almost exact.

Coherence

If we illuminate each slit with light from a different source, we find we can never produce fringes. Even using a single source such as an LED we can obtain fringes only by taking special precautions (such as putting a single narrow slit between the LED and the double slits). Using a laser, though, the simple arrangement shown above gives beautiful fringes. This is because a laser produces *coherent light*.

A beam of **coherent light**

- is nearly **monochromatic**, that is, it is a continuous stream of oscillations of a single frequency,
- has wavefronts extending across its width, as if it came from a point source.

Two sources are said to be coherent if there is a constant (not necessarily zero) phase difference between their oscillations.

If we could illuminate both slits with a laser beam accurately normal to the plane of the slits, these would act as in-phase sources because each wavefront in the beam would hit both at the same time. This is not usually achieved, but the slits will still act as *coherent*

2.5.7 Knowledge check

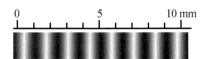

Determine the fringe spacing from the enlarged picture of the fringes with millimetre scale, and hence determine the wavelength of the light, if $a = 0.40$ mm and $D = 0.80$ m.

2.5.8 Knowledge check

In an experiment to find the speed of sound, loudspeakers each emitting a 1000 Hz note are placed at the goalposts (7.32 m apart) at one end of a football pitch. 50 m from the goal line is the halfway line. Students walk along it and agree that the quietest spots on the line are 2.3 m apart. Determine a value for the speed of sound.

◀**Stretch & challenge**

Derive the Young's slits equation by using the formula $S_2X - S_1X \approx a \sin \theta$ given in the last Stretch & challenge.

Hints: Let X be the position of the first bright fringe away from the central axis. Identify θ, and note that it is very small, so $\sin \theta \approx \tan \theta$.

2.5.9 Knowledge check

Why does the intensity of bright fringes fall off with distance from the central axis?

Hint: at still greater distances from the axis, fringes re-appear, but faintly.

sources, meaning that there will be a *constant phase difference* between them. A fringe pattern will still be seen with the same finge separation, but there probably won't be a bright fringe on the axis.

The light from an 'ordinary' source such as an LED is far from coherent. Even that from a 'coloured' LED contains a range of frequencies. This in itself wouldn't prevent us seeing a fringe pattern (actually many super-imposed patterns), but there is no fixed phase relationship between light emitted from points all over an emitting area in the order of 1 mm^2, so the LED doesn't meet the *point source* condition for producing coherent light (at least not without the single narrow slit mentioned above).

2.5.4 The diffraction grating

(a) The need for accurate measurements of wavelength

The need for accurate measurements of wavelength arises in many areas of science. For example, astronomers measure the wavelengths of light emitted and absorbed by a distant star, and from these wavelengths they can identify elements present in the star's outer layers, determine the star's line-of-sight velocity, and perhaps infer the existence of orbiting planets.

For this application, and for most others, the two-slit set-up won't do. It is impossible to measure the fringe separation accurately enough, because:

- The fringes aren't sharp: bright fringes fade gradually into dark (see diagram in Knowledge check 2.5.7).
- The brightest parts of the pattern are not as bright as if the light were concentrated on them. This matters if the light source is faint.
- The fringe separation is small.

The diffraction grating deals successfully with all three of these issues.

(b) What is a diffraction grating?

At its simplest, a diffraction grating (Fig. 2.5.11) is a flat plate, which is opaque except for thousands of straight, parallel, equally-spaced slits.

For a grating to be used with visible light, the distance, d, between the centres of adjacent slits is typically 2 or 3 μm. The makers supply the value of d.

We shall assume that light is shone normally (at right angles) on to the grating, so that the slits act as in-phase sources. Each slit is very narrow (in the order of a wavelength wide), so the diffracted wavefronts are almost semicircular, spreading right round. See Fig. 2.5.12 (a).

Instead of seeing light travelling out from the grating in all directions, we see beams of light emerging in very specific directions (Fig. 2.5.13). This is because of interference between the light from different slits.

Fig. 2.5.11 A diffraction grating

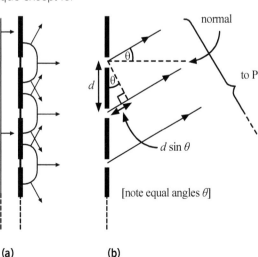

(a) (b)

Fig. 2.5.12 Diffraction grating

(c) Deriving the grating equation

Consider the light arriving at a distant point, P, from each slit. The light paths from the slits to P will be almost parallel; see Fig. 2.5.12 (b). For light going to P from adjacent slits, the *path difference* is $d \sin \theta$. To see this, use the small right-angled triangle formed by dropping a perpendicular from one slit to the 'ray' from the next slit. The condition for constructive interference at P is therefore

$$d \sin \theta_n = n\lambda \qquad \text{for} \qquad n = 0, 1, 2 \ldots$$

We can see beams emerging from the grating at these angles θ_n if we sprinkle dust in the air around the grating, or let the beams brush against a piece of paper. Note how the beams are called *zeroth order, first order, second order* and so on, according to the value of n. If a screen is put in the path of the beams, as in Fig. 2.5.14, bright spots appear on the screen.

These spots correspond to the bright fringes in Young's experiment but they are much further apart, because the slits are much closer together in the grating ($d \gg a$).

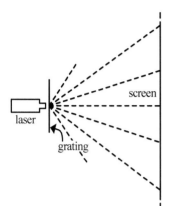

Fig. 2.5.14 Bright spot production

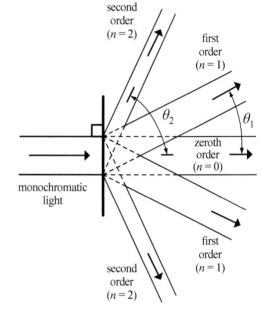

Fig. 2.5.13 Diffraction grating *orders*

> 2.5.10 **Knowledge check**

A diffraction grating has lost its label. The slit separation, d, is determined by shining a beam of light of wavelength 650 nm normally at it and noting that the second order beam emerges at 48°. Calculate the wavelength which gives a third order beam at 54° with this grating.

> 2.5.11 **Knowledge check**

In Fig. 2.5.14 the two second order spots are 180 mm apart and the distance from screen to grating is 150 mm. The centres of slits in the grating are 2.00 μm apart. Calculate:

(a) the wavelength of the laser light, and

(b) the separation of the two first order spots on the screen.

Example

A laser beam is shone normally on to a grating with 5.00×10^3 slits per metre. The angle to the normal of the second order beams is 35.1°.

(a) Calculate the wavelength of the light.

(b) Calculate the number of beams emerging from the grating.

Answer

(a) Since there are 5.00×10^5 slits per metre the distance, d, between centres of slits must be

$$d = \frac{1.00 \text{ m}}{5.00 \times 10^5} = 2.00 \times 10^{-6} \text{ m}$$

So $\lambda = \dfrac{d \sin \theta_2}{2} = \dfrac{2.00 \times 10^{-6} \text{ m} \times \sin 35.1°}{2} = 575 \times 10^{-9} \text{ m} = 575 \text{ nm}.$

(b) When $\theta = 90°$, $\sin \theta = 1$, so the path difference between light going to a distant point from adjacent slits is simply d, its greatest possible value – as should be clear without trigonometry! The number of wavelengths contained in distance d is simply

$$\frac{d}{\lambda} = \frac{2.00 \times 10^{-6} \text{ m}}{575 \times 10^{-9} \text{ m}} = 3.48.$$

There will be angles 0, θ_1, θ_2 and θ_3 corresponding to path differences of 0, λ, 2λ and 3λ but, as we have just seen, a path difference of 4λ cannot be reached for *any* angle.

So 7 beams emerge: the zeroth order and 3 orders either side.

The spots on the screen (Fig. 2.5.14) are brighter and much sharper than Young's fringes. This is because, for a diffraction grating, when $d \sin \theta = n\lambda$, light from each slit interferes constructively with light from every other slit, but for slightly larger or smaller angles, light from various non-neighbouring slits interferes destructively (See *Stretch & challenge*.) With only two slits, the interference is still almost constructive at slightly larger or smaller values of Δy than those for which $\lambda = a \, \Delta y / D$.

◢Stretch & challenge

A diffraction grating has 10 000 slits, and the slit separation is 2.000×10^{-6} m. It is used with normally incident light of wavelength 600 nm.

(a) Show that the first order beam emerges at $17.46°$ to the normal.

(b) For an angle of $17.55°$, show in terms of path difference that light from a slit will interfere destructively with light from a slit that is 100 slits away.

(c) Explain why, for the grating as a whole, there will be almost complete destructive interference at this angle.

2.5.5 Stationary waves

(a) The nature of stationary waves

The waves we have been considering so far are **progressive waves**: they progress through space (or a medium), taking energy with them. Within a confined space, a second kind of oscillation can exist, which is known as a **stationary** (or **standing**) **wave**. Three possible modes of transverse vibrations for a stretched string or wire, such as a violin or harp string, are shown in Fig. 2.5.15. In each case the string vibrates up and down: the full line shows one extreme position of the string and the pecked line shows the position half a cycle later. Fig. 2.5.16 illustrates the motion in more detail.

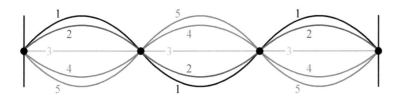

Fig. 2.5.16 Half a cycle of a stationary wire

It shows half a cycle of oscillation, starting with the string in one extreme position (1), and then at equal time intervals of $T/8$ (2, 3, 4 and 5). Following this half cycle, the string tracks back: $4 \rightarrow 3 \rightarrow 2 \rightarrow 1$ and so on. The points which are always stationary, marked with dots, are called **nodes** and the points with maximum motion, midway between the nodes, are **antinodes**.

Although, almost always, a single snapshot of a stationary wave (e.g. red line in Fig. 2.5.16) looks the same as a snapshot of a progressive wave (e.g. the full line in Fig. 2.4.11), there are major differences between progressive and stationary waves that show up if we consider what happens at various times. Think carefully about the following summary of the differences, referring to Figs 2.4.11, 2.4.12 and 2.5.16.

1. In a stationary wave, all points between a pair of neighbouring nodes oscillate in phase; points either side of a node oscillate in antiphase. In a progressive wave there is a gradual change of phase along the wave.

2. In a progressive wave, all points oscillate with the same amplitude (apart from a gradual decrease in amplitude with increasing distance from the source). In a stationary wave, the amplitude of vibration varies smoothly from zero, at the nodes, to a maximum, at the antinodes.

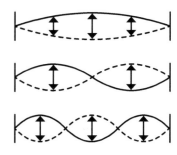

Fig. 2.5.15 Stationary waves on a string or wire

(b) The relationship between progressive and stationary waves

When two progressive waves of the same amplitude and frequency travelling in opposite directions superpose, they interfere to produce a stationary wave.

2.5.13 Knowledge check

Sketch the red and blue graphs for the instant $T/8$ after the first diagram in Fig. 2.5.17, using a copy of the same grid with vertical lines $\lambda/4$ apart. Locate the zeros and the maxima in the *resultant* wave and hence sketch it.

Comment on the positions of the zeros and on the peak displacement.

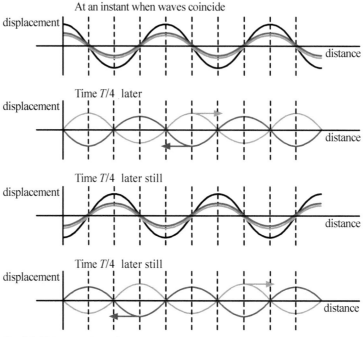

Fig. 2.5.17 Superposition of progressive waves travelling in opposite directions

Fig. 2.5.17 shows progressive waves (red and blue) interfering to produce a stationary wave (black). Check the sequence of diagrams carefully and do Knowledge check 2.5.13. Note that the nodes are a distance $\lambda/2$ apart. The antinodes are a distance $\lambda/2$ apart.

How can we *produce* progressive waves of the same amplitude and frequency travelling in opposite directions so as to superpose? An easy method is to *reflect* a progressive wave. For example, if we clamp one end of a stretched string firmly and keep sending progressive waves along the string from the other end, these waves will reflect off the clamp and the reflected waves will interfere with the 'oncoming' waves to produce a stationary wave on the string. The clamped end of the string is necessarily a node – because it can't move! The amplitude of the reflected wave may be a little less than that of the incident wave, resulting in an 'impure' stationary wave, in which the nodes are minima of amplitude rather than zeros.

(c) Harmonics

When a string is clamped at *both* ends, there must be nodes at both ends. Since neighbouring nodes are always $\lambda/2$ apart, this restricts the stationary waves to those in Fig. 2.5.15 – and its imagined continuation. So if the length of string is ℓ, we have

$$\ell = n\frac{\lambda}{2} \qquad \text{that is} \qquad \lambda = \frac{2\ell}{n} \qquad \text{in which } n = 1, 2, 3 \ldots.$$

The number n tells us the *mode of vibration* of the string, specifically the number of antinodes. When $n = 1$ the string is vibrating with no nodes except at the ends of the string. This mode is called the *first harmonic* or *fundamental mode*. When $n = 2$, we have the *second harmonic* (middle diagram in Fig. 2.5.15), when $n = 3$ we have the *third harmonic* – and so on.

Example

The vibrating section of a violin's A string has a length of 33 cm. Its fundamental frequency is 440 Hz. Calculate:

(a) The speed of transverse waves on the string.
(b) The frequencies of the second and third harmonics.

Answer

(a) The string length $= \frac{1}{2}\lambda$, $\therefore \lambda = 66$ cm.
    The wave speed $v = \lambda f = 0.66$ m $\times$ 440 Hz $= 290$ m s^{-1}

(b) For the second harmonic, the nodes are 16.5 cm apart, so λ is 33 cm. This is half that of the first harmonic, so the frequency, f_2, is double, i.e. **880 Hz**.

Similarly $f_3 = 3 \times 440$ Hz $= 1320$ Hz.

It is easy to derive a general equation for the possible vibration frequencies of a taut string. The frequency, f_n, of the nth mode is

$$f_n = \frac{v}{\lambda} = \frac{v}{2\ell / n} \qquad \text{that is} \qquad f_n = n\frac{v}{2\ell} \qquad \text{in which } n = 1, 2, 3 \dots .$$

and v is the speed of the transverse waves. This speed is a constant for a string of given mass per unit length under a given tension. We see that all modes of vibration have frequencies that are multiples of the fundamental frequency, $f_1 = v/2\ell$. In fact, $f_n = nf_1$.

(d) Stationary waves on a stretched string

A common way of investigating transverse stationary waves on strings and their harmonics is shown in Fig. 2.5.18.

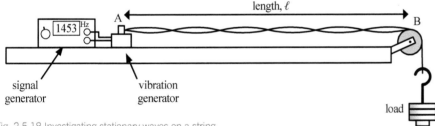

Fig. 2.5.18 Investigating stationary waves on a string

The signal generator produces alternating current of a variable frequency. The vibration generator converts these into low amplitude (typically ~1 mm) vibrations of the small peg at **A**. The load provides the required tension in the string. For a particular load and length, the frequency, f, is increased from a low value and the values of f is noted at which stable harmonics are produced. This setup may be used to verify that the frequencies of the harmonics have the mathematical relationship shown above.

(e) Stationary waves in air columns

Stationary waves can be longitudinal, too. If you blow across the open top of a tube or pipe, such as a pen top, longitudinal stationary waves are set up in the air column. Unlike waves on a string, which is fastened at both ends, the stationary wave has an antinode just beyond the open end. It is difficult to draw a longitudinal wave, so diagrams usually show the stationary wave as if it were transverse. An alternative is to mark just nodes and antinodes (as • and ↕).

The three simplest modes of vibration of the air in a pipe that is open at one end and closed at the other are shown in Fig. 2.5.19, and their frequencies worked out – check! The effective length of the pipe is ℓ, and the speed of sound in air is v. The fundamental, or first harmonic, has a frequency, f_1, of $v/4\ell$. The next highest frequency mode has a frequency of $3f_1$ and is called the third harmonic. The next highest, with a frequency of $5f_1$ is the fifth harmonic and so on. We say that pipes open at one end and closed at the other support only odd harmonics.

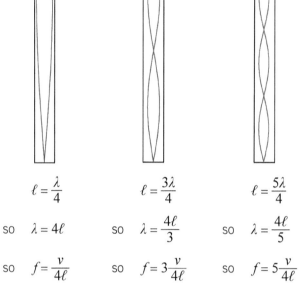

$$\ell = \frac{\lambda}{4} \qquad \ell = \frac{3\lambda}{4} \qquad \ell = \frac{5\lambda}{4}$$

so $\lambda = 4\ell$ so $\lambda = \dfrac{4\ell}{3}$ so $\lambda = \dfrac{4\ell}{5}$

so $f = \dfrac{v}{4\ell}$ so $f = 3\dfrac{v}{4\ell}$ so $f = 5\dfrac{v}{4\ell}$

Fig. 2.5.19 Harmonics in a pipe with one open end

Blowing across the mouthpiece (in a flute) or into the mouthpiece (in a trumpet) produces a wide range of frequencies, but only the ones which produce these stationary waves are selected and amplified by the instrument. This is an example of resonance, which you will meet in your year 13 course.

Example

The speed of sound in air is proportional to the square root of the kelvin temperature. A flute produces a note of frequency 440.0 Hz at 20°C. What frequency would it produce if taken outside into a temperature of 0°C?

Answer

Speed of sound at 0°C, $v_0 = v_{20}\sqrt{\dfrac{273}{293}} = 0.965 v_{20}$

Assuming that the change in the length of the flute is not significant, the wavelength of the sound waves is unchanged.

But $f = \dfrac{v}{\lambda}$, $\therefore f_0 = 0.965 f_{20} = 0.965 \times 440 = 425$ Hz.

This frequency is more than halfway down from A to G# (415 Hz), so the flute is seriously out of tune!

2.5.14 Knowledge check

'Singing tubes' are open at both ends. The diagram shows the first harmonic.

(a) Draw the 2nd and 3rd harmonics.

(b) Write the frequency of the nth harmonic, f_n, in terms of f_1.

2.5.15 Knowledge check

An organ pipe of length 50 cm is closed at one end.

(a) Calculate the frequencies of the two harmonics of the lowest frequency.

(b) What would be the effect of closing the open end of the pipe?

[Speed of sound = 340 m s⁻¹]

2.5.6 Specified practical work

(a) Determination of wavelength using Young's double slits

A microscope slide is prepared by coating it with a colloidal suspension of graphite, which is allowed to dry. Two parallel slits are scratched into the graphite using a scriber. Typically these slits are 0.2–0.3 mm wide and 0.4–0.5 mm apart (centre to centre).

To measure the wavelength of a laser, which is a monochromatic light source, the slits are set up in a darkened room, with the laser beam incident on the slits at right angles to the plane of the slits. The resulting pattern is viewed on a screen ~2 m away (Fig. 2.5.20). The distance is not critical, but the further the screen is from the slits, the more spread out the fringes will be.

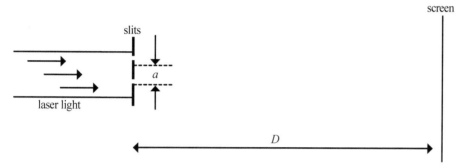

Fig. 2.5.21 Young's fringes experimental arrangement

Fig. 2.5.20 Typical fringe appearance

The wavelength, λ, of the light is given by $\lambda = \dfrac{a\Delta y}{D}$, which requires a, D and Δy to be measured.

- D is measured using a tape measure or metre rules with a percentage uncertainty of ~0.5% (1 cm in 2 m) or less.
- a is measured using a travelling microscope, with an uncertainty of ~0.01 mm, i.e. ~2%.
- Δy, the fringe separation, is measured using a mm scale. The spread of (say) 10 fringes is measured and divided by the number of fringes. Typically the spread is ~1.5 cm with an uncertainty of ~1 mm.

Notice that D, even rather crudely measured, makes the smallest contribution to the overall uncertainty in λ.

≪ **Top tip**

In the expression for λ, $\dfrac{a\Delta y}{D}$, the very small quantities, a and Δy are on the top and the large quantity, D, is on the bottom. The values of a and Δy are both < 1 mm and D > 1 m, giving λ < 1 μm, as expected.

Knowledge check 2.5.16 ◀

With the given uncertainties in D, a and Δy estimate:

(a) The percentage uncertainty in λ.

(b) The absolute uncertainty if λ ~590 nm.

(b) Determination of wavelength using a diffraction grating

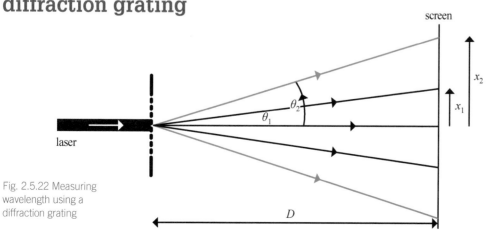

Fig. 2.5.22 Measuring wavelength using a diffraction grating

Knowledge check 2.5.17 ◀

A diffraction grating is marked '2 000 lines cm^{-1}'. Calculate the separation, d, of the slits.

The experiment is set up in the same way as for the Young's slits experiment except that a diffraction grating is used instead. With a laser, there is no need for blackout conditions as much more light is transmitted by the grating than by the slits and the spectral lines are much more sharply defined.

The diffraction grating formula is

$$d \sin \theta_n = n\lambda$$

where θ_n is the angle of the nth order spectrum and d the diffraction grating slit separation. To calculate the wavelength, d and θ_n need to be determined. The value of d is usually calculated from the **diffraction grating constant**, which is provided by the manufacturer (but see the Study point). If projecting the spectra on to a screen it is convenient to determine the θ_n values (Fig. 2.5.22) by measuring D (as with the Young's slits experiment) and x_n using a metre rule, and then using

$$\theta = \tan^{-1}\left(\frac{x}{D}\right).$$

Typically, $D \sim 2$ m and the displacements, x, are of the order of 50 cm. To reduce the uncertainty in the values of x it is sensible to measure the distance between the two first order spectra and divide by 2.

(c) Determination of the speed of sound using stationary waves

A tuning fork, F, is used to set up stationary waves in the column of air above the water surface in an open-ended transparent tube, T (see Fig. 2.5.23). The tube is raised and lowered until sounds of maximum intensity are heard at the least possible resonance length, ℓ.

The smallest distance, ℓ, for a stationary wave, is such that there is a node, N, at the surface of the water and an antinode, A, close to (but just above) the open end of the tube, as shown in Fig. 2.5.24. The distance ℓ is the measured distance of the water surface below the open end of the tube and the unknown distance e, called the *end correction*. The distance between a node and the adjacent antinode is $\lambda/4$, so we can write

$$\frac{\lambda}{4} = \ell + e.$$

The simplest method to determine the speed of sound is as follows:

- Starting tube T in its lowest position, raise the tube until the first resonance is detected with a vibrating tuning fork of known frequency, e.g. 256 Hz (middle C).
- By repeatedly raising and lowering the tube, locate the resonance position accurately.
- Measure the resonance length, ℓ, and obtain several repeat readings.
- Repeatedly with a series of tuning forks up to, e.g. 512 Hz (top C).

The analysis
For each frequency, f, $\lambda = \frac{v}{f}$, where v is the speed of sound.

Substituting for λ and rearranging: $\ell = \frac{v}{4f} - e$. So a graph of ℓ against $\frac{1}{f}$ is a straight line of gradient $\frac{v}{4}$ and intercept $-e$ on the ℓ-axis (see Fig. 2.5.25).

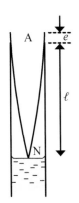

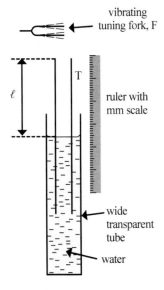
Fig. 2.5.23 Resonance apparatus

Fig. 2.5.24 1st harmonic

Variations

1. A small loudspeaker, connected to a calibrated signal generator can be used instead of the set of tuning forks.

2. The second (and third) harmonics can be located for a single frequency. This is illustrated in Fig. 2.5.26. If the length of air for the nth harmonic is ℓ_n, then $\ell_2 - \ell_1 = \dfrac{\lambda}{2}$ and $\ell_3 - \ell_2 = \dfrac{3\lambda}{4}$. If the frequency is known, the wave speed can be calculated.

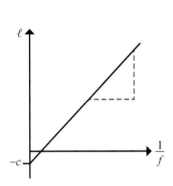

Fig. 2.5.25 Graph of ℓ against $\dfrac{1}{f}$

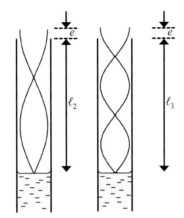

Fig. 2.5.26 2nd and 3rd harmonics

Test yourself 2.5

1. (a) State the principle of superposition.
   (b) Two water waves with the same frequency pass through the same area of the surface of a pond. One has amplitude 1.5 cm and the other 2.0 cm in this area. Explain why, in some places the amplitude of vibration of the water surface is 3.5 cm. What is the minimum amplitude of vibration of the water surface in this area?

2. Two sets of waves, arriving at a point, are coherent. Which of the following statements must be true?
   A The waves are transverse.
   B The waves have the same frequency.
   C The waves are in phase.
   D The sources of the waves oscillate in phase.
   E The phase difference between the waves is constant.

3. A signal generator, **G**, produces an oscillating electric current with a frequency of 1.70 kHz. This is fed to two small speakers, **A** and **B**, which produce sound waves of the same frequency. A microphone, **M**, is placed as shown, which receives sound waves from both speakers. [The speed of sound is 340 m s^{-1}.]

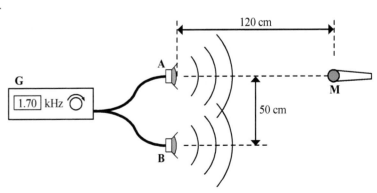

   (a) Without calculation, explain whether the two sound waves, received by **M**, are coherent.
   (b) Use Pythagoras' theorem to show that the path difference, **BM–AM**, is 10 cm.
   (c) Using their frequency calculate the wavelength of the sound waves.
   (d) Explain whether you would expect the volume of sound received by the microphone to be high or low. What is the name of the principle you have used?
   (e) The microphone is now moved 25 cm downwards in the diagram, i.e. at right angles to the dotted line between it and speaker **A**. Explain in detail what would happen to the volume of sound during the movement.

4. In an experiment to investigate stationary waves in sound, a student connects two loudspeakers to the same signal generator, which is set to produce a sine wave of **800 Hz**. She sets up the speakers facing each other and a few metres apart and uses a microphone to investigate the sound between the speakers.

(a) Explain why it is important to connect the speakers to the same signal generator.
(b) The student finds minima 0.21 m apart along the line joining the speakers. Calculate a value for the speed of sound.
(c) Explain why nodes are less distinct when much closer to one speaker than the other.

5. A string is stretched between fixed supports, **1.65 m** apart. It is set vibrating with a period of **0.016 s**, and is shown at time $t = 0$, when its displacement is a maximum.

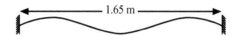

$\leftarrow$ 1.65 m $\rightarrow$

(a) Copy the diagram and add sketches of the string at $t = 0.008$ s and at $t = 0.012$ s.
(b) Show that the speed of waves on the string is approximately **70 m s^{-1}**.
(c) Describe the phase relationships between the different points on the string.
(d) Calculate the wavelength and frequency of the lowest possible frequency stationary wave on the string.

6. In an experiment to measure the speed of sound, a student uses a tuning fork of frequency **440.0 Hz**. She held the vibrating tuning fork over a tube of water which had a small hole in the side as shown, allowing the water to drain out. She measured the length, ℓ, of the air column above the water surface at which the first resonance was heard. She reasoned there was a node at the water surface and an antinode at the open end of the tube, so ℓ must represent $\frac{1}{4}\lambda$, where λ is the wavelength of the sound.

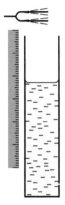

Repeating the procedure several times, she obtained the following values of ℓ using a mm scale:
185 mm, 189 mm, 190 mm, 187 mm, 189 mm.

Use the readings to determine a value for, v, the speed of sound together with its uncertainty, Δv.

7. The student in Q6 reads Section 2.5.6(c) and finds out that the antinode is not exactly at the end of the tube but a short distance (called the end correction), e, above it.

So $\ell = \dfrac{\lambda}{4} - e$. She repeated the experiment and found a second resonance with

$\ell = \dfrac{3\lambda}{4} - e$. The readings were: 579 mm, 576 mm, 577 mm, 573 mm, 575 mm.

Use the results of both experiments to find values for v and ε together with their uncertainties.

8. The diagram shows a set of plane microwaves approaching a pair of slits, whose centres are **8.0 cm** apart, made using three aluminium plates. A probe, P, is moved along the dotted line which is **50 cm** from the plane of the slits.

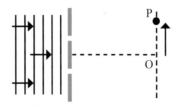

(a) When one of the slits is covered up, a signal is received at O. Name the effect which is responsible for this.
(b) When both slits are uncovered the signal received is much stronger. Explain this.
(c) As the probe is moved from O in the direction shown by the arrow, the signal initially decreases to (nearly) zero and then rises to a maximum. Without calculation, explain this observation and name the effect observed.
(d) (i) The maximum in part (c) occurs when the probe is moved 18.4 cm. Calculate a value for the wavelength of the microwaves using: (I) the Young's slits formula and (II) using Pythagoras' theorem.
(ii) Comment on your answers to part (i)

9. A narrow slit is illuminated by a red LED ($\lambda = 640$ nm) and the emergent light observed on a wall 2.00 m away. The central maximum of the diffraction pattern is measured as 2.0 cm wide. The slit is replaced by a pair of parallel slits of the same width as the first. The interference fringes are 2.0 mm apart.

   (a) Calculate the slit separation.
   (b) Estimate the number of interference fringes you would expect to see. Explain your answer.
   (c) Describe and explain what you would expect to see if:
      (i) The red LEDs were replaced by a green laser ($\lambda = 500$ nm).
      (ii) The slits are now illuminated by both LEDs but one slit is covered with a filter which lets only red light through and the other with a green filter.
      (iii) The slits are covered with polaroid filters with the directions of the two polaroids parallel.
      (iv) The slits are covered with polaroid filters with the directions of the two polaroids at right angles.

10. The speed, v, of transverse waves on a wire is given by $v = \sqrt{\dfrac{T}{\mu}}$, where T is the tension and μ is the mass per unit length. A vertical wire of mass M and length ℓ is put under tension by hanging a series of masses, m, on it. The wire is plucked so that it vibrates. Ignoring any increase in length in the wire owing to the varying tension, derive a formula relating the frequency of vibration, f, to the suspended mass, m. [Assume $M \ll m$]

11. A university physics student was given the task of identifying a mystery line in the spectrum of a sodium discharge tube. She used an accurate spectrometer and a diffraction grating as follows:

   - She measured the angles, ϕ, between the two second order spectra for four lines in the hydrogen spectrum.
   - She measured the angles between the two second order spectra of a bright line in the sodium spectrum and the mystery line in the sodium spectrum.

   Her results are in the table.

	Hydrogen lines				Sodium	
λ / nm	410.2	434.0	486.1	656.3	589.3	mystery
ϕ / °	47.92	50.90	57.54	80.71	71.39	65.46

   Evaluate whether the value of ϕ is consistent with the green line of mercury which has a wavelength of 546.1 nm.

12. A student illuminates a single slit with a laser light of wavelength 650 nm and observes this pattern on a screen which is placed 3.0 m from the slit.

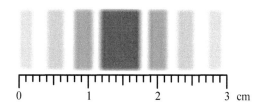

   (a) Sketch a graph of intensity against position for this light.
   (b) He replaces the laser with another of wavelength 450 nm. Sketch a graph of intensity against position for this laser.
   (c) The angular position, in radians, of the dark fringes from the centre is $\dfrac{n\lambda}{w}$, where w is the width of the slit and n takes values 1, 2, 3… By taking readings using the scale find the width of the slit.

2.6 Refraction of light

Key term

Refraction: The change of direction of travel of light (or other wave) when its speed of travel changes, e.g. when it passes from one material into another.

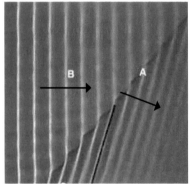

Fig. 2.6.1. Refraction of water waves in a ripple tank

All waves, such as sound, light, sea waves and seismic waves, exhibit a change of direction if they move from one material into another in which the speed of propagation is different (unless the direction of propagation is at right angles to the boundary). This effect is called refraction. The causes of the speed change depend upon the kind of wave. Some examples:

- Seismic wave speed depends upon stiffness and density of rocks.

- The speed of radio waves through the ionosphere (upper atmosphere) is affected by free-electron concentration.

- Surface water wave speed is affected by the frequency and by the depth of water – the ripple tank waves in Fig. 2.6.1 are travelling from left to right in deep water at B; they slow down at the boundary with shallower water at A, which causes them to bunch up and change direction.

- The speed of sound waves through the atmosphere depends upon the temperature.

In the case of light, we use this effect to control light in useful ways, such as making lenses for correcting defects of vision or for constructing telescopes and microscopes, or in optical fibres to transmit information. Geophysicists use the refraction of natural seismic waves to probe the structure of the Earth or artificially generated ones to locate oil and gas reservoirs.

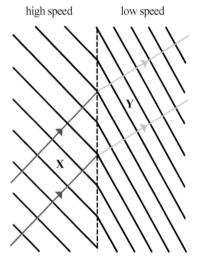

Fig. 2.6.2 Wave refraction

2.6.1 Refraction and wave properties

The wavefronts in Fig. 2.6.2 are crossing a boundary (pecked line) obliquely from left to right into a region in which they travel more slowly. The wavefronts are drawn one wavelength apart and can be thought of as wave crests. The red and green arrows show the directions of travel of the waves. These are at right angles to the line of the wavefronts and, in the case of light, the red and green lines would be light rays.

As each wavefront crosses the boundary, it slows down: the lower end of the wave slows down first so the line of the wave becomes more vertical (more nearly parallel to the boundary). Hence the direction of travel of the waves becomes more nearly horizontal, i.e. closer to right angles to the boundary.

The separation of the wavefronts along the (coloured) direction lines, is the wavelength, which clearly decreases as it crosses the boundary. What about the frequency? Every crest that passes **X** also passes **Y**. Therefore the frequency of the waves must be unchanged even though the speed changes.

Fig. 2.6.3 (a) illustrates the case of a light ray (see Study point) being refracted at a boundary. Some terms associated with refraction are introduced in the diagram and part (b) shows the relationship between the light rays and the wave model of light. Fig. 2.6.3 (b) shows the wavefronts in the narrow beam of light, which is represented by the light ray (see Study point). We can use these diagrams to derive the relationship between the angles, θ_1 and θ_2, and the wave speeds v_1 and v_2.

Study point

We usually use the concept of **rays** to illustrate the refraction of light. Light is a wave phenomenon but when the wavefronts are many orders of magnitude larger than the wavelength we can consider it to move in narrow straight line beams at right angles to the wavefronts: these are light rays.

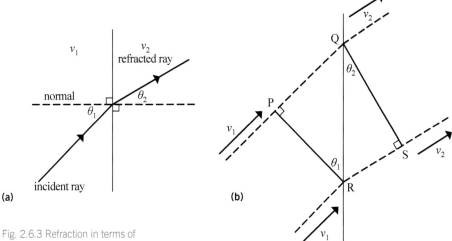

(a)　　　　　(b)

Fig. 2.6.3 Refraction in terms of
(a) rays and (b) wavefronts

PR represents a wavefront arriving at the boundary with an angle of incidence θ_1. QS represents the position of the same wavefront, a time Δt later, at the instant in which point P on the incident wavefront arrives at the boundary.

In triangle PQR: $QR = \dfrac{PQ}{\sin \theta_1}$.　　In triangle SRQ: $QR = \dfrac{RS}{\sin \theta_2}$.

$\therefore$ Equating the expressions for QR: $\dfrac{PQ}{\sin \theta_1} = \dfrac{RS}{\sin \theta_2}$

But $PQ = v_1\Delta t$ and $RS = v_2\Delta t$ $\therefore$　$\dfrac{v_1\Delta t}{\sin \theta_1} = \dfrac{v_2\Delta t}{\sin \theta_2}$

Rearranging:　　$\dfrac{\sin \theta_1}{v_1} = \dfrac{\sin \theta_2}{v_2}$ [1]

Alternatively: the ratio $\dfrac{\sin \theta_1}{\sin \theta_2} = \dfrac{v_1}{v_2}$, which is a constant. This is a form of Snell's law,

which is dealt with in Section 2.6.2

2.6.2 Refractive index

Equation [1] in Section 2.6.1 is a general relationship for all wave motions. In this section we deal only with light. The discussion also applies, in principle, to other forms of electromagnetic radiation.

For historical reasons, optical physicists discuss the ability of a material to refract light waves (rays) in terms of its **refractive index**, n, which is defined by $n = \dfrac{c}{v}$. Because light waves travel more slowly through materials than through a vacuum, the values of n have a minimum value of 1 exactly for a vacuum (by definition), i.e. $n \geq 1$. Table 2.6.1 gives the refractive indices of a range of common materials.

Material	n
vacuum	1 (exactly)
air (at 0°C)	1.000292
water	1.333*
sea water	1.343*
ice	1.31
glass	1.50–1.75
diamond	2.417
glycerine	1.473*
olive oil	1.48*
* at 293 K	

Table 2.6.1 Refractive indices

2.6.2 Knowledge check

Calculate the speed of light in: (a) water and (b) diamond.

Take $c = 3.00 \times 10^8$ m s^{-1}.

Study point

Note the following two points from the fish-tank example, which arise only because the glass surfaces are parallel:

1. The light ray in the glass makes the same angle, α, to the normal at both sides.

2. The angle α in the glass can be ignored in the calculation.

2.6.3 Knowledge check

Calculate the angle in the glass, α, in the fish-tank example.

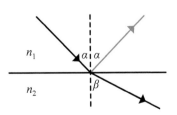

Fig. 2.6.5 Partial reflection

(a) Snell's law

We shall now write equation [1] in terms of refractive index. The refractive indices of the two materials in Fig. 2.6.3 are $n_1 = \dfrac{c}{v_1}$ and $n_2 = \dfrac{c}{v_2}$.

So equation [1] becomes: $\dfrac{n_1 \sin \theta_1}{c} = \dfrac{n_2 \sin \theta_2}{c}$, $\therefore n_1 \sin \theta_1 = n_2 \sin \theta_2$ [2]

This is **Snell's law**. Equation [2] can be written $n \sin \theta = $ constant, which also can be applied to refraction through a material with a continuously varying refractive index, e.g. the atmosphere, the density of which varies with height, temperature, humidity and so on.

◄Stretch & challenge

Many people use the qualitative term *optical density* to describe the refractive property of a material. Diamond is described as having a 'high optical density' and air has a 'low optical density'. This term should not be confused with physical density. There **can be** a rough correlation of physical density with refractive index, e.g. crown glass ($n \sim 1.5, \rho = 2.6$ g cm^{-3}) and flint glass ($n \sim 1.7, \rho = 4.2$ g cm^{-3}). However,

- the density of diamond (3.5 g cm^{-3}) is lower than that of flint glass but the refractive index is much higher
- the refractive index of perspex (1.495) is virtually equal to that of crown glass but the density is much lower (1.19 g cm^{-3}).

Example

Fig. 2.6.4 shows a light ray entering a fish tank. Use the refractive index values to calculate angle θ_w.

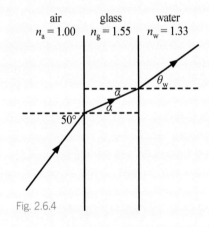

Fig. 2.6.4

Answers

From Snell's law:

$$n_a \sin 50° = n_g \sin \alpha = n_w \sin \theta_w$$

(see Study point)

$$\therefore 1.00 \sin 50° = 1.33 \sin \theta_w$$

$$\therefore \theta_w = \sin^{-1}\left(\frac{\sin 50°}{1.33}\right) = 35° \text{ (2 s.f.)}$$

2.6.3 Reflection

If a light ray strikes a smooth (polished) boundary between two media, as in Fig. 2.6.5, it is usually partially reflected and partially refracted as shown. This diagram is drawn for $n_1 > n_2$ but the same holds for $n_1 < n_2$, except that, in that case the angle of refraction, β, would be less than the angle of incidence, α. Note that the angle of reflection (i.e. the angle between the reflected ray and the normal) is equal to the angle of incidence.

The fraction of the incident power which is reflected depends upon both the angle of incidence and the refractive indices of the two materials:

- The greater the angle of incidence, the greater the power reflected.
- The greater the difference in refractive indices, the greater the power reflected.

(a) Total internal reflection

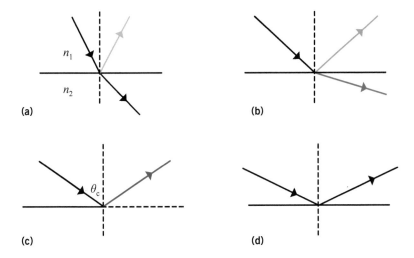

(a) (b) (c) (d)

Fig. 2.6.6 Total internal reflection

Top tip

Make sure you understand what happens to frequency and wavelength when a wave is refracted. The conditions for TIR are:

1. The light ray is incident upon a boundary with a material of lower refractive index.

2. The angle of incidence is greater than the critical angle.

The sequence of diagrams in Fig. 2.6.6 (a)–(d) illustrates qualitatively how the fraction of the reflected power varies with angle of incidence for a light ray incident upon a material **with a lower refractive index**.

- For a small angle of incidence – in (a) – the fraction reflected is small.

- As the angle of incidence increases – the sequence (a), (b), (c) – the fraction reflected increases (and the fraction transmitted decreases).

- As we reach an angle of incidence θ_c, called the **critical angle**, for which Snell's Law gives an angle of refraction of **90°**, the refracted power becomes zero, i.e. all the light is reflected. This phenomenon is called **total internal reflection**, which also applies for all angles of incidence greater than θ_c.

Study point

The critical angle for a boundary between a material of refractive index n and air (refractive index = 1.000) is given by $\sin \theta_c = \dfrac{1}{n}$.

The relationship between the critical angle, θ_c, and the refractive indices can be found by considering the limiting case in diagram (c).

Applying $n_1 \sin \theta_1 = n_2 \sin \theta_2$ with $\theta_1 = \theta_c$ and $\theta_2 = 90°$

$\rightarrow \qquad n_1 \sin \theta_c = n_2 \sin 90°$

But $\sin 90° = 1$, $\quad \therefore n_1 \sin \theta_c = n_2 \quad$ or $\quad \theta_c = \sin^{-1}\left(\dfrac{n_2}{n_1}\right)$

Knowledge check 2.6.4

A light ray, travelling in perspex of refractive index 1.495, is incident upon a boundary with air. Describe what happens if the angle of incidence is (a) 25°, (b) 35° and (c) 45°.

(b) Examples of TIR

Section 2.6.4 deals with optical fibres, which depend for their operation on total internal reflection. Here are two more everyday examples.

(i) Totally reflecting prisms

Many optical instruments, e.g. binoculars, microscopes and periscopes, use prisms to reflect light and to fold the light path. The advantage of prisms over mirrors is illustrated in Fig. 2.6.7. The multiple reflections in a mirror would cause multiple images in the instrument. In the case of the prisms, the weak partial reflections as the light enters or leaves the prism at right angles would send light back out of the instrument the way it came and wouldn't affect the final image. The glass used for prisms has refractive index in the range 1.5–1.7. You should be able to show that the critical angle is less than 45°.

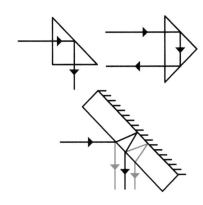

Fig. 2.6.7 Totally reflecting prisms and a glass mirror

2.6.5 Knowledge check

A light ray hits an equilateral glass prism ($n = 1.50$) with angle of incidence 50°. Calculate the angle at which it emerges from the prism.

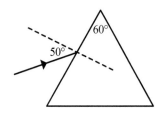

Fig. 2.6.8 Desert and highway mirages

(ii) Mirages

The reflection of light, apparently from a shimmering water surface on a hot summer road or from a non-existent lake in the baking desert, is known as a mirage (strictly an *inferior mirage*). Two examples of this are shown in Fig. 2.6.8.

This happens because the road (or desert) surface absorbs radiation from the Sun and heats up. This warms the air in contact with the surface so that there is a temperature inversion – the air temperature falls with height. The refractive index decreases as a light ray (e.g. from a car, camel or telegraph pole) approaches the surface and, if the ray travels at a glancing angle, there is enough of a difference to cause TIR.

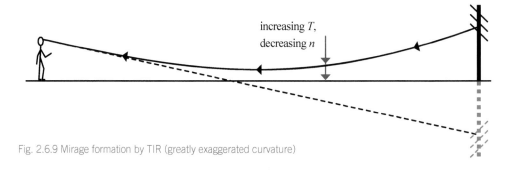

increasing T, decreasing n

Fig. 2.6.9 Mirage formation by TIR (greatly exaggerated curvature)

2.6.4 Optical fibres

Optical fibres have become ubiquitous since the 1980s. They are used for data transmission in local area networks (LANs), regional networks and long-distance networks (intercontinental). They are also used in remote imaging systems, such as internal medical examinations (endoscopy, Fig. 2.6.10) and investigating inaccessible locations such as drains and collapsed buildings (in the search for survivors).

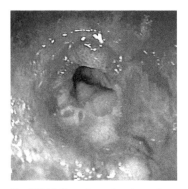

Fig. 2.6.10 Oesophageal endoscopic image showing reflux damage (Barrett's oesophagus)

core

cladding

Fig. 2.6.11 Optical fibre

This section deals with optical properties of stepped-index optical fibres. A typical optical fibre consists of a single glass thread, the central part of which (the core) carries the light signal and the outer part (the cladding) keeps the signal in the core. Around this is a protective layer of plastic (not shown in Fig. 2.6.11) called the coating. Typically the external diameter of the coating is ~ 250 μm, i.e. 0.25 mm. An optical-fibre cable can consist of hundreds of such fibres.

(a) Multimode fibres and TIR

The optical fibre relies for its operation on total internal reflection of light rays at the boundary between the core and the lower refractive index cladding. For a multimode fibre a typical core diameter is 50μm.

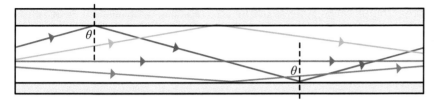

Fig. 2.6.12 Totally reflected light ray in a multimode optical fibre

All light rays that are incident upon the core–cladding boundary at angles greater than the critical angle are totally internally reflected back into the core. They then repeatedly hit opposite sides of the core at the same angle, being reflected each time and emerge from the other end of the fibre (unless they are absorbed or scattered by impurities in the glass). The optical fibre is not necessarily perfectly straight but, given that the diameter of the core is ~50 μm, any reasonable curvature of the cable will hardly affect the angles of the multiple reflections.

This **multimode** optical fibre is fine for short-distance communication or for imaging applications (endoscopy), but it runs into a problem for digital communication over long distances with rapid switching rates. To see why, consider the time taken for a signal to travel 10 km.

Example

A multimode optical fibre has a core refractive index of 1.6. Calculate the difference in the time taken for a signal to travel through 10 km of optical fibre for light rays travelling parallel to and at 20° to the axis of the fibre.

Answers

Speed of light in the core $= \dfrac{c}{n} = \dfrac{3.00 \times 10^8}{1.6} = 1.875 \times 10^8 \text{ m s}^{-1}$

$\therefore$ Time taken for parallel ray $= \dfrac{10 \times 10^3}{1.875 \times 10^8} = 53.3 \text{ μs}$

Distance travelled by signal at 20° $= \dfrac{10 \times 10^3}{\cos 20°} = 10642 \text{ m}$

$\therefore$ Time taken $= \dfrac{10642}{1.875 \times 10^8} = 56.8 \text{ μs}$

$\therefore$ Time difference = 3.5 μs.

The time difference between different paths is proportional to the transmission distance, so a 1 km fibre would give a time difference of ~0.4 μs; for a 100 m fibre Δt would be 40 ns etc. Digital data is transmitted as a series of rapidly oscillating pulses. If the transmission rate is more than 10^5 bits per second (100 kbps), the time between pulses will be less than ~10 μs, so the time difference between the arrival of the straight-through and 20° will cause the 1s and 0s of the pulses to overlap and cause the signal to be unreadable. This effect is called **multimode dispersion**. As modern data systems operate at **Gbps** rates (10^9 bits per second) this kind of optical fibre is restricted to a few metres, e.g. in a LAN. This restriction is overcome in monomode fibres.

Study point

The refractive indices of the core and cladding depend on the design, but values of 1.62 and 1.52 respectively are typical.

Knowledge check 2.6.6

Show that the critical angle for the core and cladding is approximately 70° for the n values given in the Study point.

Top tip

Note that the speed of light in an optical fibre is not c but $\dfrac{c}{n}$. This is approximately 2×10^8 m s^{-1}.

Stretch & challenge

The calculations in the example use more significant figures than justified by the data. This does not matter as this alternative approach shows. ϕ is the angle to the axis.

$t_1 = \dfrac{d}{v}; t_2 = \dfrac{d}{v \cos \phi}$

$\Delta t = \dfrac{d}{v}\left(\dfrac{1}{\cos \phi} - 1\right)$

Use the approximation that $v = 2 \times 10^8$ m s^{-1} and this equation to estimate Δt.

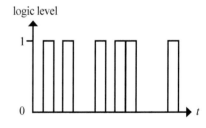

Fig. 2.6.13 Digital signal

Key terms

Multimode optical fibre: One in which light rays with a range of directions can propagate (by total internal reflection).

Multimode dispersion: Data degradation because each pulse (in a multimode fibre) travels by a range of different paths, so arrives spread out over time and may overlap neighbouring pulses.

2.6.7 Knowledge check

A telephone conversation takes place over a 3000 km monomode fibre system. Estimate the total time delay that the participants experience. State any assumption you make.

Key term

A **monomode** optical fibre: One in which light can only propagate in one direction (parallel to the axis)

(b) Monomode (single-mode) fibres

In Section 2.5 we saw that light spreads out significantly by diffraction when it passes through apertures which are similar in size to the wavelength. In these circumstances we can no longer use the light-ray model and we need to use a full wave analysis (which is beyond the scope of A level physics). For reasons which are dealt with briefly below, optical fibre communication systems use infrared radiation of wavelength (in air) of ~1.5 μm. The result of the full wave analysis is that, for core diameters of less than about 9 μm, the light waves cannot take multiple paths and they are effectively restricted to travelling parallel to the axis of the fibre. These fibres, with typical core diameters of 9 mm, are called **monomode** fibres.

Progressive signal loss due to attenuation becomes the main problem to be overcome in monomode systems. For near IR wavelengths this is mainly due to scattering by impurities in the glass of the fibre. This process is called Rayleigh scattering and becomes more serious the shorter the wavelength. This is the same effect that produces the blue of the sky – the short (blue) wavelengths are scattered more by the atmosphere than the long (red) wavelengths. Some molecules and ions in the glass (mainly OH) also selectively absorb particular wavelengths.

For long-distance communication, the signal in the optical fibre needs boosting and cleaning up by repeaters between the transmitter and receiver. The quality of current monomode fibre technology is such that repeaters need to be inserted no closer than 50 km.

2.6.5 Specified practical work

All the school laboratory methods of determining the refractive index are carried out in air. They produce results which, at best, have an estimated uncertainty of ± 0.01, so the distinction between the refractive indices of air ($n_a = 1.0003$) and a vacuum ($n = 1$ exactly by definition) can be ignored.

Measurement of the refractive index of a material

This method requires the material to be in the shape of a regular block, e.g. a semicircular or rectangular glass or perspex block. The advantage of a semicircular block is that a ray along a radius is at right angles to the curved surface, so it is not refracted at the boundary (see Fig. 2.6.14).

1. The semicircular block is set up on a piece of paper on a drawing board on which is drawn a series of lines (drawn faint) at regular angles to the straight face of the block, e.g. $5° - 40°$ in $5°$ steps.

2. The position of the ray box is adjusted so that the light ray is along one of the lines in the diagram and emerges at the centre point of the glass block.

3. Two pencil marks (P_1 and P_2) are drawn on the emergent light ray as shown in Fig. 2.6.14.

4. Steps 2 and 3 are repeated for the each of the lines from step 1.

5. The block is removed, the emergent light rays drawn in using the marks (from step 3) and angles θ_1 and θ_2 measured for each ray, using a protractor.

6. A graph is drawn of $\sin \theta_1$ against $\sin \theta_2$ and the gradient measured. This is the refractive index.

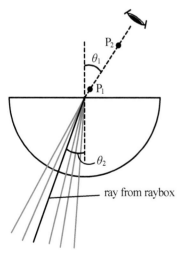

Fig. 2.6.14 Refractive index by ray tracing

Test yourself 2.6

1. A set of waves travelling at 6 m s^{-1} in water of depth 40 m has a wavelength of 10 m. It approaches a beach in a direction at $50°$ to the normal.

   (a) Calculate the frequency of the waves.
   (b) Assuming the speed of the waves is directly proportional to the square root of the depth, calculate the speed and wavelength of the waves when the water is 10 m deep.
   (c) Calculate the direction of travel of the waves at the 10 m deep point.
   (d) What direction do the waves travel in when they enter water of depth 2.5 m?

2. *The refractive index of water is 1.33.* State what this statement means.

3. A light ray is travelling in a material with refractive index n_1. It hits a boundary with a material with refractive index n_2. Under what circumstances is the light ray totally internally reflected?

4. A light ray hits an air/glass boundary with an angle of incidence of $35.0°$. The refractive index of the glass is 1.55. Calculate the angle of refraction if the incident light is travelling in (a) the air, and (b) the glass.

5. The diagram shows a light ray travelling from air into a newly created plastic. Calculate:

   (a) the refractive index of the plastic.
   (b) the angle of refraction if the light ray were incident on the boundary at $45°$ in the plastic.
   (c) the critical angle for a boundary between air and the plastic.

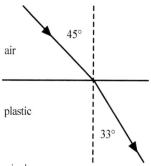

6. A light ray, travelling in air ($n = 1.00$), is incident on a boundary with glass of refractive index 1.50. The angle of incidence, θ, is varied between $0°$ and $90°$ and the angle of refraction, ϕ is measured.

   (a) Sketch a graph of $\sin \theta$ against $\sin \phi$. Label significant values on the axes.
   (b) State the significance of the gradient of the graph in part (a).
   (c) Sketch a graph of θ against ϕ. Label significant values on the axes.
   (d) State the value of the gradient at the origin of the graph in part (c).

7. The refractive indices for three common materials are as follows:

Material	barium crown glass	water	perspex	air
Refractive index	1.57	1.33	1.49	1.000 293

   (a) Calculate the speed of light in barium crown glass.
   (b) A light ray travelling in water hits a boundary with perspex. The angle of incidence is $37.3°$. Calculate the angle of refraction.
   (c) Calculate the critical angle for the boundary between water and air.
   (d) Under what circumstances will a light ray, incident upon a perspex / water boundary be totally internally reflected?
   (e) Calculate the percentage difference between the speeds of light in air and in a vacuum. Give your answer to 3 s.f.
   (f) See Stretch & challenge

◀Stretch & challenge▶

An optical range finder measures the distance of objects by the time taken for a pulse of laser light to return after bouncing off the objects. A reflector on a mountain summit is known to be approximately 25 km away. Use the answer to part (e) in Q7 to estimate the error in the measured distance if the range finder is calibrated for a vacuum rather than air.

8. The core of a multimode optical fibre has refractive index 1.580. The critical angle for the core–cladding boundary is $82.0°$. Calculate:

    (a) the refractive index of the cladding,
    (b) the speed of light in the core,
    (c) the time taken for a light ray inside the core to travel 20 km parallel to the core,
    (d) the time delay between a pulse of light travelling parallel to the core and one travelling at $8°$ to the axis of the fibre after they have travelled 20 km along the fibre. Briefly explain the significance of the answer to part (d) for data transmission.

9. The diagram shows a light ray incident on the top prism of a periscope.

    (a) Copy the diagram and complete it to show the light entering the eye.
    (b) The angle of incidence on the rear surface of the top prism is $45°$. Calculate the minimum refractive index of the material of the prism for total internal reflection to occur.

10. A light ray is incident at $45°$ at the middle of a glass block of dimensions 8.0 cm $\times$ 15 cm, and refractive index 1.50. Determine the position and direction at which the light ray emerges from the block.

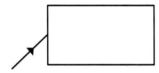

11. A light ray enters a stack of blocks with refractive indices as shown. By calculating the relevant angles in each layer, decide where total internal reflection occurs. You may assume that the blocks are long enough for this to happen.

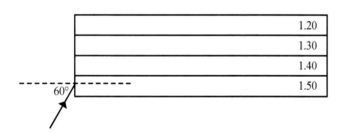

12. A student does an experiment to measure the refractive index of a semicircular perspex block.

    Her results are given in the table. θ_g is the angle in the glass and θ_a the angle in the air. All angles are measured to $\pm 0.5°$.

$\theta_g / °$	5.0	10.0	15.0	20.0	25.0	30.0	35.0
$\theta_a / °$	7.0	15.0	23.0	31.0	39.0	48.0	59.0

    Use the results to draw a graph of $\sin \theta_g$ against $\sin \theta_a$. Use the $\pm 0.5°$ uncertainties to work out the max/min values of $\sin \theta_g$ and θ_a, plot error bars and determine the refractive index of the block together with its uncertainty.

13. A horizontal light ray is incident upon the long face of a 45° isosceles prism, of refractive index 1.55, as shown. The light is refracted at the first surface and is incident upon the rear surface before emerging from the prism.

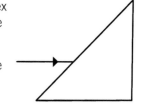

(a) Calculate the angle to the normal at which the light ray emerges following refraction at the front and rear surfaces.
(b) See Stretch & challenge.

◀Stretch & challenge▶

The light ray is incident about $\frac{1}{3}$ of the way up the long face. This light ray is partially reflected when it is incident upon the rear surface after passing though the block.

Investigate the fate of this partially reflected ray. From which face and in what direction does it emerge? Show this on a diagram. [There is no need to consider any further partial reflections.]

14. A monomode optical fibre is to be made with a core of doped silica, with a refractive index of 1.4475. It is to work with infrared of wavelength (in air) of 1.20 μm.

(a) In order to operate as a monomode fibre, the core must have a maximum diameter of 10× the wavelength in the core. Calculate the maximum core diameter.
(b) Calculate the time it takes for a signal to travel the distance between the repeater stations of 150 km.
(c) The signal is composed of a stream of 1.0×10^{11} pulses per second. Calculate the spatial separation of the these pulses in the fibre.

15. A multimode optical fibre has a core made of doped silica ($n = 1.4475$). The cladding is made of pure silica ($n = 1.4440$).

(a) Calculate:
  (i) the critical angle between the core and cladding
  (ii) the time difference between the arrival of two light rays, parallel to the axis and at the critical angle, after 10 km of travel along the fibre.
(b) A single pulse is sent along the fibre. The graph shows the difference in the pulse profile at the start and after 10 km. It also shows the detection threshold: in order for a pulse to be detected, the profile must drop below the threshold and then rise above it.

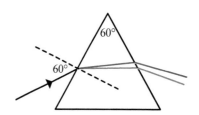

Without calculation, explain:
  (i) the change in form (width and height) of the pulse after 10 km, and
  (ii) why two close pulses will not be separately detected after 10 km (a diagram will help your answer).

16. A ray of white light, consisting of all wavelengths between 400 nm and 700 nm, hits an equilateral triangular prism with an angle of incidence of 60°. The refractive index of the prism varies with the wavelength of light. For red light of wavelength (in a vacuum) of 700 nm, the refractive index is 1.51; for violet light with $\lambda_{vac} = 400$ nm the refractive index is 1.53.

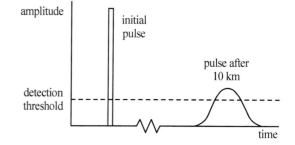

(a) Calculate the wavelengths of red and violet light in the glass.
(b) By determining the paths of red and violet light through the prism, calculate the angle between the emerging rays of red and violet light.

2.7 Photons

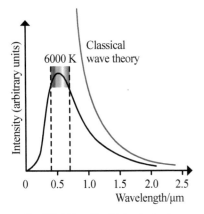

Fig. 2.7.1 The ultraviolet catastrophe

By the late nineteenth century, the wave model of light was firmly established. It explained a whole series of phenomena (see Sections 2.5 and 2.6) and the wavelength could be measured using Young's slits. However, this theory could not account for the way that light interacted with matter. According to the classical wave theory, the power radiated by a black body should get progressively higher at shorter wavelengths, leading to the conclusion that the total power radiated away across all wavelengths should be infinite! This 'prediction' was known as the *ultraviolet catastrophe* (Fig. 2.7.1). Planck successfully accounted for the shape of the black body spectrum by assuming that radiation was absorbed and emitted in discrete packets of energy, rather than continuously as a wave. These packets were of energy hf, where h is a constant called the **Planck constant**. Einstein took the idea of energy packets literally and postulated that light propagated as a stream of particles (which we now call **photons**) of energy hf and successfully accounted for another phenomenon – the photoelectric effect.

2.7.1 The photoelectric effect

If ultraviolet (UV) radiation is shone onto a negatively charged zinc plate, the plate loses its charge. This is easily demonstrated in the school laboratory using a negatively charged electroscope as in Fig. 2.7.2. The gold leaf (shown as the red line) quickly drops. A positively charged plate shows no effect, suggesting that the UV causes electrons to be given off from the zinc plate. The lack of response from a positive plate can be explained because any electrons given off are attracted back by the positive zinc. The **photoelectric effect** is not peculiar to zinc.

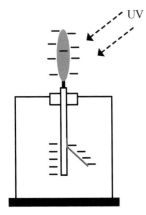

Fig. 2.7.2 Photoelectric effect demonstration

The vacuum photocell (or phototube) provides a simple way of investigating the photoelectric effect. This was originally used in camera light meters and in cinema projectors, for reading the optical soundtrack. It consists of a cylindrical cathode (K) made from a suitable metal and an anode (A) placed in front of it (Fig. 2.7.3). The whole is surrounded by an evacuated glass envelope. Incoming e-m radiation hits the surface of the cathode, which emits electrons. In the original uses, the anode is connected in a circuit with the anode positive so that liberated electrons are drawn across and out into the circuit. The current is proportional to the light intensity, enabling it to be used as a light meter. Note that, in the discussion of the photoelectric effect, the word *light* should be taken to include near infrared and ultraviolet electromagnetic radiation.

(a) Photoelectric effect experiments

If we measure the I–V characteristics of a photocell, as in Fig. 2.7.4, we notice that:

- For positive pds above a minimum level the current is independent of the pd – all the emitted electrons are collected by the anode.
- This 'plateau current' is proportional to the light intensity.
- There is a positive current for small negative values of V down to a 'stopping voltage' V_S, which has the same value for all light intensities.

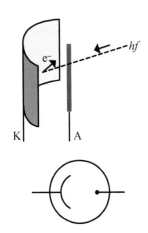

Fig. 2.7.3 The vacuum photocell

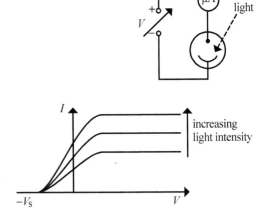

Fig. 2.7.4 Photocell characteristics

If we assume that the **photoelectrons** are emitted with a range of kinetic energies, the value of V_S enables us to measure the maximum value of this, $E_{k\,max}$. If an electron with $E_{k\,max}$ is just stopped by V_S, then from the definition of potential difference:

$$E_{k\,max} = eV_S.$$

The experiments which showed that the radiation did not behave as waves concerned the variation of $E_{k\,max}$ with the frequency, f, of the radiation. A suitable circuit is given in Fig. 2.7.5. Notice the polarity of the voltage supply.

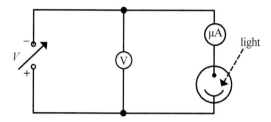

Fig. 2.7.5 Circuit for investigating the photoelectric effect

The photocell is illuminated with monochromatic radiation (i.e. consisting of a single frequency). The pd applied across the photocell is adjusted until the current just becomes zero and the value, V_S, measured. This is repeated for a range of frequencies, f, and with different metal surfaces as the cathode in the photocell.

The results of such experiments and the one in Fig. 2.7.4 are as follows:

(a) If electrons are emitted, there is no measurable time delay.
(b) For any metal there is a characteristic **threshold frequency**, f_{th}, below which no electrons are emitted, whatever the intensity of the radiation.
(c) There is a linear relationship between $E_{k\,max}$ and the frequency, with the same gradient for all metals (see Fig. 2.7.6).
(d) If electrons are emitted, $E_{k\,max}$ is independent of the radiation intensity.
(e) If electrons are emitted, the number of electrons emitted per second is proportional to the intensity of the radiation.

The results are incompatible with the idea that the energy of radiation is absorbed continuously, as it would be if light behaved as a wave. There would then be no threshold frequency. Low frequency, low intensity radiation could transfer energy gradually to electrons which would eventually build up enough energy to escape from the metal surface, after a time delay. High intensity radiation would be expected to transfer energy more quickly than low intensity and some electrons should gain more energy, making $E_{k\,max}$ higher.

(b) Einstein's explanation of the experimental results

Einstein proposed the following model to explain the experimental results:

1. Electromagnetic radiation consists of discrete packets of energy, photons, with the photon energy given by $E = hf$, where h is a constant called the **Planck constant** which has a value of 6.63×10^{-34} J s.
2. When a photon interacts with an electron in the metal surface its entire energy is transferred.
3. An electron only interacts with an individual photon – the probability of two photons interacting with an individual electron is vanishingly small.
4. There is a characteristic minimum energy, called the **work function**, ϕ, which is needed to remove an electron from a metal surface.

>>> **Study point**

The electrons emitted by the photoelectric effect are often referred to as **photoelectrons**.

Knowledge check 2.7.1

If $V_S = 0.6$ V, find the value of $E_{k\,max}$ (a) in J and (b) in eV.

($e = 1.60 \times 10^{-19}$ C.)

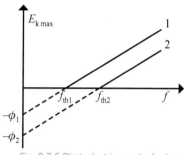

Fig. 2.7.6 Photoelectric graphs for two metals

>>> **Key term**

Work function: The minimum energy required to remove an electron from a metal surface. The work function can be expressed in either joules or **electron volts**.

2.7.2 Knowledge check

Express the work function of selenium in eV (see example).

2.7.3 Knowledge check

A selenium surface is illuminated by e-m radiation. Calculate the maximum KE of any photoelectrons if the radiation:

(a) is of frequency 2.5×10^{15} Hz

(b) is of frequency 1.0×10^{15} Hz

(c) is a mixture of these two frequencies.

2.7.4 Knowledge check

For the Sun $L = 3.85 \times 10^{26}$ W. Taking 500 nm as representative of all the radiation from the Sun, estimate (a) the number of photons emitted per second and (b) the number of photons crossing unit area per second at the distance of the Earth (150 million km).

Taken together, these postulates explain all the results (a) to (e) above. The maximum energy of the photoelectrons is given by

$$E_{k\,max} = hf - \phi,$$

which is known as **Einstein's photoelectric equation**, which is a consequence of postulates 1, 2 and 4; 3 is also involved because, if two or more photons gave their energy to one electron, it could gradually build up enough energy to escape and there should be no threshold frequency.

Metal	ϕ/eV	Metal	ϕ/eV
Al	4.1	K	2.3
Cd	4.1	Na	2.3
Cs	2.1	Sr	2.6
Ca	2.9	Ag	4.5
Mg	3.7	Au	5.1
Cu	4.7	Hg	4.5

Table 2.7.1 Miscellaneous values of the work function

Example

Calculate the threshold frequency for photoelectric emission for selenium which has a work function of 8.18×10^{-19} J.

Answer

At the threshold frequency, $E_{k\,max} = 0$, $\therefore hf_{th} = \phi$.

$$\therefore f_{th} = \frac{8.18 \times 10^{-19}\,\text{J}}{6.63 \times 10^{-34}\,\text{J s}} = 1.23 \times 10^{15}\,\text{Hz}$$

(c) Radiation intensity

We can relate the intensity of a radiation beam to the number of photons crossing an area per second.

Consider a monochromatic beam of radiation, of frequency f, crossing a surface. Let N be the number of photons crossing the surface per second.

Then the power in the radiation, $P = NE_{ph}$, where E_{ph} is the energy of an individual photon.

We can also write this as $P = Nhf = N\dfrac{hc}{\lambda}$

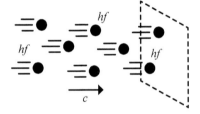

Fig. 2.7.7 Photons and intensity

We can now relate this to the inverse square law by considering a point (or spherical) source of monochromatic radiation of power P. Looking back at Fig. 1.6.7 we see that, at distance r, this radiation crosses an area of $4\pi r^2$. So the intensity, I, of the radiation (i.e. the power per unit area) is given by:

$$I = \frac{Nhf}{4\pi r^2}$$

Most sources of radiation are far from monochromatic but if the intensity, at all individual frequencies, drops off as r^{-2}, the inverse square law must apply to all radiation whatever the spectral distribution. See Section 1.6.2 (a).

(d) Is light a wave or a particle?

That is a very good question. Light exhibits the wave properties of diffraction and interference. When its speed is measured in materials, it corresponds to the wave model explanation of refraction, which was discussed in Section 2.6. Light is a part of the electromagnetic spectrum. Waves at the low frequency end of the electromagnetic spectrum are produced by oscillating electric and magnetic fields, as predicted by Maxwell's very successful theory of electromagnetism. On the other hand, the emission and absorption of light demands a particle model. But the particle properties of energy and momentum (see Section 2.7.4) are calculated using the wave-property values of frequency and wavelength.

So our picture is that light has both wave and particle properties. This is referred to as **wave–particle duality**. We shall see in Section 2.7.4 that objects that we normally think of as particles also exhibit wave properties. This whole topic is the subject of quantum mechanics.

Study point

For very long wavelengths (radio waves) wave properties dominate even when we consider emission and absorption. The higher the frequency, the more the particle properties are seen, though X-ray diffraction is used to investigate crystals.

2.7.2 The electromagnetic spectrum

A schematic diagram of the e-m spectrum is shown in Fig. 2.7.8.

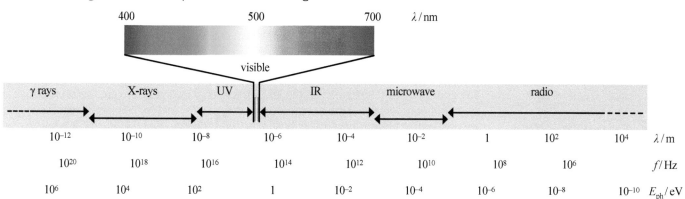

Fig. 2.7.8 The electromagnetic spectrum

Example

Show that $\lambda E_{ph} \sim 10^{-6}$ m eV.

Answer

$$E_{ph} = \frac{hc}{\lambda}$$

$$\therefore \lambda E_{ph} = hc = 6.63 \times 10^{-34} \text{ J s} \times 3.00 \times 10^8 \text{ m s}^{-1} = 1.99 \times 10^{-25} \text{ J m}$$

$$1 \text{ eV} = 1.60 \times 10^{-19} \text{ J}$$

$$\therefore \lambda E_{ph} = \frac{1.99 \times 10^{-25} \text{ J m}}{1.60 \times 10^{-19} \text{ J eV}^{-1}} = 1.24 \times 10^{-6} \text{ m eV} \sim 1 \times 10^{-6} \text{ m eV}.$$

Radiation	Typical λ / m	Typical E_{ph} /eV
γ	10^{-12}	10^6
X	10^{-10}	10^4
UV	10^{-7}	10^1
visible	5×10^{-7}	2.5
IR	10^{-5}	10^{-1}
μ-wave	10^{-2}	10^{-4}
radio	10^2	10^{-8}

Table 2.7.2 Typical wavelengths and photon energies

Many aspects of the electromagnetic spectrum are covered in Sections 1.6, 2.4 and 2.5. There are two other things you should note:

1. The boundaries between the regions are not always well defined. For example, the metastable isotope selenium-81 decays by gamma emission with a photon energy of 0.1 MeV, which is within the range of X-ray tubes (see the Medical Physics option).

2. Regions of the e-m spectrum are often subdivided (UV-A, UV-B, etc., near infrared, far infrared) with the boundary between far infrared and microwaves often called terahertz radiation (also called *submillimetre* waves by radio astronomers).

You should become familiar with typical wavelengths and photon energies of the different regions of the spectrum, especially the visible region, which is required knowledge in the specification.

Knowledge check 2.7.5

The visible spectrum may be taken as running from 400 nm (violet) to 700 nm (red) with a typical wavelength of 550 nm (yellow). Calculate photon energies (in eV) for these wavelengths.

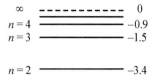

Fig. 2.7.9 Energy levels of atomic hydrogen in eV

Express the ionisation energy of atomic hydrogen in aJ.

Key terms

Ionisation: The removal of one or more electrons from an atom.

Ionisation energy of an atom: The minimum energy needed to remove an electron from the atom in its ground state.

Study point

In atoms with more than one electron, the electrons normally fill up the energy levels from the bottom. The shell with $n = 1$ can hold up to 2 electrons, $n = 2$ can hold 8, $n = 3$ can hold 18, etc. The arrangement of electrons between these energy levels gives rise to the chemical properties of the atoms.

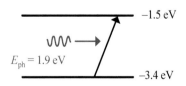

Fig. 2.7.10 Absorption transition

2.7.3 Atomic spectra

We saw in Section 1.6.4 that isolated atoms have spectra which consist of a series of wavelengths. We refer to such spectra as line (or discrete) spectra. Atomic gases both emit and absorb radiation at characteristic wavelengths. Part (d) of this section deals with the production of emission and absorption spectra using a diffraction grating. See also Section 1.6.4.

(a) Atomic energy levels

The reason why atomic gases have line spectra is to do with the way in which atoms can possess energy. We are used to thinking of systems as being able to possess any level of energy. The kinetic energy of a car could be $13\,500$ J, $13\,510$ J, $13\,511$ J, $13\,511.1$ J, etc. The microscopic world does not behave in this way. Just as light comes in lumps (called photons), atomic systems can only possess particular levels of energy.

Fig. 2.7.9 shows the energy levels (in eV) of atomic hydrogen. These values are the sum of the kinetic and potential energies of the electron in the atom (nuclear energies are ignored here). The figures on the left of the energy level diagram are the *principal quantum numbers* (n) of the energy levels – they correspond to the electron shells, which we deal with in chemistry. If an atom is in its lowest energy level (i.e. $n = 1$), it is said to be in its **ground state**. If it is in a higher energy level it is in an **excited state**. The energy level with $n = 2$ is often called the first excited state.

Notice that the energy values are negative. Consider the process of ionisation. Conventionally a stationary free electron outside the atom is shown as having zero (0) energy so an electron trapped inside the atom must be given energy to escape. Hence its total energy must be negative. The ground state has -13.6 eV of energy, so an electron must be given 13.6 eV in order to escape from a hydrogen atom in this state. This is called the ionisation energy of hydrogen.

(b) Atomic absorption spectra

In a cloud of atomic hydrogen in space, most of the atoms will be in the ground state but some will be in the first excited state (i.e. $n = 2$). The cloud will be flooded with photons of a large range of energies, which come from nearby stars. The difference in energy, ΔE between the 2nd and 1st excited states is given by:

$$\Delta E = -1.5 \text{ eV} - (-3.4 \text{ eV}) = 1.9 \text{ eV}$$

If a 1.9 eV photon hits an atom, which is in the $n = 2$ state, it may be absorbed, putting it in the higher-energy state, as shown in Fig. 2.7.10. Photons with slightly higher or lower energies (1.8 eV or 2.0 eV) will not be absorbed; so radiation which passes through the cloud will be depleted in photons with this energy, giving rise to one of the dark lines in the spectrum in Fig. 1.6.13. Using ideas from Section 2.7.1, you should be able to find the wavelength of 1.9 eV photons and identify the absorption line in the visible spectrum of atomic hydrogen to which this corresponds.

Stretch & challenge

For atoms or ions which only have one electron, e.g. H, He+, Li²⁺, Be³⁺ etc (which occur frequently in the atmosphere of stars), the energy levels (in eV) can be calculated from the simple formula:

$$(E_n \,/\, \text{eV}) = -13.6 \frac{Z^2}{n^2},$$

where Z is the proton number (atomic number). This formula also approximately gives the energy level of the innermost electron in atoms with more than one electron.

Example

Use the energy diagram for atomic hydrogen (Fig. 2.7.9) to show that only transitions between the first excited state ($n = 2$) and higher states correspond to e-m radiation in the visible part of the spectrum.

Answer

ΔE between $n = 1$ and higher energy levels is between 10.2 eV and 13.6 eV. This lies in the UV part of the spectrum. ΔE between $n = 3$ and higher energy levels is between 0.6 eV and 1.5 eV. This is in the near **IR** part of the spectrum. We have seen (main text and Stretch & challenge) that some transitions between $n = 2$ and higher lie in the visible part of the spectrum.

(c) Atomic emission spectra

In a hot cloud of atomic hydrogen, e.g. the one in Section 2.7.3 (b), which is heated by absorbing radiation from a nearby star, some of the atoms will be excited, i.e. they will be in energy levels above the ground state. This may also happen as a result of inter-atomic collisions, when some of the kinetic energy of the colliding atoms is lost. If the electron in such an atom falls into a lower energy state it gives out a photon of e-m radiation, of energy equal to the difference in energy of the two states, e.g. the electron which is promoted in Fig. 2.7.10 will subsequently return to the lower energy level, emitting a 1.9 eV photon in the process.

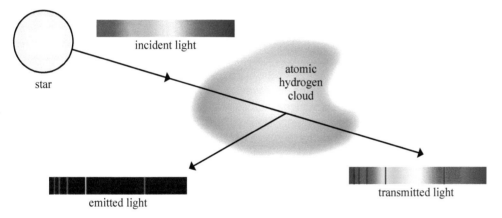

incident light

atomic hydrogen cloud

star

emitted light

transmitted light

Fig. 2.7.11 Emission and absorption by hydrogen atoms

Different observers could see emission and absorption spectra from the same object, as shown in Fig. 2.7.11. In principle the two spectra have lines at the same wavelength but in practice not all lines will appear in the absorption spectrum.

(d) Investigating and displaying atomic spectra

The simplest way to see the emission spectrum of a gas is to use a gas discharge tube. This consists of a sealed glass tube containing a low pressure gas and two high electrical terminals. When a high voltage is applied, the gas partially ionises allowing electrons to pass through it. These collide with the gas atoms and raise them to a range of excited energy states: they then drop down to the lower energy states and emit photons as they do. The discharge tube in Fig. 2.7.12 contains low pressure argon gas.

The lamp can be viewed through a diffraction grating with the results shown in Fig. 2.7.13: the central image is the zero order spectrum (with all the wavelengths at the same place); the ones on each side are the first order spectra. Alternatively, an image of

Knowledge check 2.7.7

Use the formula

$$E_n = \frac{-13.6}{n^2} \ (n = 1, 2, 3...)$$

to show that the Hδ line in the spectrum of atomic hydrogen (Fig. 1.6.13) corresponds to a transition from $n = 2$ to $n = 6$.

>> Study point

When atoms absorb photons, they become excited, i.e. they move to a higher energy level. They subsequently de-excite by emitting photons in random directions. This gives rise to the dark lines in the spectrum in the forward direction.

Knowledge check 2.7.8

Express the typical photon energies in Table 2.7.2 on page 167 in joules, to 1 s.f.

< Link >

Section 1.6.4 Line spectra.

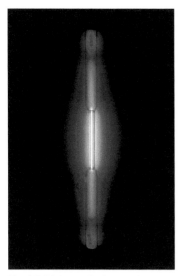

Fig. 2.7.12 Argon discharge tube

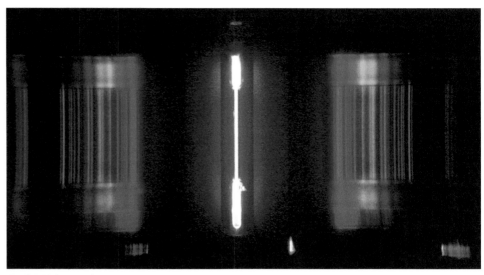

Fig. 2.7.13 Spectra from an argon discharge tube

Study point

When projecting spectra, as in Fig. 2.7.14, the wavelength of a spectral line can be calculated from the separation of the grating lines, the distance from the grating to the screen and the distance from the line to the zero order spectrum. See Section 2.5.4 and Knowledge check 2.7.9.

Knowledge check

A diffraction grating with **850** lines per cm is used to display the spectrum of a gas. A line is identified and the separation between the two first order spectra for this line is measured to be **20.0 cm**. If the screen is **2.00 m** from the grating, calculate the line's wavelength.

the discharge tube can be projected onto a screen using a lens and a diffraction grating interposed (Fig. 2.7.14).

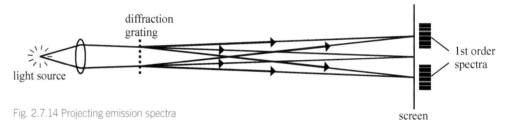

Fig. 2.7.14 Projecting emission spectra

The coloured emissions of the aurora borealis (Fig. 2.7.15) are caused by the same mechanism. Charged particles in the solar wind spiral along the Earth's magnetic field lines and enter the atmosphere at high latitudes. They collide with the atoms of the upper atmosphere (chiefly atomic oxygen and nitrogen) ionising them. The radiation is emitted as the electrons recombine with the ions and descend through the atomic energy levels.

Fig. 2.7.15 Aurora borealis: atomic oxygen emission

To display the **absorption spectrum** of a gas, a suitably shielded bright white light source with a continuous emission spectrum (e.g. a filament lamp) is used and a tube of the gas placed between the light source and the screen (usually between the source and lens). With care this method can also be used to display the absorption spectra of metals by allowing the white light to pass through a Bunsen flame in which a sample of the metal salt has been vaporised (in a flame test).

2.7.4 Wave–particle duality

(a) Electrons are waves too

Just as electromagnetic radiation can behave as both waves and particles, objects that we normally think of as particles, such as electrons, protons and even whole atoms have wave-like properties. In other words, particles exhibit diffraction and interference. The single-slit and double-slit experiments are rather difficult to arrange for electrons; although the effect was predicted in the 1920s it was not achieved until 1961. The set-up is shown schematically in Fig. 2.7.16. A stream of electrons (the red balls) is fired though a narrow slit and hits a fluorescent screen: each impact causes a bright spot.

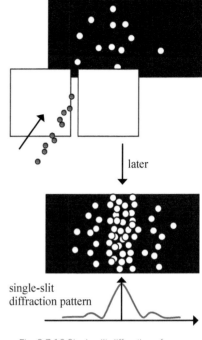

later

single-slit
diffraction pattern

Fig. 2.7.16 Single-slit diffraction of electrons

Initially the scatter of the dots appears random but gradually a pattern emerges which clearly shows similarities to the diffraction pattern of a single slit (see Section 2.5).

To achieve the double-slit pattern the individual slits must be made narrower (to spread out the diffracted electrons), as in Fig. 2.7.17. In a 2008 version of the experiment, the individual slits were only 62 nm wide and 272 nm apart! These values were chosen to suit the wavelength of the electrons – see Section 2.7.4 (c). The sequence of images in Fig. 2.7.18 indicates how the apparently random spread of electrons eventually resolves itself.

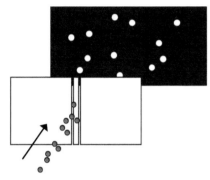

Fig. 2.7.17 Electron double-slit interference

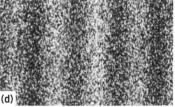

(a)   (b)   (c)   (d)

Fig. 2.7.18 The build-up of an electron interference pattern

(b) The electron diffraction tube

In the school laboratory, the wave nature of electrons is usually demonstrated using the electron diffraction tube, Fig. 2.7.19, which consists of a vacuum tube with an electron gun at one end, a graphite target (T) and a phosphor screen (P).

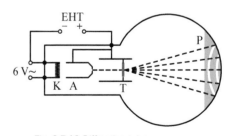

Fig. 2.7.19 Diffraction tube

- A metal-coil cathode (K) is connected to a low voltage a.c. supply which heats it up. This causes it to emit electrons by **thermionic emission**.
- The electrons are accelerated to the anode (A) by an EHT ('extremely high tension', i.e. high voltage) supply (typically $1–6$ kV).
- A beam of electrons emerges from the hole in the anode and hits the graphite target (shown as a red line)
- The beam is diffracted by the graphite, similarly to light at a diffraction grating and emerges at a series of angles to the forward direction. The electron beams hit the phosphor, producing a display of circular rings.
- If the EHT voltage is increased, the rings are observed to decrease in radius.

The diffraction pattern occurs because graphite consists of regular planes of carbon atoms hexagonally arranged, with an inter-atomic spacing of 0.142 nm. The planes are

▶▶ Study point

The electron diffraction and interference patterns are not the result of electrons interfering **with one another**. The same pattern is obtained even when the intensity of the electron beam is so low that there is only one electron at a time in the apparatus. Each electron behaves as its own wave and interferes **with itself**!

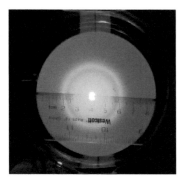

Fig. 2.7.20 Electron diffraction pattern

separated by 0.335 nm. Because the crystals in the graphite are randomly orientated, the diffracted beams at any particular angle produce circles rather than an individual dot.

(c) The wavelength of particles

The diffraction tube can be used to measure the wavelength of the electrons. All that is needed is to measure angles that the diffracted beams make with the forward direction. In Fig. 2.7.20 this is done by measuring their radii; the distance from the target to the screen also needs to be measured.

In 1924 the physicist Louis de Broglie proposed that the wavelength (λ) of a particle, such as an electron is related to its momentum (p) by the equation:

$$\lambda = \frac{h}{p},$$

where h is the Planck constant. This relationship has since been confirmed – it is consistent with both quantum and relativity theory. The following example shows how we can use the accelerating voltage to calculate the wavelength.

Top tip

The relationship $E_k = \dfrac{p^2}{2m}$ is useful when calculating the momentum of accelerated particles. You should make sure you remember it.

Example

Calculate the wavelength of 5.0 keV electrons.

Answer

5.0 keV = $5.0 \times 10^3 \times 1.6 \times 10^{-19}$ J = 8.0×10^{-16} J.

$E_k = \dfrac{p^2}{2m} \therefore p^2 = 2 \times 9.1 \times 10^{-31} \times 8.0 \times 10^{-16} = 1.46 \times 10^{-45} \rightarrow p = 3.82 \times 10^{-23}$ N s.

$\therefore$ The wavelength, $\lambda = \dfrac{h}{p} = \dfrac{6.63 \times 10^{-34}}{3.82 \times 10^{-23}} = 1.7 \times 10^{-11}$ m (17 pm)

Notice that the wavelength calculated in the example is of the same order as the inter-atomic spacing in solids and liquids (approximately 10% of the graphite inter-atomic spacing). This means that electron diffraction can be used to investigate the structure of matter. In fact X-rays, which have a similar wavelength, are more often used because of difficulties with handling charged particles and the low penetration power of electrons.

2.7.10 Knowledge check

Use the de Broglie equation to calculate the wavelength of protons which have been accelerated through 5 kV.

($m_p = 1.67 \times 10^{-27}$ kg)

(d) The momentum of photons

De Broglie's equation also applies to photons – another aspect of the duality of wave and particle descriptions. This means that if photons are absorbed or reflected by an object, their momentum changes and so, by conservation of momentum, the object suffers an equal and opposite momentum change. It is often more convenient to express the momentum of a photon in terms of the frequency (f) rather than the wavelength of the radiation:

$$\therefore p = \frac{h}{\lambda} = \frac{hf}{c}$$

2.7.11 Knowledge check

Calculate the momentum of an optical photon of frequency 600 THz.

Now hf is the photon energy, so the photon momentum is $\dfrac{E_{ph}}{c}$. One of the consequences of photon momentum is that a beam of radiation exerts a pressure on any surface which it hits. The beam of photons incident on the surface of area A, in Fig. 2.7.21, has intensity I. Thus the total energy delivered in a time Δt is given by $IA\Delta t$.

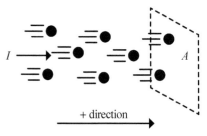

Fig. 2.7.21 Radiation pressure

∴ The momentum of the incident photons in time Δt is $\dfrac{IA\Delta t}{c}$

If the surface is totally black, so that all the momentum is absorbed, the rate of change of momentum of the photons is $-\dfrac{IA\Delta t}{c\Delta t} = -\dfrac{IA}{c}$. So by the principle of conservation of momentum, the rate of change of momentum of the surface, caused by absorbing the photons is $+\dfrac{IA}{c}$. In other words, by Newton's 2nd law (N2), the radiation exerts a force $\dfrac{IA}{c}$ on the surface, i.e. the pressure $= \dfrac{I}{c}$.

Knowledge check 2.7.12

(a) Show that the pressure exerted by the beam in Fig. 2.7.21 is $2\dfrac{I}{c}$ if the surface is perfectly reflecting.

(b) Calculate the pressure if the perfectly reflecting surface is inclined at $45°$

2.7.5 Specified practical work on light-emitting diodes

A light-emitting diode (LED) is an electronic devices constructed from a small crystal of a semiconductor material, such as gallium arsenide (**GaAs**). It conducts electricity in one direction (given by the arrow-head direction of the main part of the symbol) and, as it does so it gives out e-m radiation. Many LEDs are designed to give out close to monochromatic radiation, but some (e.g. 'white' LEDs or 'bright' LEDs) emit a broad spectrum.

>> **Study point**

LED symbol

Determination of h using LEDs

The photon energy of the emitted light is found from the pd, V_0, at which the LED has started to conduct (see Fig. 2.7.22) and give out light. Assuming that the photon energy (hf) is is equal to the electron's gain of electrical PE,

$hf = eV_0$ or, in terms of the wavelength: $\dfrac{hc}{\lambda} = eV_0$.

So, if there is a range of monochromatic LEDs available with different colours, a graph of V_0 against $\dfrac{1}{\lambda}$ (Fig. 2.7.23) is a straight line through the origin of gradient $\dfrac{hc}{e}$.

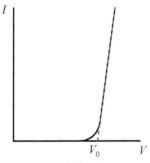
Fig. 2.7.22 LED I–V graph

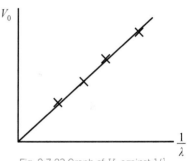
Fig. 2.7.23 Graph of V_0 against $1/\lambda$

Experimental details

1. Obtain a range of low power LEDs, with as wide a range of wavelengths as possible; determine the wavelengths using a diffraction grating as in Fig. 2.7.14. If this is not possible, note the manufacturer's values of wavelength.

2. Set up a circuit (see Study point) to determine pairs of (V, I) and plot the I–V graph for each LED.

3. Use the graphs to determine the value of V_0, as in Fig. 2.7.22, for each wavelength by drawing a best-fit straight line for the steep part of the graph.

4. Plot a graph of V_0 against $1/\lambda$ (Fig. 2.7.23) and determine the gradient, m.

5. Calculate a value for h from $m = \dfrac{hc}{e}$.

Extension to IR and UV LEDs

LEDs are available with emissions in the near IR (wavelengths 700–900 nm) and UV (wavelength ~ 390 nm). These can be used to extend the wavelength range. If using these, the wavelength cannot easily be measured in the school laboratory so the manufacturer's stated wavelength has to be relied upon.

>> **Study point**

Section 2.2.8 gives possible circuits for plotting the I–V graph. Fig. 2.7.24 is an alternative. Adjusting the variable resistor changes the LED I and V. The fixed resistor is to protect the LED from too large a current.

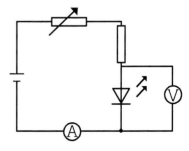
Fig. 2.7.24 I–V circuit

Test yourself 2.7

1. An LED torch has a pd of **3.0 V** and takes a current of **50 mA**.

   If the efficiency is 25%, estimate the number of visible photons it emits per second. (Take the mean wavelength to be **550 nm**.)

2. The star α Centauri has a luminosity of $1.5\,L_\odot$ where $L_\odot$ is the luminosity of the Sun. Its distance is **4.37** light years. Assuming that 70% of the star's emissions are in the visible region, estimate the number of its visible photons per second which enter the eyes of an observer on the Earth.

   Take the diameter of a dark-adapted pupil to be **7 mm**. $L_\odot = 3.85 \times 10^{26}$ W.

3. A metal has a work function of 2.4×10^{-19} J.

   (a) Express the work function in **eV**.
   (b) The metal is illuminated by monochromatic e-m radiation of photon energy 2.5 eV. State the maximum energy of the emitted electrons.
   (c) The metal is **additionally** illuminated with e-m radiation of frequency 3.0×10^{14} Hz. Explain why the maximum energy of the emitted electrons is the same as in (b).
   (d) The metal is built into a photocell and its I–V characteristic plotted as in Fig. 2.7.4. What value of stopping voltage would you expect with the 2.5 eV photons? Explain your answer.

4. Calculate the range of photon energies in the visible spectrum. Express your answer in both J and eV.

5. Helium atoms have two electrons. When a helium atom is excited, one electron remains in the lowest energy level (called 1s); the other can be raised to higher energy levels (2s, 2p, 3s etc.). The simplified diagram shows excited energy levels for helium atoms.

   A glowing gas cloud, containing helium, is heated to a high temperature by nearby stars, so that some of the atoms are in the excited state 2s. Light with photon energies up to 4 eV from other stars passes through the cloud.

   Explain, giving photon energies, what effect helium atoms have on the **visible** absorption and emission spectra of the star. There is no need to calculate wavelengths.

   (Visible photon range = 1.8 eV – 3.1 eV)

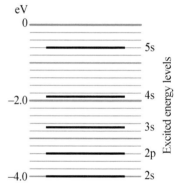

6. Electrons are given off from a heated wire, are accelerated through a pd of **500 V** and directed in a narrow beam at a graphite crystal oriented edge on. The crystal planes (separation **0.335 nm**) act as the slits of a diffraction grating. An interference pattern is observed on a fluorescent screen **30 cm** from the slits.

   (a) Explain why an interference pattern is seen on the screen.
   (b) Calculate the wavelength of the electrons.
   (c) Use the diffraction grating formula to calculate the angular position of the first and second order fringes and hence the position of these fringes on the screen.

   Electronic mass, $m_e = 9.1 \times 10^{-31}$ kg

7. A low voltage accelerator produces beams of electrons and protons by accelerating them through the same pd. Calculate the ratio of the wavelengths of the electrons and protons.

   ($m_e = 9.1 \times 10^{-31}$ kg; $m_p = 1.67 \times 10^{-27}$ kg).

8. An electron in a stationary hydrogen atom is in the second shell (see Fig. 2.7.9, $n = 2$). The electron drops down to the first shell and, in doing so, gives out a photon. By calculating the momentum of this photon, calculate the recoil speed of the hydrogen atom.
   (Mass of a hydrogen atom = 1.67×10^{-27} kg)

9. The solar constant, which is the intensity of the Sun's radiation, is 1.4 kW m^{-2} at the Earth's orbit. Calculate the area of a solar sail which is needed to produce a thrust of 1 N in the vicinity of Mars, given that the radius of the orbit of Mars is 1.5 times that of the Earth. Assume that the sail absorbs all the photons incident upon it.

10. A class of A level students undertakes the LED experiment to determine the value of the Planck constant.

    They use the manufacturer's values for the wavelengths of the LED emissions. Their results are as in the table. The estimated uncertainty in the V_0 results is $\pm 0.05 \text{ V}$.

λ / nm	420	460	540	640	660
V_0 / V	2.95	2.72	2.25	1.88	1.85

    By plotting a suitable graph, determine a value for h with its estimated uncertainty.

11. Light sometimes behaves as waves and sometimes as particles. Identify one experiment which illustrates each of these properties.

12. The kinetic energy, E_k, and momentum, p, of particles of mass m, are related by $E_k = \dfrac{p^2}{2m}$. Show that this equation is homogeneous.

13. In the electron diffraction experiment in Section 2.7.4 (b), a student notices that, when the EHT voltage is increased, the radius of the diffraction rings decreases. Explain this using the de Broglie equation.

14. State the regions of the e-m spectrum in which photons with the following energies lie.
    (a) 0.1 eV, (b) 10 keV, (c) 2.5 eV, (d) 10 eV.

15. Calculate the wavelength of a photon with the same momentum as an electron which has been accelerated through a pd of 1.0 keV. In which region of the e-m spectrum does this photon lie?

16. A student sets up the following circuit for a photoelectricity experiment.

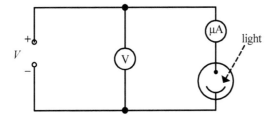

    Identify two changes that need to be made to measure the stopping voltage, V_S.

17. The graph shows how the maximum kinetic energy, $E_{k\,max}$, of the electrons emitted from a metal surface varies with the frequency, f, of the incident radiation.

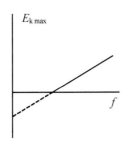

    (a) Explain how the graph is consistent with Einstein's photoelectric equation $E_{k\,max} = hf - \phi$. Identify the constants h and ϕ and state how they relate to the graph.
    (b) Explain how the values of $E_{k\,max}$ are determined by measuring the stopping voltage for the photoelectrons
    (c) A second metal has a lower work function. Explain how the graph of $E_{k\,max}$ against f relates to the given graph.

2.8 Lasers

> ## Study point

In 1962, scientists from both the USA and USSR used the time delay in laser pulses to measure the distance to the Moon. Apollo 11, 14 and 15 and Lunokhod 1 and 2 all left laser reflectors on the Moon; distance monitoring continues using them today.

> ### 2.8.1 Knowledge check

A laser sends out 10^{17} photons to the Moon per second. The beam of a laser spreads to a circle 6.5 km in diameter on the Moon. The reflector is 3 m in diameter. Assuming the reflected beam also spreads out to 6.5 km on the Earth, estimate the rate of arrival of photons in a 3 m diameter telescope.

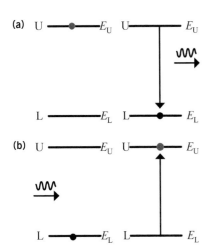

Fig. 2.8.1 (a) Spontaneous emission and (b) absorption

> ## Key term

Stimulated emission: Emission of a photon from an excited atom, triggered by a passing photon of energy equal to the energy gap between the excited state and a state of lower energy in the atom (or molecule). The emitted photon has the same frequency, phase and direction of travel and polarisation as the passing photon.

> ## Study point

Astronomers have discovered naturally occurring **masers** (**m**icrowave **a**mplification ...). See Section 2.8.5.

From the first functioning model made in 1960, lasers are now ubiquitous. Most homes have several: in DVD and CD players and optical disc drives, to say nothing of laser pointers. The first public practical use was the bar-code scanner in 1974 and since then they have been pressed into service in a variety of contexts including:

- Surgery – cutting and cauterising, minimising blood loss; also 'welding' detached retinas
- Distance measurement (see Study point), also by e.g. estate agents
- Data transmission in optical fibres and free space
- Laser printers
- Research, e.g. into laser-initiated nuclear fusion.

The usefulness of lasers arises from the fact that they produce coherent light. We refer to two light sources as coherent if they have a constant phase difference, e.g. the Young's slits, but what does it mean to say that a single source is coherent (with itself)? In fact light from a laser is coherent in two ways. It has:

1. **Spatial coherence**: different points across the width of the laser beam are in phase with one another;
2. **Temporal coherence**: monochromatic, without sudden changes of phase.

These properties allow it to be focused down to very small points (~1 mm) and to produce very short duration pulses (~1 fs). These coherent light sources are made possible by the phenomenon of stimulated emission.

2.8.1 Stimulated emission

We saw in Section 2.7 that atomic and molecular systems exist in a series of discrete energy states. For the moment, we consider just two states, U and L (upper and lower) of a system with energies, E_U and E_L. An atom or molecule can change its state:

- from U to L by spontaneously emitting a photon of frequency f where:

$$hf = E_U - E_L$$

- from L to U by absorbing the energy of a photon of the same frequency, f, as above.

These two processes, spontaneous emission and absorption, are illustrated diagrammatically in Fig. 2.8.1. The point about referring to 'spontaneous emission' rather than just 'emission' is that a second process of emission also takes place. This is known as **stimulated emission** and it was initially predicted by Albert Einstein. It is illustrated in Fig. 2.8.2.

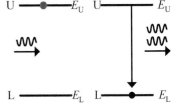

Fig. 2.8.2 Stimulated emission

In this process, an atom in the upper energy state (U) is stimulated into moving down into the lower energy state by another photon of the same energy: $hf = E_U - E_L$. In doing so it emits a second photon which is **in phase** with the first one and travelling **in the same direction**. If each of these photons now interacts with an atom in the upper state, there will be four photons: the light will be progressively amplified as it passes through the medium, giving rise to the name laser, for **L**ight **A**mplification by the **S**timulated **E**mission of **R**adiation.

Clearly for the light to be continuously amplified in this way, the photons have to carry on meeting atoms, all of which are in the upper state. Is this likely to occur naturally? The answer is, 'No!' Let's see why not.

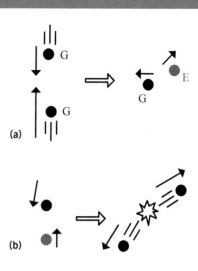

For convenience we'll consider a monatomic gas. Suppose we have a gas at room temperature (300 K) with all of the molecules being in the ground state (G). How could the molecules be put into an excited state (E)? It could happen occasionally by collision: some of the kinetic energy of the molecules is used to put one of the molecules into state E (i.e. an inelastic collision). Could this kind of collision happen repeatedly until we have more E molecules than G? No because the next time the E molecule collides, it is likely to be with a G and it is also likely to drop down again into the ground state with the energy reappearing as translational kinetic energy (we could call this a *superelastic* collision). Also the molecule is likely after a short time to lose its extra energy by spontaneous emission. And if we raised the temperature, say to 3000 K, all we'd achieve is to increase the fraction of molecules in the excited state. The 'best' we can hope for as we raise the temperature is to approach 50% of the molecules being excited because at this point the collisions would be equally likely to decrease or to raise the energy level. This is illustrated in the example below.

Example

In 1868, the Austrian physicist Ludwig Boltzmann showed how the ratio of the number of particles in two different energy states is related to the (absolute) temperature:

Let N_1 and N_2 be the number of particles in states 1 and 2 respectively and let state 2 have an energy ΔE above state 2. Then:

$$\frac{N_2}{N_1} = e^{-\Delta E/kT},$$

where $k = 1.38 \times 10^{-23}$ J K^{-1} (the Boltzmann constant) and T is the kelvin temperature.

The first excited state of sodium atoms is 2.0 eV above the ground state. In a gaseous cloud containing 10^{15} sodium atoms, estimate the number of atoms in the excited state if the temperature is: (a) 1000 K, (b) 3000 K, (c) 300 K (room temperature).

Hint: first express 2.0 eV in joules.

Fig. 2.8.3 Excitation and de-excitation processes. (a) Excitation in inelastic collision. (b) De-excitation in super elastic collision. (c) De-exitation by photon emission.

2.8.2 Achieving a population inversion

Scientists refer to a **population** of atoms (or molecules) when they mean all those with a particular property, e.g. the population of sodium atoms, the population of sodium atoms in the first excited state.

If you have worked through the Stretch & challenge, you'll have seen that the **population** of atoms at higher energy levels is normally much smaller than that at lower levels. In this case, passing photons are more likely to be absorbed than to cause stimulated emission. With energy differences in the eV range and temperatures up to 1000 K, the high-energy population is vastly outnumbered by the low-energy population. A laser can only work if the populations are the other way round, i.e. if $N_2 > N_1$. We call this situation a **population inversion**. In order to bring this about we need to find some non-thermal means of boosting the population of the upper state. We call such processes **pumping**.

It is not normally possible to achieve a population inversion with a two-state system because the upper state will usually empty as fast as it fills. So scientists work with systems of multiple energy states.

(a) Three-state laser systems

Fig. 2.8.4 is of a three-state system. The three energy states are referred to as the ground state (G), the pumped state (P) and the upper state (U).

To understand how a population inversion is achieved in multiple-state systems, you need to know two important features of transitions between energy states:

> **Key terms**

Population inversion: A situation in which a higher energy state in an atomic system is more heavily populated than a lower energy state of the same system.

Pumping: Feeding energy into the amplifying medium of a laser to achieve a population inversion.

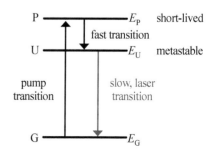

Fig. 2.8.4 Three-state system

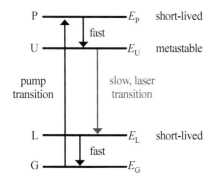

Fig. 2.8.5 Four-state system

1. Downward transitions can occur by a variety of routes, not all of which are equally likely. Laser engineers choose systems in which state P is much more likely to decay via the intermediate state U, rather than directly back to G.
2. Some energy states are very short-lived, i.e. they are populated only for a very brief period of time before decaying. Others, known as **metastable states**, last much longer before decaying. The reasons for these differences are beyond the A level course but engineers choose systems in which P is short-lived (e.g. ~ 1 ns) but U is long-lived (in atomic terms, e.g. ~ 1 ms).

If the laser medium is pumped, e.g. by flooding it with photons of energy $(E_P - E_G)$, atoms in the ground state will be raised to the pumped state and rapidly decay into the metastable upper state (partly by spontaneous emission but mainly by collisions). If the pumping is rapid enough, the population of U will exceed that of G, i.e. a population inversion will have been achieved: any spontaneous transition from U to G will produce a photon which will stimulate emissions from other atoms in state U.

Historically, a three-state laser based on a ruby laser medium was the first one to be built (see Fig. 2.8.6). However, because over half the atoms in the ground state must be pumped to achieve laser operation, three-state systems are energy hungry and inefficient. In practice, most lasers use four-state systems.

(b) Four-state laser systems

The characteristics of the pumped and upper states in a four-state laser are the same as in the three-state laser. The additional, lower, state (L) is between the upper and ground states. L is a short-lived state and rapidly decays, mainly by collisions, to G.

The advantage of this system over the three-state laser is that L is initially empty, so a population inversion between L and U is present from the first few population members in U. The short-lived nature of L ensures that a population inversion is maintained with much lower level pumping, and hence requires less energy input, than the three-state laser. This means that lasing is possible at a much lower power input.

(c) Laser inefficiency

Most of the energy input into the laser is converted into the internal energy (kinetic energy in a gas / vibrational energy in a solid) of the atoms of the amplifying medium rather than raising the energy state of the atoms themselves from G to P. Even for successful pumping events, the energy input is $(E_P - E_G)$ but the lasing output is less: $(E_U - E_L)$ for the four-state laser; $(E_U - E_G)$ for the three-state laser.

2.8.3 Laser construction

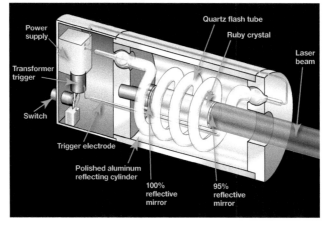

Fig. 2.8.6 The original ruby laser

The ruby laser illustrated in Fig. 2.8.6 was the first one to be produced. It is introduced here to illustrate generic aspects of lasers. As the diagram shows, it was optically pumped by the quartz 'flash tube' wrapped around the ruby (the amplifying medium). The role of the aluminium cylinder was to reflect stray pumping light back into the ruby to increase the efficiency.

Fig. 2.8.7 brings out significant features which you should study.

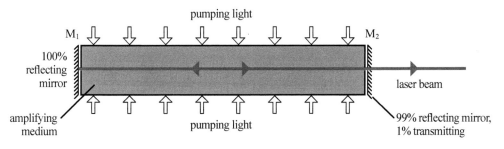

Fig. 2.8.7 Structure of an optically pumped laser

The laser operates as follows:

- The pumping radiation creates a population inversion in the amplifying medium.
- Photons of energy $E_U - E_L$ (as in Fig. 2.8.5) or $E_U - E_G$ (Fig. 2.8.4) are produced by spontaneous emission in the **amplifying medium**.
- These photons pass through the amplifying medium and produce coherent photons travelling in the same direction by stimulated emission. Each photon becomes two, four, eight, etc., in an exponential increase.
- The photons travelling parallel to the axis of the medium (see Study point) are reflected backwards and forwards, stimulating more photons and eventually escaping in the laser beam through the partly transmitting mirror.
- A dynamic equilibrium is soon established when the rate of escape of photons is equal to the rate of their production by stimulated emission (itself controlled by the rate of pumping).

2.8.4 The semiconductor diode laser

The lasers in everyday household use are almost all semiconductor laser diodes. These are constructed out of small chips of semiconductor material, often gallium arsenide (GaAs). They have many advantages:

- They are electrically pumped and operate at low voltage – some less than 2 V.
- The laser chip is very small (typically ~ 1 mm) and can be incorporated into small standard electrical packages for wiring into circuits, e.g. the laser pointer in Fig. 2.8.8.
- They are very efficient – up to 70% for infrared lasers.
- They can be mass produced cheaply.

Typical uses in the home are DVD and CD reading and writing, Blu-ray reading, optical fibre data transfer, computer scanners and printers.

Fig. 2.8.8 The laser from a laser printer (held by a crocodile clip)

The advantage of semiconductor laser diodes for optical fibres relates to the spectral purity of the output (practically only one frequency) and the coherence of the output, which allows for very rapid switching. Switching frequencies of tens of **GHz** are routinely achieved delivering very large data transmission rates.

2.8.5 Nebulium and mysterium

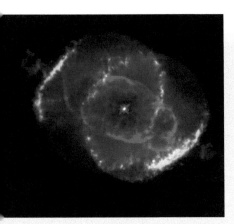

Fig. 2.8.9 Cat's Eye nebula

Helium was first identified by its emission lines in the spectrum of the Sun's corona. In 1864 the astronomer Williams Huggins studied the spectrum of the emissions from the Cat's Eye nebula. He found a group of green-blue spectral lines which had never been found in any flame test in a chemistry laboratory. He concluded that this was a previously unknown element, which he dubbed 'nebulium'. Investigations in the 20th century showed that, in fact, the lines came from O^{2+} ions (the green lines in Fig. 2.8.10) which are normally not observed because the upper levels of the transitions are metastable and normally the ions lose energy by collisions before they have had a chance to emit photons. However, in the near high vacuum conditions – better than the best vacuum produced on Earth – the collision frequency is so low that the atoms have time to emit photons with these wavelengths.

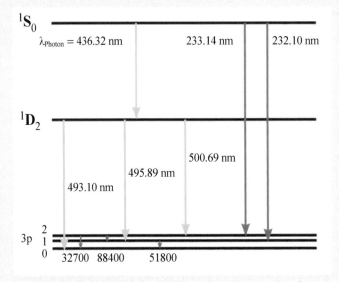

Fig. 2.8.10 'Nebulium' emissions

Calculate the energy difference between the $^1S_0$ and the $^3P_0$ energy levels in 'nebulium'.

Similarly, the molecule 'mysterium' was discovered by radio astronomers Howard Weaver and colleagues in 1963. They were looking at emissions from hydroxyl (OH) molecules in the Orion nebula, and found some lines which were very much stronger than expected (they didn't fit the known pattern of OH line strengths). They assumed that they were from an unknown molecule which they called mysterium.

The truth was that they had discovered the first natural maser – like a laser but one which emits microwaves. The emissions were from OH which was being pumped into metastable states by infrared radiation from nearby stars. Spontaneous emissions result in stimulated emissions from other OH molecules. This could not happen on Earth because, as we've seen from Knowledge check 2.8.4, a photon has to travel quite a long way before it stimulates an emission; in the high vacuum conditions of space this distance is even longer. But, to quote Douglas Adams, 'Space is big'. A hydroxide cloud is billions of km across, so there can be enough distance for lasing to occur.

Since then many space molecules have been found to exhibit maser activity, e.g. water, ammonia and hydrogen cyanide, and in 1995 the first atomic hydrogen laser was observed in the swirling disc surrounding the bright blue star MCW 349. The laser emissions are in the infrared at wavelengths corresponding to transitions between highly excited states of atomic hydrogen.

These emissions were observed because the radiation is both very intense and highly variable.

Fig. 2.8.11 Orion nebula

The radiation is much more intense than would be expected from a pure thermal emission. To emit at this intensity, just because they are hot, would require the clouds of atoms or molecules (treated as black bodies) to be at temperatures of up to 10^{15} K. But this is impossible because at these temperatures all molecules would be dissociated and all atoms completely ionised (stripped of electrons). Laser action, rather than high temperatures, must be responsible for the high intensity emission. The variability is due to the fact that the laser/maser gain is so strongly dependent (exponentially) on the length of the path in the gas cloud with appropriate conditions – that any small variation in this length produces huge changes in output. Both these properties allow astronomers to make detailed measurements of the conditions in the gas clouds.

Test yourself 2.8

1. The diagram shows two energy states, 1 and 2, in an atom.

(a) Use the principle of conservation of energy to explain why an atom in state 2 might spontaneously move to state 1, but that the opposite is not possible.

(b) State why, in a population of these atoms, more atoms will usually be in state 1 than in state 2.

(c) What is the name given to a situation in which more atoms are in state 2 than in state 1, and what is the name given to the process by which this is achieved?

2. The diagram shows a system with energy levels 1–4. It is used as the basis of a 4-level laser.

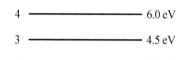

(a) Use the information on the diagram to explain the terms, ground state, population inversion, laser transition and pumping, referring to energy levels by their numbers.

(b) Explain why levels 2 and 4 need to be short-lived energy states but that level 3 needs to be long-lived.

(c) An atom moves through the levels, $1 \rightarrow 4 \rightarrow 3 \rightarrow 2 \rightarrow 1$, including the laser transition. Calculate the efficiency of this process in terms of producing laser radiation.

3. In a laser, the laser transition is from the upper state (U) to the lower state (L). Explain why there needs to be a population inversion between these states for light amplification to occur.

4. A photon causes the production of a second photon by stimulated emission. Compare the properties of the two photons.

5. Explain why at least half the atoms must be pumped from the ground state in order for a population inversion to exist between the upper (U) and ground (G) states of a **three-state** laser system.

The remaining questions in this section are all based around the energy diagram of a helium–neon laser and the following short passage describing the laser.

The helium–neon laser

Real lasers have more complicated energy levels than in the textbook notes. The He-Ne laser has a mixture of the two gases helium and neon as the amplifying medium. The two gases at low pressure are sealed into a glass tube and an electric current passed through them. Electrons colliding with the helium atoms excite them to the 2 ^{1}s and 2 ^{3}s energy states (don't worry about the names of these energy states). These two energy levels happen to be almost the same as the 4s and 5s energy levels in neon, so energy can easily be transferred from helium to neon atoms by inelastic interatomic collisions.

The 4s and 5s states of neon are *metastable* and this results in laser emissions down to the 4p and 3p states. The wavelengths of the laser emissions are 633 nm, 1.15 μm and 3.39 μm (shown as red arrows).

The 3p and 4p states are very short-lived. An atom in 3p decays very rapidly, by spontaneous emission of wavelength 600 nm (green arrow), to the 3s state. The 3s atom de-energises by collision (usually with atoms in the container walls) to the ground state. The 4p state also decays rapidly by spontaneous emission (not shown) to 3s.

A data book gives the energy of the helium 2 ^{1}s energy state as 20.65 eV above the ground state.

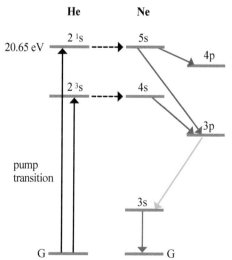

6. Explain what is meant by: (a) an inelastic collision and (b) a metastable state.

7. A neon atom in the 5s state decays to the 3p state by emitting a photon. This photon causes stimulated emission in another neon atom.

   (a) Draw a diagram to illustrate the stimulated emission event.
   (b) How do the properties of the second photon compare with the first?

8. Why is important that the 3p state in neon is much shorter-lived than the 5s for establishing a population inversion?

9. Identify which of the laser transitions produce the 633 nm, 1.15 μm and 3.39 μm emissions. You may assume that the separations of the energy states above the 3s state are drawn roughly to scale. Identify the region of the e-m spectrum in which each of these emissions lies.

10. Use your answers to Q9 and the wavelength of the 3p→3s transition to calculate the energies of the excited neon states above the ground state. Give your answers in both J and eV.

11. Calculate the wavelength of the spontaneous 4p→3s transition. In which region of the e-m spectrum does this lie?

12. Describe in your own words the process of energy transfer between the He 2 ^{1}s state and the neon 5s state.

13. The diagram shows an electron, with a kinetic energy of 5.3×10^{-18} J, approaching a helium atom (in a He–Ne laser) in the ground state. It collides with one of the electrons in the helium atom and promotes is to the 2 ^{1}s energy state (shown as a dotted line).

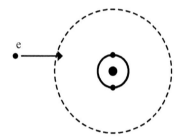

   (a) Calculate the pd through which the incident electron has been accelerated.
   (b) Express the energy of the incident electron in eV.
   (c) Sketch a diagram showing the atom and electron, with their energies, after the collision.
   (d) State the minimum accelerating voltage needed to enable an electron to excite a helium atom to 2 ^{1}s.

14. A He–Ne laser operates at 1.2 kV. Calculate the number of helium atoms an electron can excite to the 2 ^{1}s state.

Unit 2 equations

The WJEC Data Booklet contains the equations you may need to use in the examination. The symbols in the equations are not identified in the Data Booklet: they are standard symbols and you are expected to recognise them. The equations below are the ones needed for Unit 2.

Equation	Description
$I = \dfrac{\Delta Q}{\Delta t}$	Current, I, is the flow of charge, ΔQ, divided by time interval, Δt
$I = nAve$	Current, I, in a wire related to the flow of charges: n = number density of free electrons, A = cross-sectional area v = drift velocity of free electrons, e = electronic charge
$R = \dfrac{V}{I}$	Definition of resistance, R. V = pd across the conductor, I = current in the conductor.
$P = IV = I^2R = \dfrac{V^2}{R}$	Power, P, dissipated in a component. Other symbols as above.
$R = \dfrac{\rho\ell}{A}$	Definition of resistivity, ρ. ℓ = length of conductor, R = resistance, A = cross-sectional area.
$V = E - Ir$	pd, V, across a power supply. E = emf, r = internal resistance I = current
$\dfrac{V}{V_{total}}$ or $\left[\dfrac{V_{OUT}}{V_{IN}}\right] = \dfrac{R}{R_{total}}$	Potential divider formula, V = output voltage, R = output resistance.
$T = \dfrac{1}{f}$	Relationship between period, T, and frequency, f.
$c = f\lambda$	c = wave speed, f = frequency, λ = wavelength
$\lambda = \dfrac{a\Delta y}{D}$	Young's slits formula: a = separation of slit centres, Δy = fringe separation, D = slits-screen distance
$d\sin\theta = n\lambda$	Diffraction grating formula: d = line separation, n = spectrum order, θ = angle of spectrum, λ = wavelength
$n = \dfrac{c}{v}$	n = refractive index of a material, c = speed of light in a vacuum, v = speed of light in the material
$n_1 v_1 = n_2 v_2$	Relationship between the refractive index, n, and speed of light, v, in two materials 1 and 2.
$n_1 \sin\theta_1 = n_2 \sin\theta_2$	Relationship between the refractive index, n, and the angle, θ, to the normal for a light ray travelling between materials 1 and 2.
$n_1 \sin\theta_C = n_2$	Relationship between the critical angle, θ_C, and the refractive indices for a light ray in material 1 incident on material 2.
$E_{k\,max} = hf - \phi$	Photoelectric equation. $E_{k\,max}$ = maximum KE of emitted electrons h = the Planck constant, f = frequency, ϕ = work function
$p = \dfrac{h}{\lambda}$	de Broglie's equation: p = particle momentum, h = the Planck constant, λ = particle wavelength

Exam practice questions

Unit 2

1. **(a)** Two 3.30 Ω resistors are connected in series across a battery of emf 4.80 V as shown. The voltmeter reads 4.33 V.

    Diagram 1

    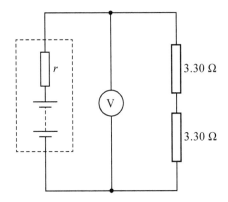

    (i)   State, in terms of work or energy, what is meant by the emf of a battery.   **[2]**
    (ii) Show that the internal resistance, r, of the battery is approximately 0.7 Ω.   **[2]**

    (iii) When the resistors are connected in parallel as shown below, the voltmeter reads 3.35 V.

    Diagram 2

    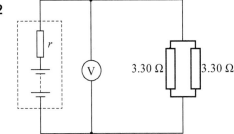

    (I)   **Without further calculation**, explain why you would expect the reading to be lower, now that there is a lower resistance across the cell terminals.   **[2]**
    (II) Calculate the number of electrons entering **either** of the resistors (shown in Diagram 2) per **minute**.   **[2]**

    (b) The heating element (a coil of wire) in an electric heater dissipates energy at a rate of 1.00 kW, when connected across the 230 V mains supply. Calculate:

    (i)   the resistance of the coil;   **[2]**
    (ii) the energy dissipated per hour, giving your answer in megajoules (MJ).   **[1]**

    (c) For each megajoule of heat from an electric heater, approximately 0.08 m^3 of gas would have to be burned in a gas-fired electricity power station. For each megajoule of heat from a domestic gas fire or boiler, approximately 0.03 m^3 of gas is burned. Discuss whether the use of electric heaters in houses should be discouraged. Calculations are not required.   **[3]**

    (Total 14 marks)
    [*WJEC AS Physics Unit 2 2018 Q1*]

2 (a) A progressive wave is travelling from left to right along a stretched string at a speed of 0.40 m s^{-1}. The diagram shows the string at time $t = 0$.

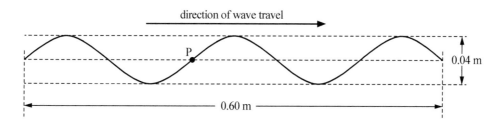

direction of wave travel

P

0.04 m

0.60 m

Carefully sketch, on a copy of the grid below, a displacement–time graph for point **P** on the string between $t = 0$ and $t = 1$ s.

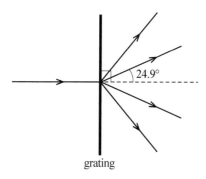

displacement / m

time / s

[3]

(b) The distance between the centres of the slits in a diffraction grating is 1500 nm. Monochromatic light is shone normally on to the grating.

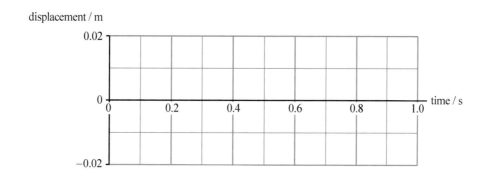

24.9°

grating

(i) First order beams emerge at angles of 24.9° to the normal (see diagram). Calculate the wavelength of the light. **[2]**

(ii) Explain **in terms of path difference** why the **second** order beams emerge from the diffraction grating at 57.4° to the normal. You will need to copy and add to the diagram (which shows two adjacent slits in the grating). **[3]**

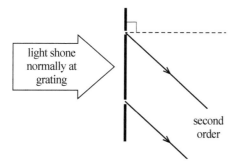

light shone normally at grating

second order

(Total 8 marks)

[WJEC AS Physics Unit 2 2018 Q4]

3 (a) (i) A narrow beam of light enters a glass block of quarter-circle cross-section, as shown. The refractive index of the glass is 1.60.

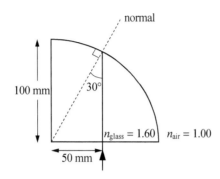

Calculate the angle of refraction into the air at the curved surface and carefully sketch the refracted beam on the diagram. [Use of a protractor is not required.] **[3]**

(ii) Calculate the largest distance, x, along the bottom face of the block, at which the beam can enter the block normally, for it to emerge from the curved face. You should refer to the angle θ in your calculation. **[3]**

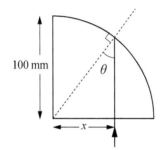

(b) Explain how a multimode (thick) glass fibre transmits light, and why rapid streams of data cannot be transmitted successfully through long lengths of multimode fibre. **[6 QER]**

(Total 12 marks)
[*WJEC AS Physics Unit 2 2018 Q7*]

4 (a) (i) Light of frequency less than ϕ/h cannot eject electrons from a surface of work function ϕ, even if the light intensity is increased. Explain this in terms of photons. **[3]**

(ii) The emitting surface in a vacuum photocell is known to be made of one of the metals listed below (with their work functions).

Metal	caesium	potassium	barium	calcium	zinc
Work function / 10^{-19} J	3.12	3.68	4.03	4.59	5.81

The photocell is included in the circuit shown, and illuminated with light of frequency 6.59×10^{14} Hz.

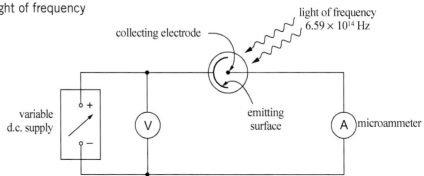

With zero pd applied, the microammeter indicates a current. At some pd between 0 V and 0.35 V the microammeter reading drops to zero.

Determine which metal the emitting surface is made of, giving your reasoning clearly. **[4]**

(b) Rachel varies the pd across a light-emitting diode (LED) and notes the value, V, for which she can just see light from the LED. She also notes the frequency, f, of the light, as supplied by the LED's makers. She does the same for three other LEDs and plots V against f (below).

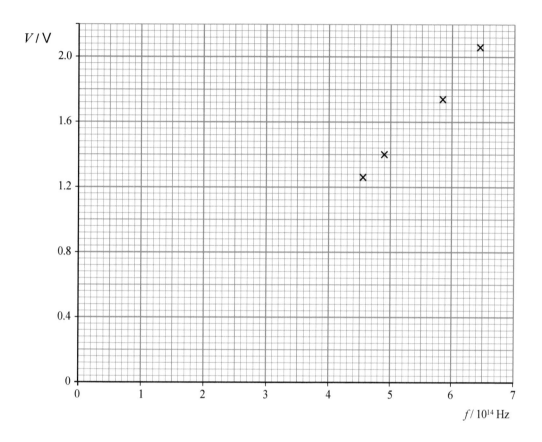

It has been suggested that V and f are related by the equation:

$$V = \frac{h}{e}f$$

(i) On **a copy of the graph**, draw the line of best fit. **[1]**

(ii) Discuss the extent to which the graph supports an equation of this form. **[2]**

(iii) Determine the gradient of the graph, and hence a value for h to an appropriate number of significant figures. Assume that the equation predicts $\frac{\Delta V}{\Delta f}$ correctly. Show your working clearly. **[3]**

(Total 13 marks)
[*WJEC AS Physics Unit 2 2018 Q8*]

Chapter 3

Practical skills

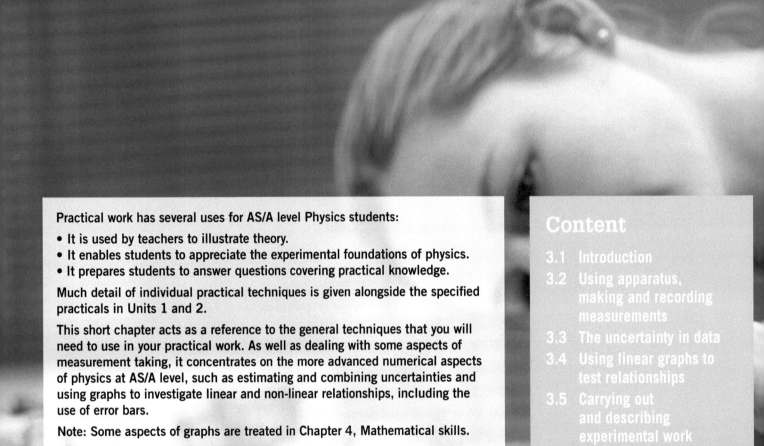

Practical work has several uses for AS/A level Physics students:

• It is used by teachers to illustrate theory.
• It enables students to appreciate the experimental foundations of physics.
• It prepares students to answer questions covering practical knowledge.

Much detail of individual practical techniques is given alongside the specified practicals in Units 1 and 2.

This short chapter acts as a reference to the general techniques that you will need to use in your practical work. As well as dealing with some aspects of measurement taking, it concentrates on the more advanced numerical aspects of physics at AS/A level, such as estimating and combining uncertainties and using graphs to investigate linear and non-linear relationships, including the use of error bars.

Note: Some aspects of graphs are treated in Chapter 4, Mathematical skills.

Content

3.1 Introduction

3.2 Using apparatus, making and recording measurements

3.3 The uncertainty in data

3.4 Using linear graphs to test relationships

3.5 Carrying out and describing experimental work

3.1 Introduction

The predictions of physics theories are tested experimentally. The results of experimental work are used creatively by physicists to propose new theories. Additionally, engineering and other science-based activities, such as medicine, rely on experimental techniques from physics to investigate phenomena and characterise materials. The WJEC GCE Physics specification contains a set of specified practical work, which is designed as an introduction to techniques of measurement, analysis and evaluation which physicists use. You should be prepared to answer questions which test both your knowledge of the specified practical work and your ability to apply practical experience in novel situations. (Unit 5 of the A level consists of a practical examination and a data analysis test.)

3.2 Using apparatus, making and recording measurements

The list of general apparatus used in A level physics includes: distance-measuring equipment such as metre rules, digital callipers, micrometers and (possibly) travelling microscopes; electrical meters, mainly ammeters and voltmeters; digital balances; timers, such as digital stopwatches and light-gate activated timers; thermometers or temperature probes; liquid volume measuring apparatus such as measuring cylinders; signal generators, oscilloscopes, radiation detectors and counters.

3.2.1 Dealing with zero errors

Apparatus should be checked to ensure that it reads zero when it should do. Some equipment can be zeroed, e.g. analogue electric meters. If a piece of equipment cannot be zeroed the initial reading should be subtracted from each reading taken, e.g. a digital calliper reads 0.02 mm with the jaws closed and 0.34 mm when reading the diameter of a wire: the wire diameter should be reported as $0.34 - 0.02 = 0.32$ mm.

3.2.2 Avoiding parallax errors

Parallax errors affect readings from analogue instruments such as metre rules or electrical meters with pointers and scales. In Fig. 3.1 it is clear that only the middle reading (in red) is correct. The position to be measured should be as close to the scale as possible and the eye should be at right angles to the scale. In 'bouncing ball' experiments the eye should be at the anticipated height of bounce, the approximate value of which can be determined by trial runs.

> **>> Study point**
>
> WJEC has produced a set of guidance notes for practical work in GCE Physics, which is available to download from the website.

> **< Top tip**
>
> The ends of metre rules are often damaged. Readings should be taken from a convenient point, e.g. 100 mm away from the end.

> **< Top tip**
>
> To find the diameter of a wire, it is good practice to measure the diameter at various points along its length and repeat each measurement at right angles in case the wire isn't accurately circular in cross section.

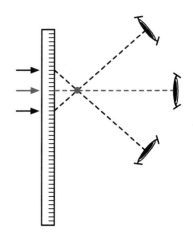

Fig. 3.1 Parallax errors

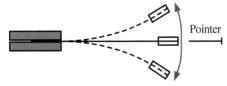

Fig. 3.2 Fiducial mark

3.2.3 Using a fiducial mark

In many experiments, it is useful to have a fixed mark from which to take measurements. This is particularly useful in oscillation experiments. The oscillations should be timed when the object crosses the central point. This is where it is moving most rapidly. Also, if the oscillations get smaller because of damping, after a few oscillations the object might not reach a pointer placed at the extreme positions.

3.2.4 Resolution

Always record the **resolution** of any instrument you use.

For a digital instrument, this is 1 in the least significant figure in the display. The resolution of the ammeter in Fig. 3.3 is 0.1 mA

For an analogue instrument, e.g. the voltmeter in Fig. 3.3, the resolution should be taken as the interval between the smallest graduations. In this case 0.1 V.

Fig. 3.3 Resolution of digital and analogue instruments

3.2.5 Recording and displaying data

The standard way of displaying systematic data is to use a table. Table 3.1 shows the points you should consider when constructing a table.

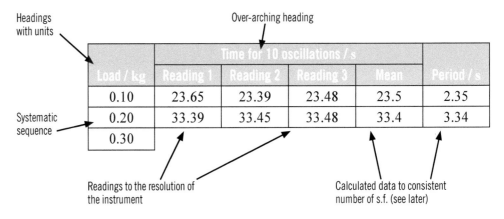

Load / kg	Time for 10 oscillations / s				Period / s
	Reading 1	Reading 2	Reading 3	Mean	
0.10	23.65	23.39	23.48	23.5	2.35
0.20	33.39	33.45	33.48	33.4	3.34
0.30					

Table 3.1 Tabulation

The table should also include values calculated from the raw experimental data. In Table 3.1, this includes the mean value of the time for 10 oscillations and the calculated period. In an electrical experiment, it could include calculated resistance or power. In many cases, such as when error bars are to be plotted (see Section 3.3.3), it is helpful to include uncertainty values in the table (see Sections 3.3.1 and 3.3.2).

3.2.6 Number of significant figures in data

A stated value of a quantity without an indication of uncertainty is of limited use because we cannot be sure how precise it is. For example, what does a current of stated value 53 mA mean? With no other information, we take 53 mA to mean 'somewhere between $52.5\ldots$ mA and $53.4\ldots$ mA'. This means that the range of uncertainty is ± 0.5 mA, which is roughly 1 part in 100 (or 1%).

If we calculate an electrical resistance, from a single pair of values of current and pd, we need to decide how precise the answer is, i.e. how many figures to give in the answer. Suppose the pd is 25.52 V and the current is 53 mA. The value of resistance, R is calculated as follows:

$$R = \frac{V}{I} = \frac{25.52 \text{ V}}{0.053 \text{ A}} = 481.509... \ \Omega$$

But the current is only known to 1 part in 100. If the actual value of the current were 0.05270 A the calculated resistance would be $484.250...$ Ω. These two answers are different if we use more than two significant figures, which is the precision of the current, the less precise datum, so we give R as 480 Ω (2 sf).

If we can estimate the uncertainty in the data we can give a better answer to this problem. This is dealt with in the next section.

3.3 The uncertainty in data

No experimental value is known with unlimited precision. A resistor with a marked value of 22 kΩ cannot be taken to have a resistance of $22.00000...$ kΩ. Experimental results are quoted together with what is called the **absolute uncertainty**. The resistance could be quoted as 21.6 ± 0.5 kΩ, suggesting that the best estimate is that the resistance, in kΩ, lies between 21.1 and 22.1.

This section introduces the methods employed in A level Physics in estimating uncertainties.

3.3.1 Estimating the best value and absolute uncertainty

(a) Uncertainty from a single measurement
The instrument resolution should be used as the uncertainty estimate. For example, when using a metre rule, the uncertainty should normally be given as ± 0.001 m (± 1 mm). It may be possible to estimate readings to better than this, e.g. to half the resolution, but it should be remembered that all length readings are the difference between the readings at each end and hence have two uncertainties associated with them.

(b) Best value and uncertainty from several measurements
The first thing to be done is to examine the set of measurements to decide whether they are all valid. Sometimes an examination reveals an **outlier**, which is a result that is very different from the others. For example, consider the following readings of the time for 10 oscillations of a pendulum:

Time/s: 6.37, 6.28, 6.38, 6.29, 6.95, 6.33, 6.31, 6.41, 6.25, 6.37

The circled reading, 6.95 s, is an outlier. The safest way to proceed is to ignore this result when calculating the mean and uncertainty (see Knowledge check 3.3).

Fig. 3.4 shows the results of 5 length measurements, $x_1, x_2, ..., x_5$, each made to the nearest 0.5 mm. The clustering gives no reason to reject any value. The value to be reported is the **arithmetic mean** of the readings.

$$\langle x \rangle = \frac{x_1 + x_2 + ... \, x_5}{5} = \frac{68.25 + 68.70 + 69.05 + 69.40 + 69.50}{5} = 68.98 \text{ cm}$$

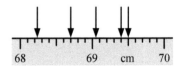

A current, I, is measured 4 times with the following results in mA: 36.7, 37.2, 36.6, 37.0. Give the reported value of I together with its uncertainty.

Calculate the precision of the wavelength of the microwaves in Section 3.3.1(c).

▶▶ **Study point**

If we are calculating ρ from $R = \frac{\rho\ell}{A}$, we manipulate the equation to give:

$\rho = \frac{RA}{\ell}$. Then $p_\rho = p_R + p_A + p_\ell$.

Note that this is not a rearrangement of the equation in Section 3.3.2(a).

Uncertainties always add!

The absolute uncertainty Δx is estimated by dividing the spread of values by 2, in this case:

$$\Delta x = \frac{x_{max} - x_{min}}{2} = \frac{69.50 - 68.25}{2} = 0.625 \text{ cm}$$

Because Δx is only an estimate of the uncertainty, we report it to 1 s.f. only. In this case, we report Δx as 0.6 cm and we quote the value of x as $x = \langle x \rangle \pm \Delta x$, which in this case is 69.0 ± 0.6 cm.

Note that we report the best value of x to the same decimal place as the uncertainty. In this case, the uncertainty is in the 1st decimal place, so we report x to 1 d.p.

(c) Precision, percentage uncertainty, fractional uncertainty

The absolute uncertainty in a quantity does not, by itself, indicate the level of precision with which a value has been determined. For example, an uncertainty in wavelength, $\Delta\lambda$ of 10 nm would be very precise for a microwave ($\lambda \sim 1$ cm) but unimpressive for UVB ($\lambda \sim 100$ nm) and very imprecise for X-rays ($\lambda \sim 0.1$–10 nm). The **precision**, p, is defined as follows:

$$\text{Precision}, p = \frac{\text{absolute uncertainty}}{\text{mean value}} (\times 100\%) = \frac{\Delta x}{x} (\times 100\%)$$

It is also referred to as the percentage (or fractional) uncertainty.

As with absolute uncertainty, the precision is normally given to 1 s.f.

Example

Calculate the percentage uncertainty in the length readings in Fig. 3.4.

Answer

From the working in Section 3.3.1(b), the length is 69.0 ± 0.6 cm

$$\therefore p = \frac{\Delta x}{\langle x \rangle} = \frac{0.625}{68.98} \times 100\% = 0.9\%$$

Note that it is good practice to use the unrounded data in calculating the value of p.

For the UVB radiation mentioned above, $p = \frac{\Delta\lambda}{\lambda} = \frac{10}{100} = 0.1 = 10\%$.

It is a matter of taste whether the fraction (0.1) or the percentage (10%) is quoted. The precision is useful when combining uncertainties, which is covered in the next section.

3.3.2 Finding the uncertainty in calculated quantities

(a) Multiplying and dividing

Many quantities in physics are found by multiplying and dividing others, e.g.

$$\text{speed} = \frac{\text{distance}}{\text{time}}; \qquad \text{power}, P = IV; \qquad \text{resistance}, R = \frac{\rho\ell}{A}$$

The uncertainty in the quantity to be calculated (speed, power or resistance) is found by combining the uncertainties in the known quantities. For multiplications and divisions the precisions (fractional or percentage uncertainties) add to give the precision in the answer. So, in the resistance equation:

$$p_R = p_\rho + p_\ell + p_A$$

To calculate the absolute uncertainty in the resistance, we then use:

$$\Delta R = p_R R.$$

Example

The pd, V, across a component is (5.35 ± 0.02) V; the current, I, is (25.3 ± 0.8) mA. Calculate the resistance of the component together with its absolute uncertainty and report the value correctly.

Answer

1st step: Calculate the best value of R: $R = \dfrac{V}{I} = \dfrac{5.35 \text{ V}}{0.0253 \text{ A}} = 211.46\ \Omega$

2nd step: precision in the voltage, $p_V = \dfrac{0.02}{5.35} = 0.0037 \ [= 0.37\%]$

                  precision in the current, $p_I = \dfrac{0.8}{25.3} = 0.0316 \ [= 3.16\%]$

3rd step: Add the precisions: $p_R = p_V + p_I = 0.0037 + 0.0316 = 0.0353$

4th step: Calculate ΔR: $\Delta R = R\, p_R = 211.46 \times 0.0353 = 7.46\ \Omega = 7\ \Omega$ (1 s.f.)

5th step: $\therefore$ Reported $R = (211 \pm 7)\ \Omega$

Note: In the example, p_V is much less that p_I so that it could really be ignored. In fact ignoring p_V still gives us $\Delta R = 7\ \Omega$ (1 s.f.)!

(b) Multiplying or dividing by an integer

One good way of increasing the level of precision in a measurement is to take a single measurement of multiple identical quantities together. A good example is measuring the period of an oscillation. It is good practice to time 10 or 20 oscillations and divide the time by the number of oscillations.

Let's take another example. Suppose we determine the thickness, d, of a stack of 10 identical microscope slides to be 1.32 ± 0.02 cm. See Fig. 3.5.

$d = 1.32 \pm 0.02$ cm

Fig. 3.5 Thickness of a stack of microscope slides

The thickness, t, of a single microscope slide is given by:

$$t = \frac{(1.32 \pm 0.02)\text{ cm}}{10} = (0.132 \pm 0.002)\text{ cm}$$

The **precision** in t is the same as the **precision** in d. In this case:

$$p = \frac{0.02\text{ cm}}{1.32\text{ cm}} = \frac{0.002\text{ cm}}{0.132\text{ cm}} = 0.015 = 1.5\%\ (2\%\text{ to 1 s.f.})$$

Of course this is just a special case of the rule in Section 3.3.2(a). How is this? Let's make it more general and suppose we have n microscope slides:

$$t = \frac{1}{n}d \qquad \therefore\ p_t = p_n + p_d$$

But there is no uncertainty in n, so $p_n = 0$ and therefore $p_t = p_d$.

> **Top tip**

It is easier to work with fractional rather than percentage uncertainties when combining them. In the example the percentage uncertainties are 0.37 and 3.16 giving a total of 3.53. To convert this to a value of ΔR requires a division by 100, which is not needed if fractional uncertainties are used.

> **Top tip**

When doing uncertainty calculations, keep 2 or 3 s.f. during your working and only reduce to 1 s.f. in the answer.

>> **Study point**

Had we measured the thickness of a single slide to within ± 0.02 cm, p would be 10× larger.

>> **Study point**

Constants such as 2 and π also have zero uncertainty, so the percentage uncertainty in the circumference of a circle $(= 2\pi r)$ is the same as that for the radius.

3.5 Knowledge check

Calculate the absolute uncertainty in the period for loads of 0.10 kg and 0.20 kg in Table 3.1. Hence comment on the number of d.p. in the data in the period column.

3.6 Knowledge check

Calculate the acceleration in Section 3.3.2 (c) if the time taken was 4.0 ± 0.1 s.

3.7 Knowledge check

The diameter of a sphere is measured as (2.00 ± 0.01) mm. Calculate its volume, $V \pm \Delta V$.

Use the formula $V = \frac{4}{3}\pi r^3$ and remember that $\frac{4}{3}\pi$ has zero uncertainty.

Example

Calculate the percentage uncertainty in the period for a load of 0.10 kg in Table 3.1.

Answer

The absolute uncertainty in 10 oscillations $= \dfrac{T_{max} - T_{min}}{2}$

$$= \dfrac{23.65 - 23.39}{2} = 0.13 \text{ s}$$

$\therefore$ For 10 oscillations, $p = \dfrac{0.13}{23.5} \times 100\% = 0.55\%$ (= 0.6% to 1 s.f.)

$\therefore$ For 1 oscillation, $p = 0.6\%$

(c) Adding and subtracting quantities

When combining by addition or subtraction, the **absolute uncertainties add**. For example, if a car accelerates from (12.0 ± 0.2) m s^{-1} to (20.5 ± 0.2) m s^{-1}, the change in velocity is $(20.5 - 12.0) \pm (0.2 + 0.2) = (8.5 \pm 0.4)$ m s^{-1}. Notice that subtracting quantities tends to produce a result with a much larger percentage uncertainty. In this case the precisions in the velocities are 1.7% and 1.0% but the precision in the change in velocity is 5% (1 s.f.)!

(d) Powers

We often need to square or find the square root of quantities in calculations. If we remember that $A^2 = A \times A$, then we can apply the rule from part (a). So:

$$p(A^2) = 2p_A.$$

We can generalise from this: $p(A^n) = np_A$ and $p(\sqrt[n]{A}) = \frac{1}{n}p_A$. As an example, the area of a circle is given by πr^2; π has no uncertainty; so, if we know the radius to a precision of 1%, the precision in the area is $2 \times 1\% = 2\%$.

3.3.3 Uncertainties and graphs

(a) Error bars

Consider an investigation into the variation of acceleration of a model rocket with height above the ground. At a height of 1.9 m, the acceleration is measured to be (32 ± 4) m s^{-2}. We plot this information as in the red barred vertical line in Fig. 3.6(a). The line, called an error bar, is at 1.9 m and extends from 28 to 36 m s^{-2}. The best value of acceleration, 32 m s^{-2}, has no special significance apart from being the centre of the error bar. The horizontal ends of the bar are not significant and serve only to draw attention to

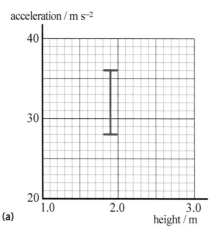

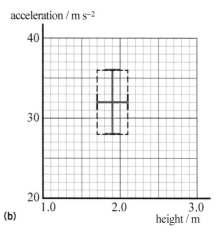

Fig. 3.6 Error bars   **(a)**   **(b)**

the extent of the error bar. When we come to draw the best-fit graph, we expect it to pass through the bar.

If the height of the rocket was only known to within 0.2 m we would also plot a horizontal error bar to represent this, as in Fig. 3.6 (b). If there were no vertical error bars, the best-fit line would pass through the horizontal error bar. If there are uncertainties in both the x and y directions the best-fit graph would pass through the dotted box, which encloses both error bars. In fact, a sensible way of plotting the uncertainties in this case is just to draw the 'error box': this is not conventional but it is perfectly acceptable.

(b) Best-fit straight lines and error bars

We use the plotted error bars to:

- decide whether the results are consistent with a linear relationship
- determine the relationship (with an estimate of the uncertainties) between the variables.

Consider the set of results plotted in Fig. 3.7. It is possible to draw a straight line through all the error bars, so the results are consistent with a linear relationship between y and x. The lines drawn are the extremes – they represent the steepest and least steep lines it is possible to draw through all the error bars.

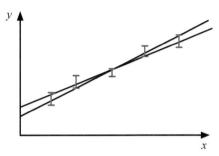

Fig. 3.7 Graphs and error bars

Maths tip

The equation of a straight line graph of y against x is $y = mx + c$, where m is the gradient and c the intercept on the y axis. See Section 4.5.4.

We can use these extreme lines to find the best values of the gradient and intercept together with their uncertainties. If the extreme gradients and intercepts are m_1, m_2 and c_1, c_2 (see Maths tip), then:

$$m = \frac{m_1 + m_2}{2} \pm \frac{m_1 - m_2}{2} \text{ and } c = \frac{c_1 + c_2}{2} \pm \frac{c_1 - c_2}{2}$$

i.e. we take the arithmetic means of the extremes of gradient and intercept to be the best values, with the absolute uncertainties being half the ranges. So the equation between y and x is written:

$$y = \left(\frac{m_1 + m_2}{2} \pm \frac{m_1 - m_2}{2} \right) x + \left(\frac{c_1 + c_2}{2} \pm \frac{c_1 - c_2}{2} \right)$$

In Fig. 3.8, the relationship $y \propto x$ is consistent with the error bars because it is possible to draw a straight line through the origin and the error bars. This should not be assumed, however; the extreme graphs should be drawn as shown and the best values of m and c together with their uncertainties reported. It is not possible to draw a straight line through the error bars in Fig. 3.9.

Knowledge check 3.8

In a graph of V/V against I/A, the extreme graphs have the following gradients and intercepts:

Gradient: -0.165, -0.169

Intercept: 9.05, 9.17

Write the equation between V and I as in Section 3.3.3 (b).

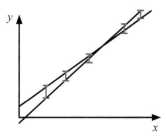

Fig. 3.8 Proportionality

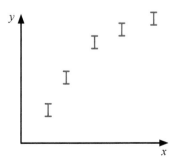

Fig. 3.9 Data inconsistent with a linear relationship

3.4 Using linear graphs to test relationships

This involves comparing the expected relationship between the variables with the straight-line equation, $y = mx + c$, where m and c are constants, the gradient and y-intercept respectively.

3.4.1 For a linear relationship

3.9 Knowledge check

What are the gradient and intercept of a graph of v against t for constant acceleration?

We shall examine this using a typical Year 12 example, the relationship between terminal pd, V, and current, I, for a power supply: $V = E - Ir$. It is advisable to rearrange the equation so that it is easy to compare with $y = mx + c$.

Rearranging the equation:

Compare with

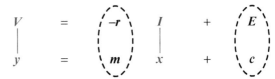

3.10 Knowledge check

State the values of r and E from Knowledge check 3.8.

The variables in the two equations and the correspondence between them are indicated in red. If the internal resistance, r, and emf, E, of the supply are constant, a graph of V against I is a straight line with gradient $-r$ and intercept E on the V-axis. Thus the relationship is tested and (assuming the relationship holds) r and E determined.

3.4.2 For non-linear relationships

Maths tip

The label of the v^2 axis should be $(v \, / \, \mathrm{m \, s^{-1}})^2$.

Many non-linear relationships can be plotted to give a linear graph by a careful choice of variables. Some are rather straightforward; e.g. $v^2 = u^2 + 2ax$, for a constant acceleration. If we plot v^2 against x (rather than v against x), the graph should be a straight line of gradient $2a$ and intercept u^2 on the v^2-axis.

Other relationships are more difficult to linearise. For example $V = \dfrac{ER}{R+r}$, where E and r

3.11 Knowledge check

Show that $V = \dfrac{ER}{R+r}$ can be rearranged to give $\dfrac{1}{V} = \dfrac{r}{ER} + \dfrac{1}{E}$.

are constants as before, is not easy but it can be rearranged to give $\dfrac{1}{V} = \dfrac{r}{ER} + \dfrac{1}{E}$.

This suggests that a graph of $\dfrac{1}{V}$ against $\dfrac{1}{R}$ should be a straight line of gradient $\dfrac{r}{E}$

and intercept $\dfrac{1}{E}$ on the $\dfrac{1}{V}$-axis.

3.5 Carrying out and describing experimental work

3.12 Knowledge check

The resonance frequency, f, for a pipe which is open at one end, relates to its length, ℓ, by the equation

$\ell + e = \dfrac{c}{4f}$, where c is the speed of sound waves and e a constant (called the 'end correction'), relating to the diameter of the pipe. What graph, relating f and ℓ, should be plotted to verify the relationship and how should c and e be determined?

Experimental work in physics has two main purposes:

1. To test relationships between variables

2. To determine the value of a physical quantity.

In practice these two purposes often overlap. For example, if the purpose is to determine the presumed-constant acceleration of an object, it is sensible to demonstrate that it really is constant, e.g. by plotting a graph of v^2 against x (as in Section 1.3.9) or v against t, showing that the results are consistent with a constant acceleration and using the gradient to determine its value. This procedure has the additional advantage that it allows for a more accurate estimation of the uncertainty in the result.

Physics teachers have additional reasons for carrying out practical work: demonstrating a phenomenon or setting up an event which can be discussed in terms of physical laws. The purpose of a particular piece of experimental work should always be clearly stated.

3.5.1 Identifying variables

All the quantities that need to be measured in experimental work are referred to as variables. This is the case whether or not their values actually change during an investigation – they have the potential to vary. An investigation often involves looking at the effect of varying one or more variables, the **independent variables**, on another variable, the **dependent variable**. Some variables, the **control variables**, need to be kept the same so that the effect of varying the independent variables is isolated. When planning an investigation, all the key variables need to be identified.

Often, when plotting graphs, the dependent variable is plotted on the vertical axis (the ordinate or y-axis) with the independent variable on the horizontal axis (the abscissa or x-axis). This is not always convenient, for example when investigating the resistance of a wire, it doesn't matter whether the current or pd is on the y-axis: if the current is on the x-axis, the gradient is the resistance; current on the y-axis makes the resistance equal to the reciprocal of the gradient.

Unlike living systems, the objects under investigation usually behave in a relatively simple way and we can describe the effect of one variable on another in terms of algebraic or trigonometrical functions.

3.5.2 Measuring variables

The measuring technique and the instruments should be chosen which: (a) measure the relevant variables, (b) have appropriate precision and accuracy, (c) are as convenient as possible and (d) give reproducible results. Some measurements do not need be very precise – when measuring the volume of a long (~ 5 m) piece of thin (~ 1 mm) wire, there is no point in trying to make the length measurement to better than 1 cm uncertainty. Why not? See the Study point. On the other hand, there is every justification for improving the precision of the diameter determination.

3.5.3 Look out for systematic uncertainties

Uncertainties in measurement which have the effect of always making the result too large or always too small are called systematic uncertainties. They are sometimes easy to spot: the relevant distance between the shielded radioactive source and the G-M tube 'hot spot' is unknown but will always differ from the distance x by the same amount, ε.

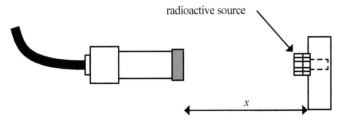

Fig. 3.10 Systematic uncertainty

> **Study point**
>
> The distinction between independent and dependent variables is not always useful. In the equation $v^2 = u^2 + 2ax$, we could measure the velocity at a set of displacements (making v the dependent variable) or the displacements at a set of velocities.

> **Study point**
>
> Some variables are almost self-controlling; e.g. when using a wire from a single reel to investigate the effect of length of a wire on resistance, the diameter and composition of the wire are automatically kept the same.

> **Study point**
>
> Consider a wire of approximate diameter 1 mm and length 5 m: If the diameter, D, is measured to ± 0.01 mm that is a precision of 1%. The volume depends upon D^2, which has a precision of 2%. A 1 cm uncertainty in the length means a precision of 0.2% which is negligible.

3.13 Knowledge check

From the graph of $\dfrac{1}{\sqrt{C}}$ against x opposite:

(a) State the values of the gradient and intercept.

(b) Explain how ε can be determined.

3.14 Knowledge check

The electromagnet in the experiment to determine g in Section 1.2.5 doesn't lose its magnetisation instantly, so hangs on to the ball for an unknown time, τ.

The fall height, h, and true time of fall, T, are related by

$h = \dfrac{1}{2} gT^2$.

Explain how to determine g and τ.

Carefully designed graphs are useful in trapping this kind of uncertainty.

If the inverse square law is being investigated, the count rate, C, is expected to be given by: $C = \dfrac{k}{d^2}$, where $d = x + \varepsilon$.

The equation can be rearranged to give $\dfrac{1}{\sqrt{C}} = \dfrac{x + \varepsilon}{\sqrt{k}}$, so a graph of $\dfrac{1}{\sqrt{C}}$ against x should be a straight line, if the inverse square law is obeyed. (See Knowledge check 3.13.) When selecting the graph to plot, the variable with the systematic uncertainty should be linear (i.e. in this case we plot x not x^2). Knowledge check 3.14 gives another example.

3.5.4 Experimental plans

Your report should be succinct but sufficiently detailed to allow another A level Physics student to follow it. Whether you use a numbered list of steps or bullet points is a matter of taste but the report should be clearly sequential. Details of standard items of equipment and their operation need not be given, other than to explain the precautions to take into account random and systematic uncertainties (e.g. measuring the diameter of a wire at several points and in directions at $90°$).

Your plan should include:

- the variables involved, including the controlled variables
- the method of varying the independent variable, if appropriate, and maintaining the controlled variables at constant values, if this is not obvious
- a clear sequence of steps for obtaining values of the variables
- the instruments used for making the measurements, including their resolution
- whether repeats will be taken and how this is planned for
- the method for analysing the results, including theoretical justification.

The method of analysis should be clear. For example, a statement of the graph to be drawn (such as velocity^2 against distance), how this relates to any algebraic equation and how features of the graph (linearity, gradient, intercept) will be used.

Example of stating the method of analysis

When determining the Young modulus, E, of the material of a wire, the diameter, D, and initial length, ℓ_0, are measured and a graph of force, F against extension, $\Delta\ell$, plotted.

$$F = \frac{\pi D^2 E}{4\ell_0} \, \Delta\ell, \text{ so the gradient of the graph is } \frac{\pi D^2 E}{4\ell_0}.$$

The gradient of the graph is measured and E calculated from $E = \dfrac{4\ell_0}{\pi D^2} \times \text{gradient}$.

Not all experiments involve testing the relationship between variables; the following example is one which does not.

Example

(a) Write a plan for determining the density of the material of a steel sphere of approximate diameter 3 cm.

(b) Give details of how the percentage uncertainty in the value of density will be estimated.

Answer

(a) The plan:

1. Determine the mass, M, of the steel sphere using an electronic balance with resolution ± 0.01 g.

2. Measure the diameter, d, of the sphere along five different diameters using a pair of digital callipers with resolution 0.01 mm.

3. Calculate the density, ρ, using $\rho = \dfrac{M}{\frac{4}{3}\pi r^3}$, where r is radius of the sphere, calculated as half the mean value of d.

(b) The absolute uncertainty in M is taken to be the resolution of the balance; the absolute uncertainty in the diameter is half the spread of the results for d. The percentage uncertainty in density, p_p, is calculated using $p_p = p_M + 3p_d$.

Knowledge check 3.15

What would be the absolute uncertainty in d, if all the values of diameter in the example were the same?

Knowledge check 3.16

In the example, the following values were obtained:

$M = 68.49$ g; $d = (25.46 \pm 0.02)$ mm

What values of density and uncertainty should be reported?

3.5.5 Evaluating experimental work

The final stage of an investigation involves looking at the outcome and determining to what extent it has been successful. This may involve deciding:

- whether, and to what extent, the results are consistent with the theoretical predictions
- whether the uncertainty in the final measured result is acceptable
- what changes could be made to the procedure to improve the outcome
- how the investigation could be developed to test further aspects.

Test yourself 3

1. An experiment is done to investigate the variation of resistance with length of cylindrical metal wire. The resistance is measured directly using a multimeter and length using a metre rule.

    (a) State (i) the independent variable, (ii) the dependent variable, (iii) two controlled variables.
    (b) Give one potential systematic uncertainty and explain how it can be taken into account.

2. You are provided with several metal spheres with different diameters. Plan an experiment to investigate whether they are made of the same material. You are provided with an electronic balance and a pair of digital callipers.

3. Bethan is asked to find the volume of a glass slide. She is given a set of 10 identical glass slides, a metre rule (with a mm scale) and a pair of digital callipers (±0.01 mm) which measures up to 15 cm.

    She uses the following method:

    - Put all 10 slides end to end along the metre rule and measure the total length.
    - Put all 10 sides side to side along the metre rule and measure the total width.
    - Stack the 10 slides and measure the height using the callipers.
    - Use the resolution of the instruments to estimate the uncertainty.

Her results were: Total length = 75.3 cm; Total width = 24.6 cm; Total thickness = 1.113 cm

(a) Which measurement had the greatest percentage uncertainty? Justify your answer.
(b) Calculate the volume of a single slide together with its absolute uncertainty.
(c) Eirian suggests that the uncertainty would be reduced by using the callipers to measure the total width of 5 slides. Evaluate this suggestion.

4. Paul is given a squash ball and an inclined plane. He decides to investigate the acceleration of the ball rolling down the plane – to show that it is constant and to measure the acceleration. His only other equipment is a stopwatch and a metre rule. He decides to time the ball for a series of different distances, to use the equation $x = ut + \frac{1}{2}at^2$ and to plot a suitable graph.

(a) Write a plan for this investigation, including how the results will be analysed.
(b) Paul reads that the acceleration, a, of a ball on a plane with a slope, θ, to the horizontal is given by $a = g \sin \theta$. He sets the plane with $\theta = (8.0 \pm 0.5)°$ and determines the time taken for the ball to accelerate from rest for a distance of (1.000 ± 0.001) m to be (1.50 ± 0.05) s. Evaluate whether Paul's results are consistent with the equation.

5. Some students measure the speed of a train for a period of 10 seconds. The following is a plot of their results with error bars.

(a) Explain how the results are consistent with the equation $v = u + at$, where a is the constant acceleration.
(b) Use the plotted error bars to determine the initial velocity, u, and the acceleration, a, together with their absolute uncertainties.
(c) Another student makes a measurement, not using the data above, and determines that the displacement of the train after 10 seconds is 95 ± 3 m. Evaluate whether this is consistent with the plotted data.

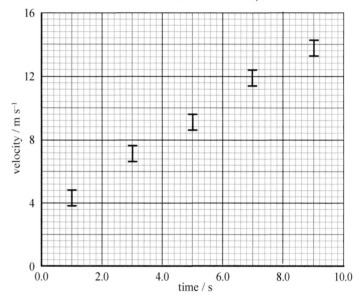

6. Jewellers often polish items of jewellery by putting them into a tumbler – a rotating horizontal cylinder – together with *tumbling shot* and water, for about 24 hours. The tumbling shot consists of small (a few mm) pieces of polished stainless steel in a variety of shapes. A group of students used the following method to measure the density of the shot:

- Place a measuring cylinder containing water on to an electronic balance and note the measuring cylinder and balance readings.
- Add a quantity of tumbling shot to the measuring cylinder and note the measuring cylinder and balance readings.
- Calculate the density of the shot using $\text{density} = \dfrac{\text{mass}}{\text{volume}}$.
- Use the resolution of the measuring cylinder and the balance to estimate the uncertainty.

Results

Instrument	Initial reading	Final reading	Resolution
Measuring cylinder	23.0 cm³	48.0 cm³	0.5 cm³
Balance	82.63 g	262.88 g	0.01 g

(a) Explaining your reasoning, use the results to find a value for the density of the shot along with its absolute uncertainty.
(b) The students noticed that their result was lower than the expected value of 7.9 g cm^{-3}. One of them suggested that air bubbles could have been trapped between the pieces of shot. Evaluate whether this could account for the lower value and suggest how this difficulty could be overcome.

Chapter 4

Mathematical skills

In many ways, mathematics is the language of physics. Physicists develop their models and theories in mathematical terms; they make quantitative predictions about the behaviour of the world; they subject their experiments to mathematical analysis; even the extent to which physicists and engineers are uncertain is expressed mathematically.

As with Chapter 3, this short section of the book is not meant as a substitute for a detailed mathematical course. It provides, rather, a ready reference to the mathematical techniques you will employ in AS Physics: use of indices and standard form; algebraic manipulation; angles and trig ratios; graphs.

The topic of Graphs is split between this chapter and Chapter 3. Graphs are sometimes plotted directly from equations; at other times experimental results are plotted, to which best-fit or theoretical lines are added. The final section on graphs in this chapter deals mainly (but not exclusively) with aspects of graphs you will meet in non-experimental work.

Content

4.1 Introduction

Chapters 1 and 2 include mathematical treatment of physics ideas. Basic GCSE level mathematics is assumed. This chapter has two purposes:

- It serves as a quick reference for the necessary skills.
- It introduces some new material, e.g. radians and the use of the 'Δ' symbol, which is more conveniently dealt with away from the main text.

4.2 Arithmetic and numerical computation

4.2.1 Indices

(a) Basic rules

- $a^n = a \times a \times a \ldots$ (n times) – where n is a positive integer, i.e. 1, 2, 3 … e.g. $5^3 = 5 \times 5 \times 5 = 125$.
- $a^{-n} = \dfrac{1}{a^n}$, e.g. $10^{-2} = \dfrac{1}{10^2} = \dfrac{1}{10 \times 10} = \dfrac{1}{100} = 0.01$
- $a^{\frac{1}{n}} = \sqrt[n]{a}$, e.g. $16^{\frac{1}{2}} = \sqrt{16} = 4$ and $81^{\frac{1}{4}} = \sqrt[4]{81} = 3$
- $a^x \times a^y = a^{(x+y)}$, e.g. $10^2 \times 10^3 = 10^5$
- $\dfrac{a^x}{a^y} = a^x \times a^{-y} = a^{(x-y)}$, e.g. $\dfrac{4^2}{4^{0.5}} = 4^{1.5}$
- $(a^x)^y = a^{xy}$, e.g. $(5^4)^{0.5} = 5^{4 \times 0.5} = 5^2$

(b) Indices and units

At GCSE we wrote the units of velocity and acceleration as **m/s** and **m/s²** respectively. At A level and higher, the units are written **m s⁻¹** and **m s⁻²**.

When multiplying or dividing two quantities, the units are treated in the same way; e.g. a car accelerates at 1.5 m s⁻² for a time of 10 s. The increase in velocity is 15 m s⁻¹. See the Study point for the working.

(c) Standard form

There are two ways of expressing the values of large or small quantities: **standard form** and **SI multipliers**. You need to deal with both and to convert between them. Dealing with standard form first, look at the following examples:

- The quantity $350\,000$ m s⁻¹ is written as 3.5×10^5 m s⁻¹ in standard form.
- The quantity $0.000005\,6$ A is written as 5.6×10^{-6} A in standard form.

Note that to qualify as standard form the number before the multiplication sign is between **1.00** and **9.99**. This doesn't mean that, e.g. 0.35×10^6 m s⁻¹ is not a valid way of writing the same speed as in the first example – it's just not standard form.

In order to express a number in standard form, we make use of the fact that

$10^1 = 10;\ 10^2 = 100;\ 10^3 = 1\,000;\ \ldots\ 10^6 = 1\,000\,000$; etc.

$10^{-1} = 0.1;\ 10^{-2} = 0.01;\ 10^{-3} = 0.001;\ \ldots\ 10^{-6} = 0.000001$; etc.

4.1 Knowledge check

What are

(a) $y^2 \times y$? (b) $y^{-2} \times y$? (c) $y\,(y^2 + 2a)$

Study point

For acceleration of 1.5 m s⁻² for 10 s:

Vel inc. $= 1.5$ m s⁻² $\times$ 10 s
$= 1.5 \times 10$ (m s⁻² $\times$ s)
$= 15$ m s⁻¹.

Examples

$5 \times 10^{-3} = 5 \times 0.001$
$= 0.005$
and $5.57 \times 10^{-3} = 0.00557$
$5 \times 10^3 = 5 \times 1000$
$= 5000$
and $5.57 \times 10^3 = 5570$

So, multiplying a number by 10^n involves moving the decimal point n places to the right.

Hence $2.536 \times 10^2 = 253.6$

Similarly, multiplying a number by 10^{-n} involves moving the decimal point n places to the left.

Hence $9372.8 \times 10^{-3} = 9.3728$

Example
Convert the following numbers to standard form: (a) **820000** (b) **0.00365**

Answer
(a) The answer will be 8.2×10^n. To find n, imagine moving the decimal point in **8.2** to the right, until we get **820000**.

8.200000, that's 5 hops, so $n = 5$, i.e. $820000 = 8.2 \times 10^5$.

(b) The answer will be 3.65×10^{-n}. To find n, imagine moving the decimal point in **3.65** to the left, until we get **0.003 65**.

00003.65, that's 3 hops, so $n = 3$, i.e. $0.00365 = 3.65 \times 10^{-3}$.

One big advantage of using standard form is that it gives clarity in the number of significant figures.

Consider, for example, the distance **5100 m**. Without other information, the number of significant figures could be two, three or four. But:

5.1×10^3 m has two significant figures
5.10×10^3 m has three significant figures
5.100×10^3 m has four significant figures.

See Calculator check.

(d) SI multipliers
The following table of SI multipliers is given in the Data Booklet.

Multiple	Prefix	Symbol	Multiple	Prefix	Symbol
10^{-18}	atto	a	10^3	kilo	k
10^{-15}	femto	f	10^6	mega	M
10^{-12}	pico	p	10^9	giga	G
10^{-9}	nano	n	10^{12}	tera	T
10^{-6}	micro	μ	10^{15}	peta	P
10^{-3}	milli	m	10^{18}	exa	E
10^{-2}	centi	c	10^{21}	zetta	Z

Table 4.1 SI multipliers

The multiplier is essentially an abbreviated way of writing $\times 10^n$. For example 53 μA: From the table, μ means $\times 10^{-6}$, so 53 μA is equivalent to 53×10^{-6} A, which is 5.3×10^{-5} A in standard form – but see Top tip. Usually the number before the multiplier is in the range of 1 to 999.

Calculator check

To check that you are entering numbers correctly into your calculator, try:

$(5 \times 10^{-6}) \times (3 \times 10^7)$.

If your answer is 150 you are doing it right.

Top tip

Enter 53 μA into your calculator by keying it in as

53×10^{-6}

You are less likely to make a mistake than if you mentally convert it to 5.3×10^{-5} first.

Knowledge check 4.2

A resistor has a voltage of 300 mV across it and passes a current of 50 μA. Calculate its resistance.

4.2.2 Fractions, ratios and percentages

The fraction $\frac{a}{b}$ means a divided by b. The ratio of a to b is also $\frac{a}{b}$. If $a = 4$ and $b = 10$, the ratio of a to b is $\frac{4}{10} = \frac{2}{5} = 0.4 = 40\%$. These are just different ways of writing the same thing.

Top tip

If $a = 3$ and $b = 25$ the ratio of a to b is $\frac{3}{25} = 0.12 = 12\%$. It can be written as 3 : 25 but this is best avoided.

We can express a fraction as a **percentage** by multiplying by 100; so

$\frac{7}{20} = \frac{7}{20} \times 100\% = 35\%$. You should know the following percentage equivalents:

$\frac{1}{10} = 10\%$; $\frac{2}{10} = 20\%$ ... $\frac{9}{10} = 90\%$; $1 = 100\%$; $\frac{1}{4} = 25\%$; $\frac{1}{3} = 33.3\%$; $\frac{1}{2} = 50\%$;

$\frac{3}{4} = 75\%$; $2 = 200\%$; $2.5 = 250\%$ ...

4.2.3 Expressing angles in radians

For most purposes, physicists express angles in the everyday unit of *degree* (°). You will be familiar with the use of angles expressed in degrees in geometry and trigonometry. For small angles, the subdivisions of minutes (′) and second (″) are used. Like the subdivision of the hour, there are **60** minutes in a degree and **60** seconds in a minute. For clarity in written text, scientists often write these units as arcmin and arcsec respectively. Astronomers in particular often express 'angular separation between close stars' in milli arcsec.

Top tip

Remember the following degree / radian equivalents:

$30° = \frac{\pi}{6}$	$135° = \frac{3}{4}\pi$
$45° = \frac{\pi}{4}$	$180° = \pi$
$60° = \frac{\pi}{3}$	$270° = \frac{3}{2}\pi$
$90° = \frac{\pi}{2}$	$360° = 2\pi$
$120° = \frac{2}{3}\pi$	$540° = 3\pi$

Your calculator probably has a special button ⬛ for converting from angles expressed in degrees, minutes and seconds into decimals of degrees.

For example, to enter 53° 27′ 6″, key in

53 ⬛ 27 ⬛ 36 ⬛

The display should be **53.46**

For many aspects of motion, especially rotations and oscillations, it is mathematically much more convenient to use a different measure of angle – the radian (rad). This is defined in the following way.

Angles in radians

Fig. 4.1 shows an angle, θ, between two radii of a circle and the arc between them.

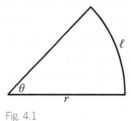

Fig. 4.1

θ, in radians, is given by $\qquad \theta = \frac{\ell}{r}$

Notice that the calculated value of θ is independent of ℓ and r because if the radius is increased, the arc length, ℓ, will increase in the same proportion and the ratio will be unchanged.

To understand how to convert between degrees and radians, let θ be a complete revolution, i.e. $360°$. In this case the arc length, ℓ, would be the circumference of a circle, i.e. $2\pi r$.

Hence $\theta\,/\text{rad} = \dfrac{2\pi r}{r} = 2\pi$

Hence $360° = 2\pi\,\text{rad}$, $180° = \pi\,\text{rad}$, etc.

So, the conversion is: $\theta\,/\text{rad} = \theta\,/° \times \dfrac{\pi}{180}$

Example

Convert

(a) $120°$ to radians

(b) 15.0 rad to degrees.

Answer

(a) $120° = 120 \times \dfrac{\pi}{180}\ \text{rad} = 2.1\ \text{rad}$ (2 s.f.)

(b) $15.0\ \text{rad} = 15.0 \times \dfrac{180}{\pi}° = 859°$ (3 s.f.)

Knowledge check **4.3**

Convert:

(a) 1.0 rad to degrees

(b) $37°$ to radians.

Top tip

Angles in radians can be left as multiples of π, e.g. $\theta = 2.5\pi$. It is then assumed that the angle is in radians.

Fig. 4.2 illustrates the **small angle approximation** which is only valid if angles are expressed in radians.

The red line is the arc of a circle with radius r.

By definition $\sin\theta = \dfrac{f}{r}$; θ (in rad) $= \dfrac{g}{r}$; $\tan\theta = \dfrac{h}{r}$

The three lengths, f, g and h, are very close, with $f < g < h$ and, as $\theta \longrightarrow 0$ the ratios $\dfrac{f}{g}$ and $\dfrac{g}{h} \longrightarrow 1$. So for small angles we can write $\sin\theta \approx \theta \approx \tan\theta$. For many purposes, angles < 0.1 rad [$\sim 6°$] can be considered small.

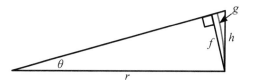

Fig. 4.2 Small angles

4.3 Algebra

4.3.1 Symbols

(a) Less than (<) and greater than (>)

$a < b$ means 'a is less than b'; similarly, $x > y$ means 'x is greater than y'.

Examples: $10 > 5$; $5 \times 10^6 < 2 \times 10^7$. Care should be taken with negative numbers, e.g. $-10 > -20$.

With care, these can be used in the same way as the '=' sign (meaning 'is equal to'), e.g. $p > q$ then: $p + x > q + x$ and $10p > 10q$ but $-5p < -5q$ (!)

Top tip

You should know the meaning of the following symbols and how to use them:

$=, <, \ll, \gg, >, \propto, \approx, \Delta$

Study point

If c is the speed of light in a vacuum, v_a the speed of sound in air and v_c the speed of sound in carbon dioxide:

$v_a > v_c$ and $c \gg v_a$

(b) Much less than (<<) and much greater than (>>)

These can be used in the same way as < and >. More often they are just used to express that there is a big difference in value without undertaking any algebra.

(c) Direct and inverse proportion

$y \propto x$ ('y is directly proportional to x', or just 'y is proportional to x') means that the ratio $\frac{y}{x}$ is constant, i.e. if x is multiplied by any number (e.g. 2, 3 or π), y is multiplied by the same number.

Sometimes in Physics we find that when one variable is doubled, the other one halves. The product, xy, of the variables is constant. In these cases we say that 'y is inversely proportional to x' and write it

$$y \propto \frac{1}{x}.$$

Similarly, y can be proportional to x^2 ($y \propto x^2$), or y can be inversely proportional to x^2 ($y \propto \frac{1}{x^2}$). See the box for examples.

Top tip

If $y \propto x$, a graph of y against x is a straight line through the origin. If we re-write $y \propto x$ as $y = kx$, then k is the gradient of the graph.

Study point

Proportion relationships can also be written as follows:

$y \propto x \quad \longrightarrow \quad y = kx$

$y \propto \frac{1}{x} \quad \longrightarrow \quad y = \frac{k}{x}$

where k is a constant.

Examples of proportion

Hooke's Law	$F \propto x$
Waves	$\lambda \propto \frac{1}{f}$
Kinetic energy	$E_k \propto v^2$
Gravity	$F \propto \frac{1}{r^2}$

4.4 Knowledge check

Identify the constant k in each of the examples above and give its unit.

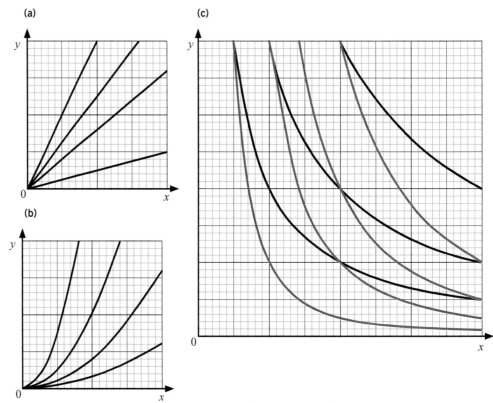

Fig. 4.3 Proportion graphs: (a) $y \propto x$, (b) $y \propto x^2$, (c) $y \propto \frac{1}{x}$ (black) and $y \propto \frac{1}{x^2}$ (red)

(d) Delta (Δ)

Δ means 'change in', i.e. Δv means 'change in velocity'. If a length, x, changes from $x_1 = 2.5$ cm to $x_2 = 7.6$ cm:

$$\Delta x = x_2 - x_1 = 7.6 \text{ cm} - 2.5 \text{ cm} = 5.1 \text{ cm}$$

We always subtract the first value from the second value. So, if $x_2 < x_1$, then Δx will be negative ($\Delta x < 0$).

Study point

The mean resultant force, F, on an object is defined as the rate of change of momentum, p of the object, i.e. $F = \frac{\Delta p}{\Delta t}$

4.3.2 Manipulating equations

To find the value of a quantity, an equation must be manipulated to make the quantity into the subject, e.g. find a if $10 = 3 + 2a^2$.

In every part of the manipulation, we must perform the same arithmetic operation on the two sides of the equation. Examples:

- Add or subtract the same quantity
- Multiply or divide by the same quantity
- Square both sides or take the square root of both sides.

In our example:

- Subtract 3 from both sides $\longrightarrow$ $7 = 2a^2$
- Divide both sides by 2 $\longrightarrow$ $3.5 = a^2$
- Take the square root of both sides $\longrightarrow$ $a = \pm\sqrt{3.5} = \pm1.87$ (3 s.f.)

Top tip

When solving an equation, e.g. finding x from $v^2 = u^2 + 2ax$, you might find it easier to substitute the numbers into the equation before manipulating the equation.

Knowledge check 4.5

Find x if $7x + 16 = 49$.

4.3.3 Evaluating expressions to solve equations

When evaluating expressions, such as $16 + (6 \times 5^2 - 29)^{0.5}$ the sequence of actions is summarised by the mnemonic BODMAS – bracket, **orders**, division, multiplication, addition, subtraction. With the above expression this works as follows:

- to evaluate the *bracket*, work through the ODMAS sequence within the bracket:
$5^2 = 25; 6 \times 25 = 150; 150 - 29 = 121$

- order: $121^{0.5} = \sqrt{121} = 11$

- addition: $16 + 11 = 27$ (answer)

Another example: $\frac{3}{8} + 4.2$. The $\frac{3}{8}$ term means 3 divided by 8, so start with that:

3 divided by 8 is 0.375. Then add 4.2 to give 4.575.

Study point

The word 'orders' in BODMAS means powers or indices, e.g. 2^3, $\sqrt{40} = 40^{0.5}$.

Knowledge check 4.6

Evaluate the following:

(a) $8 - (3 + \sqrt{2})^2$, (b) $\frac{24}{8 + 2^2}$, (c) $\frac{1}{3} + \frac{2}{5}$

4.3.4 Quadratic equations

If an unknown quantity, x, satisfies the equation, $ax^2 + bx + c = 0$, where a, b and c are constants, the solutions are:

$$x = \frac{-b \pm \sqrt{b^2 - 4ac}}{2a}$$

In A level Physics, the most common type of question which requires this formula arises in finding an unknown time from $x = ut + \frac{1}{2}at^2$. Here, t is the unknown quantity and the constants are x, u and $\frac{1}{2}a$.

Knowledge check 4.7

Evaluate the following:

If $x = 24$ m, $u = 4$ m s^{-1} and $a = 3$ m s^{-2}.

(a) From $x = ut + \frac{1}{2}at^2$, show that $1.5t^2 + 4t - 24 = 0$.

(b) Solve the equation using the quadratic formula.

4.4 Geometry and trigonometry

4.4.1 Angles in geometrical figures

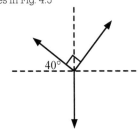

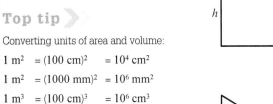

Fig. 4.5

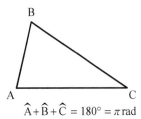

$$\hat{A} + \hat{B} + \hat{C} = 180° = \pi\,\text{rad}$$

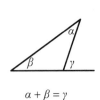

$$\alpha + \beta = \gamma$$

$$\theta + \phi = 180° = \pi\,\text{rad}$$

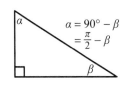

$$\alpha = 90° - \beta = \frac{\pi}{2} - \beta$$

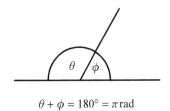

$$\alpha + \beta + \gamma + \delta + \varepsilon = 360° = 2\pi\,\text{rad}$$

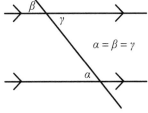

$$\alpha = \beta = \gamma$$

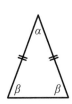

Fig. 4.4 Angles in geometrical figures

4.4.2 Areas and volumes of geometrical figures

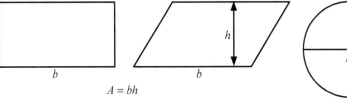

$$A = bh$$

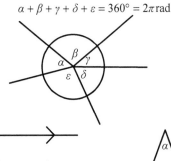

$$A = \pi r^2 = \frac{\pi D^2}{4}$$

Circumference $= \pi D = 2\pi r$

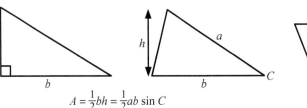

$$A = \frac{1}{2}bh = \frac{1}{2}ab \sin C$$

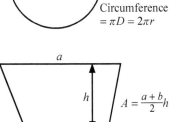

$$A = \frac{a+b}{2}h$$

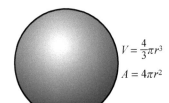

$$V = \frac{4}{3}\pi r^3$$

$$A = 4\pi r^2$$

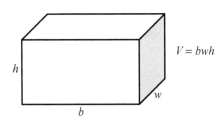

$$V = bwh$$

Fig. 4.6 Areas and volumes

The volumes of wires can be found by multiplying the cross-sectional area by the length.

4.4.3 Right-angled triangles

Pythagoras' theorem: $a^2 + b^2 = c^2$,

where c is the long side, the hypotenuse, of the right-angled triangle.

e.g. if $a = 8$ cm and $b = 15$ cm

$c = \sqrt{8^2 + 15^2} = \sqrt{64 + 225} + \sqrt{289} = 17$ cm.

A well known *Pythagorean triple* is 3, 4, 5. Others are 5, 12, 13 and 5, 24, 25.

For angles from $0 - 90°$ [$0 - \frac{\pi}{2}$ rad] the trig ratios, sine, cosine and tangent, are defined as follows:

Sine: $\sin \theta = \dfrac{\text{opposite}}{\text{hypotenuse}} = \dfrac{a}{c}$

Cosine: $\cos \theta = \dfrac{\text{adjacent}}{\text{hypotenuse}} = \dfrac{b}{c}$

Tangent: $\tan \theta = \dfrac{\text{opposite}}{\text{adjacent}} = \dfrac{a}{b}$

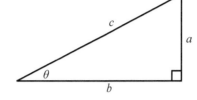
Fig. 4.9 Trig ratios

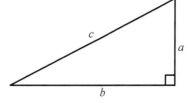
Fig. 4.8 Pythagoras theorem

4.4.4 Trig ratios for angles outside the range 0–90°

Fig. 4.10 shows the sine and cosine functions for angles which cannot be found in a right-angled triangle, i.e. $> 90°$ ($\frac{\pi}{2}$ rad) and negative values.

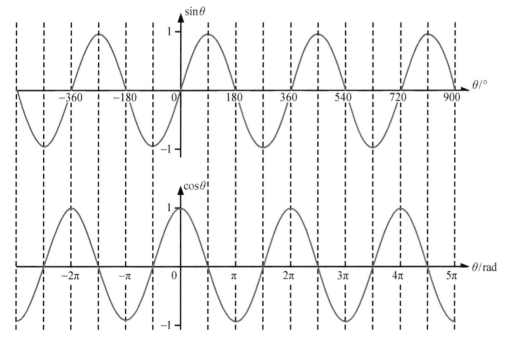

Fig. 4.10 Sine and cosine graphs

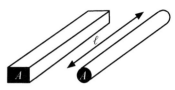

A is called the *cross-sectional area*, c.s.a.

The volume, $V = A\ell$

For the cylinder, $V = \pi r^2 \ell$

Fig. 4.7

Top tip

Know the following values of trig ratios:

θ /°	$\sin \theta$	$\cos \theta$	$\tan \theta$
0	0	1	0
30	$\dfrac{1}{2}$	$\dfrac{\sqrt{3}}{2}$	$\dfrac{1}{\sqrt{3}}$
45	$\dfrac{1}{\sqrt{2}}$	$\dfrac{1}{\sqrt{2}}$	1
60	$\dfrac{\sqrt{3}}{2}$	$\dfrac{1}{2}$	$\sqrt{3}$
90	1	0	∞

Study point

$\cos \theta = \dfrac{x}{v}$

$\therefore x = v \cos \theta$

Similarly,
$y = v \sin \theta$.

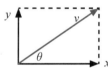

Study point

We have used different units in the $\sin \theta$ and $\cos \theta$ graphs. This does not matter because $180° = \pi$ radians, etc. The labels could be swapped or both could use the same.

Top tip

Learn the graphs and use them when finding angles when you know the value of $\sin \theta$ or $\cos \theta$, e.g. if $\cos \theta = 0.5$, the calculator will give $\theta = 60°$. From the graph, other possible values of θ are: $-300°$, $-60°$, $300°$, $420°$, …

Knowledge check 4.10

What are the possible values of θ if $\sin \theta = 0.5$? Give your answer in ° and in **rad**.

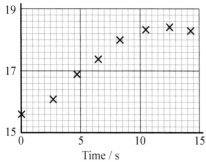

Fig. 4.11 Good practice in graphs

4.5 Graphs

4.5.1 Graph preparation and point plotting

Fig. 4.11 illustrates the following rules which you should always follow when graph plotting:

1. The axes are both labelled clearly with the quantity being plotted.
2. The scales increase in equal steps and don't include factors which make the scales difficult to read, e.g. multiples of 3 or 7.
3. The scales are chosen so that the points occupy at least half the grid in both directions.
4. The units of the quantities plotted are included: the standard method to adopt is <quantity> / <unit>. If the quantity is raised to a power, e.g. v^2 then the label should be $(v \,/\, \text{m s}^{-1})^2$.
5. The points are clearly plotted, with the centre of the cross representing the position of the point.

'Plotting a graph' requires a suitable line to be drawn in addition to setting up the axes, scales and plotting the points. Chapter 3 deals with handling experimental data including decisions on drawing the most appropriate line, e.g. best-fit straight line or best-fit curve.

4.5.2 Determination of rate of change from a graph

The mean rate of change of a quantity y with respect to x is defined by:

$$\text{Mean rate of change} = \frac{\Delta y}{\Delta x}.$$

For example, if the gravitational potential energy of an object increases by $500\ \text{J}$ when it is raised by $20\ \text{m}$, the mean rate of change of potential energy with respect to height is $25\ \text{J m}^{-1}$.

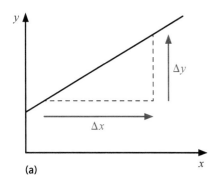

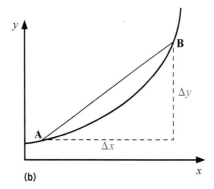

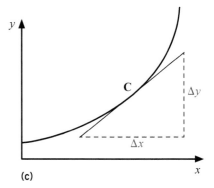

Fig. 4.12 Calculating the rate of change

The relationships between the rate of change and the graph are illustrated in Fig. 4.12. For a straight line variation of y with x (Fig. 4.12 (a)), the rate of change of y with respect to x is the gradient of the graph.

For a curved graph, the mean rate of change between **A** and **B** (Fig. 4.12 (b)) is the gradient of the chord joining **A** and **B**. The instantaneous rate of change of y with respect to x at **C** (Fig. 4.12 (c)) is the gradient of the *tangent* to the graph at **C**. In principle it doesn't matter how large Δx is in finding the slopes but the larger Δx is the more accurately you will be able to determine the rate of change.

4.5.3 Determination of the 'area' under a graph

Graphs presented in A level Physics are often idealised and drawn as a series of linear sections. In this case the method of calculating the area between the graph and the horizontal axis is to divide it into triangles and rectangles (or trapeziums). For instance, consider the velocity–time graph in Fig. 4.13(a). To find the total displacement between 0 and 15 seconds, we could divide the area into the three sections as shown in Fig. 4.13(b).

>> **Study point**

We should really talk about the 'area' between the graph and the horizontal axis – and if the graph is below the axis, the area is negative.

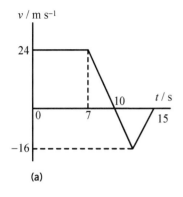

(a)

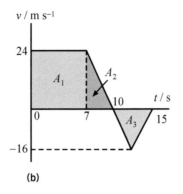

(b)

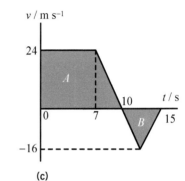

(c)

Fig. 4.13 finding the area under a v–t graph

Area A_1 (rectangle) $= 7\ \text{s} \times 24\ \text{m s}^{-1} = 168\ \text{m}$

Area A_2 (triangle) $= \frac{1}{2} \times 3\ \text{s} \times 24\ \text{m s}^{-1} = 36\ \text{m}$

Area A_3 (triangle) $= \frac{1}{2} \times 5\ \text{s} \times (-16)\ \text{m s}^{-1} = -40\ \text{m}$

∴ Total displacement $= (168 + 36 - 40)\ \text{m} = 164\ \text{m}$

Alternatively, we could divide up the area into a trapezium and a triangle as in Fig. 4.13(c). Do Knowledge check 4.12 to show we get the same answer.

If the graph is curved we estimate the area underneath it using one of three methods – which to use is up to you! Consider the force–displacement curve in Fig. 4.14. We shall calculate the total work done by estimating the area under the curve.

Knowledge check 4.12

Calculate the area of trapezium A in Fig. 4.13(c) and hence show that this result agrees with the calculation in the main text.

1. Square counting: The area of each 1 cm square is $5.0\ \text{N} \times 0.1\ \text{m} = 0.5\ \text{J}$. Treating $< \frac{1}{2}$ a square as 0 and $> \frac{1}{2}$ a square as 1, there are 12 (or 13) such squares ⟶ $12 \times 0.5 = 6.0\ \text{J}$ (13 squares gives 6.5 J). You should confirm this by identifying all the squares which are $> \frac{1}{2}$

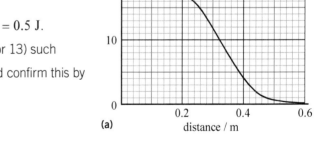

(a) distance / m

2. Divide the graph into equal trapeziums, as in Fig. 4.14(b), find the area of each and add to find the total area. For example

$A_3 = \frac{1}{2}(17 + 11.5) \times 0.1 = 1.425\ \text{J}$

See Knowledge check 4.14.

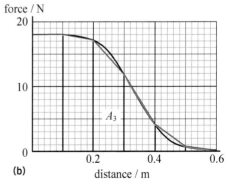

(b) distance / m

Fig. 4.14 (a)–(b) Finding the area under a curved graph

4.13 **Knowledge check**

What work do you get if you count up the 2 mm squares in Fig. 4.14 (a)? Each is worth $1.0 \text{ N} \times 0.02 \text{ m} = 0.02 \text{ J}$

4.14 **Knowledge check**

Find the area of all the trapeziums in Fig. 4.14 (b) and show that the work done is approximately 6.0 J.

3. Draw a straight line which cuts the curve into two so that the area above the line is (by eye) equal to the area below the line. In Fig. 4.14 (c):

$$A_1 = 18 \times 0.08 = 1.44 \text{ J}$$

$$A_2 = \frac{1}{2} \times 18 \times 0.52 = 4.68 \text{ J}$$

$$\therefore \text{ Area } = 6.1 \text{ J (2 s.f.)}$$

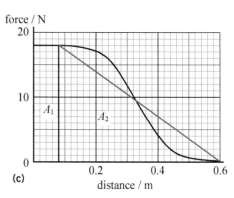

Fig. 4.14 (c) Finding the area under a curved graph

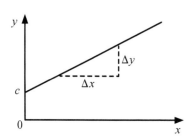

Fig.4.15 Linear graph

4.5.4 Linear graphs

(a) Equation of a straight-line graph

The graph in Fig. 4.15 represents a linear relationship, i.e. it is a straight line. The equation of the graph is:

$$y = mx + c, \text{ where } m = \frac{\Delta y}{\Delta x}.$$

If the graph passes through the origin (0,0), i.e. $c = 0$, the equation becomes $y = mx$. In this case we say that y is directly proportional to x, $y \propto x$. This is also covered in Section 4.3.1(c).

The relationships $v = u + at$ and $V = E - Ir$ are examples of linear relationships.

- A graph of v against t for constant acceleration is a straight line with gradient a and intercept u on the v axis.

- A graph of V against I for a power supply with a constant internal resistance is a straight line with gradient $-r$ and intercept E on the V axis.

Chapter 3 includes a section on plotting straight line graphs from experimental data including cases where the relationship is non-linear.

(b) Find the equation from the graph

Method 1: If the horizontal axis (the x-axis) of the graph goes back to 0 as in Fig. 4.15, draw a large triangle as shown, measure Δx and Δy, calculate m, read off c from the intercept on the axis.

Method 2: If the data points are all a long way from the origin, as in Fig. 4.16, (e.g. if the values of x are in the range $120-150$) then:

1. Locate two well-separated points on the graph (x_1, y_1) and (x_2, y_2).

2. Calculate the gradient from $m = \dfrac{y_2 - y_1}{x_2 - x_1}$.

3. Substitute m and x_1 and y_1 (or x_2 and y_2) into $y = mx + c$ to calculate c.

Method 3: As method 2 but after identifying the points, write the equation as:

$$\frac{y - y_1}{x - x_1} = \frac{y_2 - y_1}{x_2 - x_1}.$$

Insert the values and rearrange into the form $y = mx + c$.

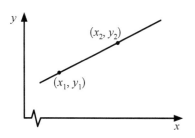

Fig.4.16 Graph remote from the origin

4.15 **Knowledge check**

If $v = 55 \text{ m s}^{-1}$ at $t = 25.0 \text{ s}$ and 34 m s^{-1} at 27.3 s, find the relationship between v and t in the form $v = u + at$. State the values of u and a.

Test yourself 4

It is not possible in this textbook to give exercises covering the many different mathematical skills you need to demonstrate in physics. If you have difficulty with any of these questions, it is suggested that you consult a mathematics textbook and your physics teacher for further advice.

1. Evaluate the following **without** using your calculator.

(a) $3^2 \times 2^3$ (b) $5^3 + 3^4$ (c) $3^2 \times 2^{-3}$ (d) $100^{1/2} \times 5^{-2}$ (e) $\dfrac{4^{16}}{4^{14}}$ (f) $(3 \times 2)^3$

(g) $\sqrt[3]{8^2}$ (h) 20^4 (i) $\sqrt{25^3}$ (j) $(10^3)^2$ (k) $10^2 \times 10^3$ (l) $10^2 + 10^3$

(m) 0.5^{-3} (n) 0.1^3 (o) $\dfrac{0.4^3}{0.1^3}$ (p) $2 \times (1 + 4^{-1})$ (q) $\dfrac{3.0 \times 10^8}{5.0 \times 10^{14}}$ (answer in standard form)

(r) $\sqrt{15^2 - 8 \times 13}$ (s) $6.6 \times 10^{-34} \times 5.0 \times 10^{15}$ (t) $\dfrac{6.6 \times 10^{-34}}{3.3 \times 10^{-7}}$ (u) $\dfrac{6.6 \times 10^{-34} \times 5.0 \times 10^{14}}{3.0 \times 10^8}$

2. Use your calculator to evaluate the following:

(a) $12^{0.6}$ (b) $5^{0.5} + 5^{1.2}$ (c) $0.5^{-2.3}$ (d) $\dfrac{\sqrt{8} - 3^2}{\sqrt[3]{10}}$ (e) $\dfrac{3.0 \times 10^8}{7.2 \times 10^7}$ (f) $\sqrt{15^2 - 2 \times 3 \times 10}$

(g) $\dfrac{1.99 \times 10^{30}}{1.67 \times 10^{-27}}$ (h) $6.63 \times 10^{-34} \times 7.50 \times 10^{14} \times 4.05 \times 10^{-19}$

3. Express the following as percentages.

(a) $\dfrac{5.0}{7.0}$ (b) $\dfrac{0.15}{3.72}$ (c) 0.64 (d) 0.005 (e) $\dfrac{3.0 \times 10^8}{240 \times 0.26}$ (f) $\dfrac{5.0 \times 9.81 \times 3.5}{24 \times 8.0}$ (g) $\dfrac{1.5 \times 10^8}{2.1 \times 10^{11}}$

4. Convert the following angles from degrees to radians.

(a) $90°$ (b) $35°$ (c) $0.15°$ (d) $415°$ (e) $(2.5 \times 10^{-6})°$

5. Express the following angles from radians to degrees.

(a) 0.10 rad   (b) $\dfrac{1}{24}\pi$ rad   (c) 0.1π rad   (d) 8.3×10^{-9} rad   (e) 0.157 rad

6. Solve the following inequalities, i.e. write the answer as $x <$ or $x >$ a number.

(a) $3x > 15$ (b) $2x < -5$ (c) $3 > x + 1$ (d) $5 > 2x - 1$ (e) $x + 2 > 3 - x$

(f) $\dfrac{x}{x-1} > 3$ (g) $-3 > -x + 2$ (h) $\dfrac{6}{3x-5} > 3$ (i) $\dfrac{3}{x} - 2 > 1$ (j) $\dfrac{3}{2x} - \dfrac{2}{3x} < 1$

7. The following table contains two values of each of 6 variables, u, v, w, x, y and z.

u	v	w	x	y	z
72	5	2	25	12	0.65
8	15	18	75	4	1.95

(a) When $v = 5$, $x = 25$ and when $v = 15$, $x = 75$. Show that this is consistent with x and v being directly proportional.

(b) The data are consistent with one other variable being directly proportional to x. Identify this variable.

(c) Explain which one of the variables might be inversely proportional to x.

(d) One of the variables is proportional to x^2 and one is inversely proportional to x^2. Identify these.

(e) Explain briefly why, even if these values are accurate, more data are needed to test these suggested relationships.

(f) A student suggests that the variables v and w might be linearly related, i.e. $v = aw + b$, where a and b are constants. If this is correct, find the values of a and b, and give the value of w when $v = 10$.

8. Calculate the values of the given angles and lengths in the following figures:

(a)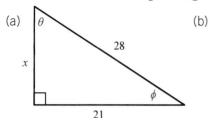

(b)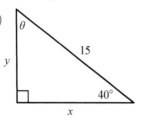

(c)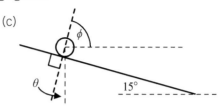

(d)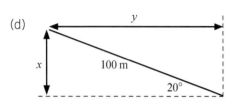

(e)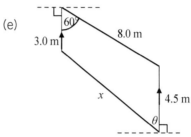

(f)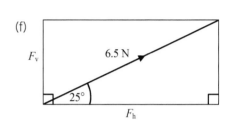

9. Find the values of the angle, θ, in the range 0–360° which satisfy the following equations: (a) $\sin \theta = 0.60$
(b) $\cos \theta = 0.80$ (c) $\sin \theta = -0.30$ (d) $\cos \theta = -0.70$ Express your answers to the nearest degree.

10. (a) $0° \le \theta \le 90°$. (i) If $\sin \theta = 0.40$, calculate $\cos \theta$ and $\tan \theta$.   (ii) If $\cos \theta = 0.30$, calculate $\sin \theta$ and $\tan \theta$.
  (iii) If $\tan \theta = 0.30$, calculate $\sin \theta$ and $\cos \theta$.

(b) $0° \le \theta \le 360°$. Use the answers in part (a) to answer the following questions.
  (i) If $\sin \theta = 0.40$, what are the possible values of $\cos \theta$ and $\tan \theta$?   (ii) If $\cos \theta = 0.30$, what are the possible values of $\sin \theta$ and $\tan \theta$?   (iii) If $\tan \theta = 0.30$, what are the possible values of $\sin \theta$ and $\cos \theta$?

11. The displacement–time graph is of a car decelerating to rest.

(a) State the feature of a displacement–time graph which is the velocity of motion.
(b) Explain at what time the car stops moving.
(c) Calculate the mean velocity of the car between 2.0 s and 5.5 s.
(d) Calculate the velocity of the car at 3.5 s.
(e) Calculate the initial velocity of the car.
(f) Explain how you could use the graph to determine whether the resultant force on the car during deceleration was constant.

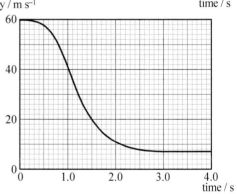

12. The velocity–time graph is for a skydiver from the time at which the parachute is deployed.

(a) Identify the features of a velocity–time graph which are the acceleration and displacement.
(b) State the terminal velocity of the diver before and after the parachute is deployed.
(c) Calculate the mean rate of change of velocity between 0 and 3.0 s.
(d) Calculate the maximum deceleration of the skydiver.
(e) Estimate the displacement of the skydiver from the moment the parachute is deployed until the lower terminal velocity is achieved.

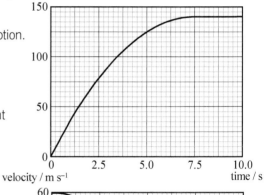

13. A graph of v against t had a gradient of 1.26 m s^{-2} and passes through the point $(v, t) = (8.6 \text{ m s}^{-1}, 2.4 \text{ s})$.

(a) Find the equation of the graph.   (b) State the initial velocity, i.e. the velocity at $t = 0$.

14. When a battery delivers a current of 12.0 mA to a circuit, the terminal p.d. is 7.53 V. When $I = 15.0$ mA, its terminal p.d. is 6.92 V.

(a) V, is related to the current, I, by an equation of the form: $V = aI + b$. Determine the values of a and b.
(b) Comparing your answer to (a) with the equation $V = E - Ir$, state the values of the emf, E, and internal resistance, r, of the supply.

Knowledge check answers

Unit 1

1.1

1.1.1 (a) $8a$ (b) $18a^2$ (c) $2\dfrac{a}{b}$ (d) $36a^2$

1.1.2 $V = \ell bh$

∴ $[V] = [\ell][b][h] = $ m m m $ = $ m^3

1.1.3 Pa s $= $ N m^{-2} s $= $ kg m s^{-2}

m^{-2} s $= $ kg m^{-1} s^{-1}

(same as the example)

1.1.4 $[x] = $ m; $[ut] = $ m s^{-1} s $= $ m;

$[\tfrac{1}{2}at^2] = $ m s^{-2}. s$^2 = $ m.

∴ **The** two terms on the RHS have the same units, so can be added together; and the LHS has the same units. ∴ The equation is homogeneous.

1.1.5 139 N, 30.2° anticlockwise from 120 N.

1.1.6 $F_v = 75$ N; $F_h = 130$ N

1.1.7 $B \sin \theta$

1.1.8 Mass $= 7.9$ g cm^{-3} × (10 cm × 5 cm × 4 cm) $= 1580$ g

1.1.9 (Wheelbarrow):

weight $= $ CM; lift $= $ ACM

(Spanner):

Force $= $ ACM; friction on nut $= $ CM

1.1.10 Total weight $= (2 + 3 + 5.5 + 10)$ kg × 9.8 N kg^{-1} $= 201$ N

∴ For equilibrium, upward force by pivot on plank ∼200 N

1.1.11 $6 \times 60 = 3 \times 90 + 1 \times 50 + 2(100 - d)$;

∴ $360 = 520 - 2d$; ∴ $2d = 160$;

∴ $d = 80$ cm

1.1.12 **A**: $40F = 3 \times 10 + 1 \times 50 + 2d$;

∴ $20F = 40 + d$ [1]

B: $60F = 3 \times 90 + 1 \times 50 + 2(100 - d)$;

∴ $30F = 260 - d$ [2]

Then, e.g., solve [1] and [2] for F by adding to eliminate $d \rightarrow 50F = 300$, etc.

1.1.13 By all methods: $F = 11.7$ N at 31.0° to downward vertical

1.1.14 $F = 66.1$ N, $\theta = 61.9°$ to horizontal

1.2

1.2.1 10.8 m s^{-1}

1.2.2 0−3 s; 6−8 s (0 velocity); 9.6−12 s

1.2.3 0−3 s constant forward velocity; 3−6 s decelerating; 6−8 s stationary; 8−9.6 s accelerating backwards; 9.6−12 s constant velocity back

1.2.4 3.2 m s^{-1} to the right

1.2.5 (a) 2.5 m s^{-1} East

(b) 2.5 m s^{-1} South

(c) 2.5 m s^{-1} West

1.2.6 (a) 32 m on a bearing of 219°

(b) 1.8 m s^{-1} on a bearing of 219°

1.2.7 (a) 2.0 m s^{-1} on a bearing of 310°

(b) 20 s (2 sf)

1.2.8 20−30 s (BC): uniform acceleration from 10 to 20 m s^{-1}

30−54 s (CD): uniform velocity of 20 m s^{-1}

54−68 s (DE): uniform deceleration to rest from 20 m s^{-1}

1.2.9 0.4 m s^{-2}

1.2.10 Area 0−20 s (trapezium) $=$

$\tfrac{1}{2}(15 + 20) \times 10 = 175$ m, etc.

1.2.11 72 m s^{-1} at 56.3° to horizontal

1.2.12 (a) $10^2 = 26^2 - 2 \times 1.2x$; ∴ $2.4x = 576$;

∴ $x = 240$ m, etc.

(b) $10 = 26 - 1.2t$; ∴ $1.2t = 16$;

∴ $t = 13.3$ s

(c) $240 = 26t - 0.6t^2$;

∴ $0.6t^2 - 26t + 240 = 0$;

∴ $t = \dfrac{26 \pm \sqrt{676 - 576}}{1.2}$

$= 13.3$ s or 30.0 s

1.2.13 63.4 m s^{-1}

1.2.14 (2nd part) Because the acceleration is different (and not constant) in the powered stage.

1.2.15 (a) 26.0 m s^{-1}; 15 m s^{-1}

(b) 26.0 m s^{-1}; −34.1 m s^{-1}

(c) 42.9 m s^{-1} at 52.7° below horizontal

(d) Horizontal: 130 m; vertical −47.6 m (i.e. 47.6 m below starting point)

1.2.16 $t = 2.89$ s. $v_v = -8.3$ m s^{-1} → $v = 35.6$ m s^{-1} at 13.5° below horizontal

1.2.17 (a) $t \sim 0.32$ s

(b) 6% (uncertainty in height is negligible)

1.3

1.3.1 (a) 20 000 kg m s^{-1} (or 20 kN s)

(b) 1.8×10^{29} N s

1.3.2 $p_1 = 0.15 \times 6 = 0.9$ kg cm s^{-1}

$p_2 = 0.45 \times 2 = 0.9$ kg cm s^{-1} $= p_1$

1.3.3 (a) $KE_1 = 2.7 \times 10^{-4}$ J;

$KE_2 = 0.5 \times 0.3 \times 0.032 = 1.35 \times 10^{-4}$ J

∴ 50% of KE lost

(b) 67% of KE lost

1.3.4 $1000 \times (-6) + 4000 \times 2 = 5000\,v$;

∴ $5000v = 2000$; ∴ $v = 0.4$ m s^{-1}

1.3.5 KE_1 as in 1.3.3;

$KE_2 = 0.5 \times 0.15 \times (-0.02)^2 + 0.5 \times 0.3 \times 0.04^2 = 2.7 \times 10^{-4}$ J

1.3.6 (a) 3.0 J (b) 0.060 J (c) Ball gains 0.98 (98%) of the energy.

1.3.7 0.983 (98.3%)

1.3.8 $[F] = \dfrac{[\Delta p]}{[\Delta t]}$; ∴ N $= \dfrac{[p]}{s}$; ∴ $[p] = $ N s

1.3.9 (a) 1.4 kN East (bearing 90°)

(b) 0.82 kN East (bearing 90°)

(c) 4 kN West (bearing 270°)

1.3.10 (a) F on sand to right because Δp to right

(b) F on belt to left by N3

1.3.11 $[c_d] = \dfrac{[F_d]}{[\rho][v^2][A]} = \dfrac{\text{kg m s}^{-2}}{\text{kg m}^{-3} \text{ m}^2 \text{ s}^{-2} \text{ m}^2}$

$= \dfrac{\text{kg m s}^{-2}}{\text{kg m s}^{-2}}$, i.e. no units

1.3.12 mg: gravitational force of brick on the Earth (vertically upwards)

N: normal contact force of brick on the plane (downwards ⊥ slope)

F: frictional force of the brick on the slope (down the slope)

1.3.13 1.0(4) m s^{-2} down the slope

1.3.14 72.0 kg [706 N]; 0

1.3.15 Taking a spread-eagled skydiver, estimating $A = 1$ m^2; $m = 90$ kg; → $v_{term} \sim 40$ m s^{-1}

1.3.16 Smaller area → lower drag force. Only if v is larger will we get $F_d = mg$ and, hence, terminal velocity.

1.3.17 (a) 0.49 m s^{-1} (b) 0.20 m s^{-2}

1.3.18 dependent: a independent: m
controlled: x, M

1.3.19 The same 'pulling mass', m, is used.

1.4

1.4.1 Initially random vibrational energy of molecules in the rope and winch.

1.4.2 614 J

1.4.3 $a = \dfrac{F}{m} = \dfrac{1200}{800} = 1.5 \text{ m s}^{-2}$

$\therefore \sqrt{15^2 + 2 \times 1.5 \times 250} = 31.2 \text{ m s}^{-1}$

1.4.4 (a) 540 kJ (b) 5.0 kJ

1.4.5 (a) 16.7 kJ (b) 19.8 m s^{-1}
(c) 62 N

1.4.6 1 kW h = 3.6 MJ

1.4.7 (a) 450 kJ (b) 0.125 kW h

1.4.8 4.0×10^{16} J [= 40 PJ]

1.4.9 (a) unit of the 0.3 term = $\dfrac{[F_d]}{[\rho][v^2]} =$

$\dfrac{\text{kg m s}^{-2}}{\text{kg m}^{-3}\text{ m}^2\text{ s}^{-2}} = \text{m}^2$. QED

(b) 6.1 kW

1.5

1.5.1 23.5 N m^{-1}

1.5.2 200 μm

1.5.3 0.1 J

1.5.4 $\dfrac{1}{2}\sigma\varepsilon = \dfrac{1}{2}\sigma\dfrac{\sigma}{E} = \dfrac{1}{2}\dfrac{\sigma^2}{E}$;

$\dfrac{1}{2}\sigma\varepsilon = \dfrac{1}{2}\varepsilon E\varepsilon = \dfrac{1}{2}\varepsilon^2 E$ QED

1.5.5

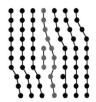

1.5.6 Thickness halves

1.5.7 $\varepsilon_{\text{rubber}} \sim 10\ 000 \times \varepsilon_{\text{steel}}$

1.6

1.6.1 The redness indicates that αTau has a higher wavelength of peak emission than α Cen. Hence, by Wien's law, αTau has a lower temperature.

1.6.2 (a) ~8 mW m^{-2}
(b) 80 μW m^{-2}
(c) 8 nW m^{-2}

1.6.3 $L_B = 4L_A$

1.6.4 $L = I \times 4\pi d^2 = 42.8 \times 10^{-9} \times 4 \times \pi \times (9.5 \times 10^{16})^2 = 4.9 \times 10^{27}$ W

1.6.5 1.4×10^9 m [1.4 million km]

1.6.6 Diameter = $9.34 \times 10^{-3} \times 150 \times 10^6$ km ~ 1.4 million km

1.6.7 $T = 3700$ K; $r \sim 6.0$ million km

1.6.8 A – O$_2$; B – O$_2$; C – Hα; D – NaI;
E – FeI; F – Hβ

1.7

1.7.1 $E = \dfrac{8.20 \times 10^{-14}\text{ J}}{1.60 \times 10^{-19}\text{ J(eV)}^{-1}} = 5.12$ keV

1.7.2 $f = 1.47 \times 10^{20}$ Hz; $\lambda = 2.03$ pm;
$p = 3.3 \times 10^{-22}$ N s

1.7.3 $^{12}_{6}$C : 50%; $^{56}_{26}$Fe : 54%; $^{197}_{79}$Au : 60% neutrons

1.7.4 Neutron decay:
excess energy = 939.6 – (938.3 + 0.5)
= 0.8 MeV

Proton decay: $m_p < m_n + m_e (+ m_\nu)$ $\therefore$ Not enough energy

1.7.5 Q: LHS = 1 + (−1) = 0 = 0 + 0 = RHS
L: LHS = 0 + 1 = 1 = 0 + 1 = RHS

1.7.6 (a) Weak interaction. We know this because:
■ neutrinos are involved
■ change of quark flavour
(U: 1 → 0; D: −1 → 0)

(b) Baryon number, B: 0 → 0
Lepton number, L: 0 → (−1) + 1 = 0
Charge, Q: 1 → 1 + 0 = 1

Unit 2

2.1

2.1.1 9.4×10^{20}

2.1.2 2×10^{10} electrons are transferred to the surface of the polythene rod.

2.1.3 (a) 8640 C, (b) 4.8 hours

2.1.4 v is inversely proportional to A, hence to diameter^2. Hence 4× the drift velocity = 0.48 mm s^{-1}.

2.2

2.2.1 30 MV

2.2.2 (a) 720 mW (0.72 W) (b) 216 C
(c) 1.3 kJ

2.2.3 $t = \dfrac{VQ}{P}$

2.2.4 5.0 V

2.2.5 3.6 MΩ

2.2.6 2.6 mA (2 sf)

2.2.7 (a) 2 Ω (b) 11 Ω

2.2.8 The constant pd produces a (constant) acceleration of the free electrons between collisions. When the electrons collide with the ions, their velocity is randomised, so their mean velocity, which is the constant drift velocity, is half the gain in velocity between collisions.

2.2.9 5 A

2.2.10 1060 Ω; 9.9 m

2.2.11 1.34×10^{-6} m^2; 6.3 mΩ

2.2.12 14.3×, 3000°C

2.2.13 Higher pd gives greater energy of free electrons when they collide with ions. Hence the temperature increases, so the random speed of the free electrons increases and the time between collisions decreases.

2.2.14 When the pd across a filament lamp is increased the filament's resistance increases, so the current increases less than proportionally to the pd.

2.2.15 Gradient = 8.29 [Ω m^{-1}];
diameter = 0.28 mm.

2.3

2.3.1 (a) 0.20 A, 0.50 A (b) 3.6 V, 3.6 V
(c) 12 Ω, 18 Ω, 8.0 Ω

2.3.2 8.3 Ω

2.3.3 (a) 4 Ω
(b) 3 parallel resistors in series with one resistor.

2.3.4 $x = 0.5$ A, $y = 0.4$ A

2.3.5 1%

2.3.6 Roughly 30°C

2.3.7 The brighter the light, the lower the output voltage.

2.3.8 (a) 0.27 A
(b) 0.41 W (total) [0.37 W exported]

2.3.9 (a) 0.56 Ω (b) 1.25 V

2.3.10 Equations, using $I = \dfrac{E}{R + r}$:

$0.88 = \dfrac{E}{1.25 + r}$ and $0.28 = \dfrac{E}{5.0 + r}$

→ $E = 1.54$ V; $r = 0.50$ Ω

2.3.11 (a) $I = 1.75$ A, $V = 3.49$ V;
(b) $I = 0.58$ A, $V = 1.16$ V

2.3.12 (a) R and r form a potential divider:

$V = \dfrac{ER}{R + r}$ $P = \dfrac{V^2}{R}$,

$\therefore P = \dfrac{(ER)^2}{R(R + r)^2} = \dfrac{E^2R}{(R + r)^2}$

(b)

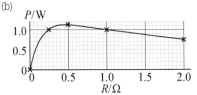

(c) Max power when $R = r$.

2.3.13 (a) 5.9 V; 120 Ω

(b) 0.018 A; 3.7 V

2.3.14 $E = \dfrac{1}{\text{gradient}}; r = \dfrac{\text{intercept}}{\text{gradient}}$

2.3.15 $E = \dfrac{1}{\text{intercept}}; r = \dfrac{\text{gradient}}{\text{intercept}}$

2.4

2.4.1 e.g. radiation from the Sun; electricity

2.4.2 $n = 0.5$

2.4.3 ρ may be density as increasing it would reduce particle accelerations.

Unit check: Rearranging: $\rho = \dfrac{1.40p}{v^2}$.

Assuming that 1.40 is dimensionless:

$[\rho] = \dfrac{[p]}{[v]^2} = \dfrac{\text{kg m}^{-1}\,\text{s}^{-2}}{(\text{m s}^{-1})^2}$

$= \text{kg m}^{-3}$, which is the unit of density

2.4.4 To start with the brightness is maximum. The brightness decreases to minimum and back up to maximum as the second filter rotates through 90° and on to 180°. This is repeated for the next 180° of rotation.

2.4.5 (a) 0.156 s

(b) 1.6 Hz

2.4.6 The horizontal axis represents time, so the peak-to-peak interval cannot be a length. He should have said that this represents the period.

The amplitude is the maximum displacement, so 'maximum amplitude' is a tautology. He should have said that A was the amplitude.

2.4.7 (a) $\frac{1}{4}\lambda$ (b) $1\frac{3}{4}\lambda$.

2.4.8 0.6 m

2.4.9 $\lambda/4 \sim 8$ cm, so this is possible.

2.5

2.5.1 Approximately 2×

2.5.2 The low frequencies have long wavelengths which spread out a lot by diffraction through the open window. The higher frequencies, which are important for understanding do not diffract to the same extent. Taking $f > 3$ kHz (for consonants), this gives $\lambda < 0.1$ m. Doorways and window gaps ~ 1 m, so diffraction spreading small.

2.5.3 The string still has kinetic energy – just to the left of the centre line, the string is moving upwards; just to the right it is moving downwards.

2.5.4 They do not affect each other.

2.5.5 $\lambda = 3.9$ mm
$S_1R = 31.0$ mm; $S_2R = 35.0$ mm;
$S_2R - S_1R = 4.0$ mm
∴ reasonable agreement

2.5.6 $S_1U - S_2U = 2\lambda$; $S_1V - S_2V = 2.5\lambda$

2.5.7 Fringe spacing = 1.34 mm; $\lambda = 670$ nm.

2.5.8 337 m s⁻¹ [340 m s⁻¹ to 2 sf]

2.5.9 The intensity of the **diffracted** beams from the slits decreases with angle from the centre.

2.5.10 470 nm (2 sf)

2.5.11 (a) 510 nm (2 sf) (b) 80 mm (2 sf)

2.5.12

2.5.13

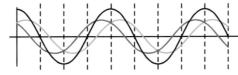

2.5.14 (a) 2nd harmonic

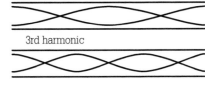

3rd harmonic

(b) $f_n = nf_1$.

2.5.15 (a) $f_1 = 170$ Hz; $f_2 = 510$ Hz

(b) With closed-end pipe: $f_1 = 340$ Hz and $f_2 = 680$ Hz

2.5.16 (a) % uncertainty $\sim 9\%$

(b) absolute uncertainty $\sim \pm 50$ nm

2.5.17 $d = 5$ μm

2.5.18 (a) $d = 2.75$ μm

(b) grating constant = 3.63×10^5 m⁻¹
= 3630 cm⁻¹.

2.6

2.6.1 (a) deep = 2.5 cm, shallow = 1.5 cm

(b) 23°

2.6.2 (a) 2.25×10^8 m s⁻¹

(b) 1.24×10^8 m s⁻¹

2.6.3 29.6°

2.6.4 (a) Ray emerges at 39°; weak reflected ray at 25°

(b) Weaker refracted ray at 59°, stronger reflected ray at 35°

(c) Total internal reflection – strong reflected ray at 45°

2.6.5 47.2°

2.6.6 $\sin c = \dfrac{1.52}{1.62}$. ∴ $c = \sin^{-1}\dfrac{1.52}{1.62} = 69.8°$

2.6.7 0.03 s, assuming the glass has a refractive index of 1.5.

2.7

2.7.1 $E_{k\,\text{max}} = 9.6 \times 10^{-20}$ J = 0.6 eV

2.7.2 5.11 eV

2.7.3 (a) 5.2 eV = 8.4×10^{-19} J

(b) no electrons emitted

(c) same as (a).

2.7.4 (a) 9.7×10^{44} photons s⁻¹

(b) 3.4×10^{21} m⁻²

2.7.5 3.1 eV (violet); 1.8 eV (red); 2.3 eV (yellow)

2.7.6 2.18 aJ

2.7.7 $E = 13.6 \left(\dfrac{1}{2^2} - \dfrac{1}{6^2}\right) = 3.02$ eV

$= 4.84 \times 10^{-19}$ J

∴ $\lambda = \dfrac{hc}{E} = \dfrac{6.63 \times 10^{-34} \times 3.0 \times 10^8}{4.84 \times 10^{-19}}$ m

$= 411$ nm.

In Fig. 1.6.12, Hδ has $\lambda = 410$ nm.

2.7.8 2×10^{-13}, 2×10^{-15}, 2×10^{-18}, 4×10^{-19}, 2×10^{-20}, 2×10^{-23}, 2×10^{-27}

2.7.9 590 nm (2 sf)

2.7.10 410 fm (2 sf)

2.7.11 1.3×10^{-27} N s.

2.7.12 (a) The momentum of the incident photons in $\Delta t = \dfrac{IA\Delta t}{c}$

The momentum of the reflected photons in $\Delta t = -\dfrac{IA\Delta t}{c}$

∴ The momentum change of the incident photons in $\Delta t = -2\dfrac{IA\Delta t}{c}$

∴ The force exerted by the photons on the plane =

$\dfrac{\Delta p}{\Delta t} = 2\dfrac{IA}{c}$ by N2 and N3

∴ The pressure exerted by the photons on the plane $= 2\dfrac{I}{c}$

(b) pressure $= \dfrac{I}{c}$

2.8

2.8.1 4500 s⁻¹

2.8.2 (a) 2.5 eV (0.4 aJ)

(b) 1.8 eV (0.29 aJ)

2.8.3 0.72 (= 72%)

2.8.4 For a laser operating with a constant beam, the rate of production of photons must equal their rate of loss. Hence a photon must travel ~ 280 lengths of the medium before stimulating an emission. This is 140 m in this case.

2.8.5 (a) Energy input per pumping event =
1.80 eV = 2.88×10^{-19} J

Photon energy $= \dfrac{hc}{\lambda} = 2.46 \times 10^{-19}$ J

∴ Max efficiency =

$\dfrac{2.46 \times 10^{-19}\,\text{J}}{2.88 \times 10^{-19}\,\text{J}} \times 100\% = 85.4\%$

$< 86\%$

(b) 1.0×10^{17} photons s⁻¹.

2.8.6 8.59×10^{-19} J (5.37 eV)

Practical skills

3.1 478 Ω (3 sf)

3.2 6.33 ± 0.08 s

3.3 36.9 ± 0.3 mA

3.4 $p = 1.0 \times 10^{-6} = 1.0 \times 10^{-4}$ %

3.5 0.1 kg: $\Delta T = 0.013$ s [0.01 s to 1 sf]

0.2 kg: $\Delta T = 0.0045$ s [0.005 s to 1 sf]

Hence the periods could have been expressed to 3 decimal places, especially for 0.2 kg.

3.6 2.13 ± 0.15 m s^{-2} or 2.1 ± 0.2 m s^{-2}

3.7 4.19 ± 0.06 mm^3

3.8 $V = -(0.167 \pm 0.002)I + (9.11 \pm 0.06)$

3.9 Gradient = acceleration, a; intercept = initial velocity, u

3.10 $r = 0.167 \pm 0.002$ Ω; $E = 9.11 \pm 0.06$ V

3.11 Inverting the equation: $\dfrac{1}{V} = \dfrac{R+r}{ER} = \dfrac{R}{ER} + \dfrac{r}{ER} = \dfrac{r}{ER} + \dfrac{1}{E}$ QED

3.12 Graph of ℓ against f^{-1}. e is minus the intercept on the ℓ axis and c is 4× the gradient.

3.13 (a) Gradient = $k^{-0.5}$; intercept = $ek^{-0.5}$

(b) $e = \dfrac{\text{gradient}}{\text{intercept}}$

3.14 If t is the measured time; $t = T + \tau$.

Hence $h = \dfrac{1}{2}g\,(t - \tau)^2$, which can be rearranged to give

$t = \sqrt{\dfrac{2h}{g}} + \tau$

A graph of t against $\sqrt{h}$ is plotted which should be a straight line of gradient $m = \sqrt{\dfrac{2}{g}}$.

Hence $g = \dfrac{2}{m^2}$ and τ is the intercept on the t axis.

3.15 ±0.01 mm

3.16 (7.93 ± 0.02) g cm^{-3} / (7.93 ± 0.02) × 10^3 kg m^{-3}

Maths skills

4.1 (a) y^3 (b) y^{-1} (c) $y^3 + 2ay$

4.2 6.0 kΩ

4.3 (a) 57.3° (57° to 2 sf) (b) 0.65 rad

4.4 k – the spring [or stiffness] constant; N m^{-1}

v or c – the wave speed; m s^{-1}

$\dfrac{1}{2}m$ where m is the mass; kg

[In the A2 spec]: GMm – (no name); N m^2

4.5 4.7 (to 2 sf)

4.6 (a) −11.5 (3 s.f.) (b) 2 (c) $\dfrac{11}{15}$ = 0.73 (2 s.f.)

4.7 (a) Substituting: $24 = 4t + 1.5t^2 \rightarrow 1.5t^2 + 4t - 24 = 0$

(b) $t = \dfrac{-4 \pm \sqrt{16 + 4 \times 1.5 \times 24}}{3} = \dfrac{-4 \pm \sqrt{160}}{3} = -5.55$ s or 2.88 s (to 3 s.f.)

4.8

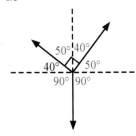
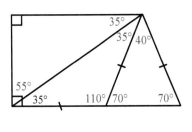

4.9 1.0×10^6 m [1.0×10^3 km]

4.10 θ in the range [−360°, 360°] : −330°, −210°, 30°, 150°

θ in the range [−2π, 2π] : $-\dfrac{11}{6}\pi, -\dfrac{7}{6}\pi, \dfrac{1}{6}\pi, \dfrac{5}{6}\pi$

(or −5.76 rad, −3.67 rad, 0.52 rad, 2.62 rad)

4.11 (a) acceleration (b) velocity

4.12 Trapezium area = $\dfrac{1}{2} \times (7\text{ s} + 10\text{ s}) \times 24$ m s^{-1} = 204 m.

This area = $A_1 + A_2$ = 168 m + 36 m = 204 m

4.13 301 squares $\rightarrow$ 6.02 J

4.14 Total area = 1.8 + 1.75 + 1.425 + 0.775 + 0.225 + 0.025 = 6.0 J

4.15 $v = 283 + (-9.13)t$; $u = 283$ m s^{-1}; $a = -9.13$ m s^{-2}

Test yourself answers

Unit 1

1.1

1
- Moments can be calculated about any point, not just a pivot.
- The distance is not specified clearly enough.

The moment of a force about a point is the force multiplied by the perpendicular distance from the line of action of the force to the point.

2 J s = kg m^2 s^{-1}

3 Direction omitted: acceleration is a vector.

4 Taking moments about left hand end:

CM = 2.5 N × (L/2), ACM = 1.0 N × L,

so there is a resultant moment and rod can't be in equilibrium.

5 (a) 77 N (b) 64 N (2 sf)

6 (a) 13.0 N at 22.6° to horizontal

(b) 2.6 m s^{-2} at 22.6° to horizontal

7 (a) 2.72 g cm^{-3} or 2.72 × 10^3 kg m^{-3}

(b) 3.21 cm because

$$\frac{0.01}{3.21} > \frac{0.01}{4.75} > \frac{0.01}{10.30}$$

(c) p_ρ = 0.62%; Δρ = ± 0.016 kg m^{-3}; so
ρ = (2.72 ± 0.02) × 10^3 kg m^{-3}

(d) $p_m = \frac{0.01}{4.27} \times 100\%$ =

0.0023% << 0.62%

so assumption reasonable.

8 9.5 × 10^{-11} m^3

9 (a) [p] = kg m^{-1} s^{-2}, [g] = m s^{-2}, [ρ] = kg m^{-3}

(b) [p] = [p_A] = kg m^{-1} s^{-2},
[$g\rho d$] = m s^{-2} kg m^{-3} m = kg m^{-1} s^{-2}

so homogeneous

(c) 354 kPa

10 (a) 72 Nm clockwise, 36 Nm clockwise

(b) 90 N

(c) Upward force = sum of downward forces. F$_1$ = 140 N

11 (a) [G] = [F][d^2][$M_1 M_2$]$^{-1}$ = N m^2 kg^{-2}

(b) N = kg m s^{-2}

∴ [G] = kg m s^{-2} m^2 kg^{-2} = kg^{-1} m^3 s^{-2}

12 34 kg

1.2

1 (a) 5.0 m s^{-1}

(b) 4.4 m s^{-1}

(c) 4.7 m s^{-1} [≠ 0.5 (5.0 + 4.4)]

(d) 0.53 m s^{-1} downwards

(e) 18.9 m s^{-1} upwards

2 47.41 km s^{-1}

3 ⟨a⟩ = 5.7 m s^{-2} SE [135°]

4 (a) 12.0 m North [0°]

(b) 12.6 s (c) 9.5 m s^{-1} North [0°]

(d) 30 m s^{-1} West [270°]

(e) 2.4 m s^{-2} West [270°]

(f) 13.5 m s^{-1} NE [45°]

(g) 3.4 m s^{-2} NW [315°]

5 (a) 1.0 m s^{-2}

(b) 250 m

(c) [Gradient of tangent] = −0.60 m s^{-2}

(d) Distance est = 355 m. ⟨v⟩ = 8.9 m s^{-1}

6 [v] = [u] = m s^{-1}; [at] = m s^{-2} × s = m s^{-1}

But [x] = m, [u] = m s^{-1},

[at^2] = m s^{-2} × s^2 = m

7 Total displacement = 8325 m;

total time = 270 s

∴ Mean velocity = 30.8 m s^{-1}

8 (a) +25 m s^{-1}, +15 m s^{-1}, +5 m s^{-1}

(b) 3.5 s (c) 7.0 s

9 (a) 123 m,

(b) 49.1 m s^{-1},

(c) 57.5 m s^{-1} at 59° to horizontal

10 (a) 31.9 m

(b) 2.54 s

(c) 43.3 m s^{-1} horizontal

(d) 220 m

11 h against t^2 : best fit straight line has gradient 4.25 (m s^{-2}) and intercept −0.045 (m) → g = 8.5 m s^{-2}

√h against t : best fit straight line has gradient 2.21 (m$^{0.5}$ s^{-1}) and intercept −0.086 (m$^{0.5}$) → g = 9.77 m s^{-2} and $τ \sim$ 0.04 s

13 (a) v_1 = 26 m s^{-1} (right), 15 m s^{-1} (up)

v_2 = −10 m s^{-1} (right), 17 m s^{-1} (up)

(b) ($v_1 + v_2$) = 36 m s^{-1} at 4° to the right above dotted line

(c) ($v_2 - v_1$) = 36 m s^{-1} at 4° to the left above dotted line

1.3

1 (a) 800 kg m s^{-1} (N s) to the right

(b) 200 N s to the right

(c) 580 N s at 31°

2 7750 J

3 Let m be mass of a rider. Initial momentum = 2m × 6 = 12m; final momentum = 3m × 4 = 12m. Equal initial and final momenta, ∴ as predicted

4 33%

5 (a) 28 kN s in the direction of the first car

(b) 11.2 m s^{-1} (in the direction of the first car), assuming no resultant horizontal external force acts

6 5.0 m s^{-1}

7 (a) 2.8 × 10^5 m s^{-1}

(b) 3.4 × 10^{-13} J [~2.1 MeV]

8 A 17 m s^{-1}; B 18 m s^{-1} both in reverse directions to their original ones

9 Change of momentum of football = [(−25) − 30] × 0.45 = −24.75 N s. ∴ Mean force exerted by the wall on the ball is

$\frac{-24.75}{0.04}$ = −620 N [2.s.f] by N2,

i.e. −620 N in the original direction of the ball

∴ By N3 the mean force exerted by the ball on the wall = − (−620 N) = 620 N in the ball's original direction

10 (a) $α_{max}$ = 11.0 m s^{-2}; $α_{min}$ = 4.5 m s^{-2},

(b) 8.4 m s^{-2} at 22.6° to the 12 N force.

11 (a) 4.9 × 10^4 m s^{-2}

(b) Assuming

$⟨F_{res}⟩ = \frac{1}{2}F_{max} \rightarrow$ 186 m s^{-1}

[constant $F_{res} \rightarrow$ 260 m s^{-1}]

12 (a) Momentum has both magnitude and direction.

(b) A body's rate of change of momentum is proportional to the resultant force acting on it, and is in the direction of the resultant force.

13. For W, upward force of skydiver on the Earth; gravitational

For F_d, downward force of skydiver on air; contact force (electromagnetic)

14 (a) Without air resistance

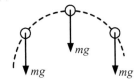

(b) With air resistance

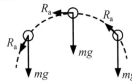

15 (a) Graph of x/t against t is a straight line, $\therefore$ constant acceleration

(b) Gradient = 14.4 (cm s^{-2}) $\therefore$ acceleration = 28.8 cm s^{-2}

(c) 0.17 kg.

16

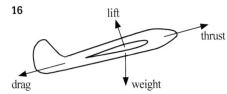

17

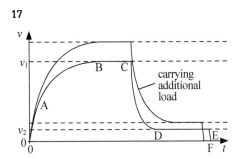

Terminal velocities v_1 and v_2 are higher, as greater resistive forces are needed to balance the greater weight. It will take rather longer to approach terminal velocity each time. Because of the greater speed the skydiver will reach the ground earlier.

1.4

1 (a) 500 kJ   (b) 5.0 kN

2 (a) 0.5 J

(b) 5 m s^{-1} assuming all PE transferred to KE in sphere

3 (a) 1.62 m s^{-1}

(b) 42.9° assuming no loss in energy, i.e. same height as originally (0.134 m)

4 The work done *by a force* is force × distance moved by force *in the direction of the force*. Omission of second italicised phrase makes the student's definition at best ambiguous.

5 1.69 m

---

6 (a) 84.9 m s^{-1}

(b) Height = 92 m, range = 640 m

7 (a) Direction of thrust relative to displacement of satellite.

(b) If thrust in same direction as displacement, then 3.5 MJ + 0.6 MJ = 4.1 MJ.

If opposite, 3.5 MJ – 0.6 MJ = 2.9 MJ.

8. (a) (i) 785 J; (ii) 491 J; (iii) 196 J; (iv) 0

(b) (i) 0; (ii) 294 J; (iii) 589 J; (iv) 785 J

(c) Using (b)(iv), v = 28 m s^{-1}. Using $v^2 = 2gx$, v = 28 m s^{-1}. First method assumes zero energy loss through air resistance; second assumes acceleration unaffected by air resistance.

9. (a) Tension per unit extension,

(b) $[k] = [F].[x]^{-1}$ = kg m s^{-2} m^{-1} = kg s^{-2}

10 (a) $E_k = \frac{1}{2}mv^2 = \frac{1}{2}\frac{(mv)^2}{m} = \frac{p^2}{2m}$

(b) m_A = 7.8 kg; v_A = 3.2 m s^{-1} North
m_B = 1.0 kg; v_B = 10.0 m s^{-1} North

11

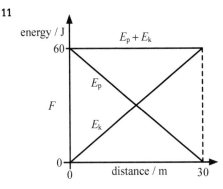

1.5

1 (a) 24.5 N m^{-1}

(b) 0.078 J

(c) 0.157 J

2 Extension = 16.0 cm

(a) 12.25 N m^{-1} (b) 0.157 J

(c) 0.304 J

3 (a) 0.885 m s^{-1}

(b) 16.0 cm

(c) 9.81 m s^{-2}

4 (a) end-to-end k = 12 N m^{-1}. Total extension double for any given load

(b) side-to-side k = 48 N m^{-1}. Total load double for any given extension

5 F_{max} = 650 N, $\Delta\ell$ = 0.14 mm

6 (a) 58 900 N   (b) 295 kJ

(c) 7.4 m   (d) 540 m

7 Any surface scratch will close up and not be under tension until the glass is stretched so that the compression in the surface layer is overcome.

---

8 $T_A = T_B$; $\Delta\ell_B = 2\Delta\ell_A$; $\sigma_B = \sigma_A$; $\varepsilon_B = 2\varepsilon_A$; Energy$_B$ = 2×Energy$_A$.

9 (a) ε has no unit, being the ratio of two lengths, so $[E] = [\sigma]$ = N m^{-2}

(b) $[\sigma]$ = N m^{-2} = kg m^{-1} s^{-2}

10 $T_X = T_Y$; $\sigma_X/\sigma_Y = \frac{1}{2}$; $\varepsilon_X/\varepsilon_Y = \frac{1}{3}$; $U_X/U_Y = \frac{1}{6}$

11 (a) (i) 0.285 mm is well outside the range of the other readings and is likely to be a mistake.

(ii) Wire might not be perfectly round, but $\pi d^2/4$ will be close to true cross-sectional area if d is mean 'diameter'.

(iii) Mean diameter = $\frac{1}{7}$ × (sum of allowed readings) = 0.2744 mm

$p_d = \frac{0.2744 - 0.273}{2 \times 0.274} \times 100\%$

= 0.73%

So $A = \pi\left(\frac{0.2744}{2}\right)$mm^2 ± 1.46%

= (0.0594 ± 0.009) mm^2

(b) (i) Elastic limit reached when load = 100 MPa × 0.0591 mm^2 = 6 N, so sensible to keep load below 5 N for determining Young modulus.

(ii) (105 ± 4) GPa

(c) The % uncertainty in length will indeed remain insignificant, but that in the extension will double because the extension will halve while its absolute uncertainty will stay the same. Thus the uncertainty in E will increase.

(d) There will indeed be a smaller % uncertainty in the extension, but this benefit will be (exactly) cancelled by the greater % uncertainty in d^2.

12 (a) *Elastic limit*: the maximum stress for which the specimen returns to its original length when the stress is removed. If exceeded, the specimen would not return to its original length when the load was removed.

(b) Diameter, d, and length, ℓ of the wire. Mass, m, of load used to extend the wire by extension x. She would calculate the Young modulus thus:

$$E = \frac{4mg\ell}{\pi d^2 x}$$

(c) A straight line graph of m against x will confirm that the wire remained Hookean. If readings taken during unloading lie on the same line it will confirm elastic behaviour. Drawing the best straight line and using the equation below averages E values from different m, x pairs.

$$E = \frac{4g\ell}{\pi d^2} \times \text{graph gradient}$$

13 (a) At B molecules are straightening (by bond rotation) requiring little force (except to overcome thermal agitation tending to return molecules to shorter, more probable shapes). At C most molecules are almost straight and bonds have to be stretched, needing more force.

(b) Hysteresis

(c) Less work is done by the band on the external restraints when the band contracts than is done by the restraints on the band when it expands. This 'missing work' results in an increase in the band's random thermal energy.

1.6

1 (a) 1.2×10^{25} W

(b) 120 nm

2 $\lambda_{WD} = \frac{1}{4} \lambda_\odot$

$L_{WD} = \frac{256}{10^4} L_\odot \sim \frac{1}{40} L_\odot$

3 Using Wien's law $\lambda_{max} \sim$

(a) 0.3 mm   (b) 3 μm

(c) 30 nm   (d) 300 pm

This suggests

(a) microwave   (b) (far) IR

(c) (far) UV   (d) X-ray

4 The peak intensity is much less so the vertical scale needs to be expanded. The wavelengths of significant intensity extend to much larger values so this scale needs to be compressed.

5 $L_{red\ giant} = 10^6 \times L_{red\ dwarf}$. ∴ At same distance the red giant appears $10^6 \times$ as bright. To reduce the brightness of the red giant by a factor of 10^4 (so it only appears $10^2 \times$ as bright) it needs to be moved $10^2 \times$ the distance. So the red giant is 100 times as far away as the red dwarf.

6 Nearly all the hydrogen outside stars has been used up (or blown away by radiation pressure).

21 cm emission is only from neutral hydrogen atoms.

7 Visible light, of wavelength 0.4–0.7 μm, is strongly scattered by interstellar dust. IR radiation, especially far IR, has a longer wavelength than the size of the dust particles so is not scattered by them and can therefore better penetrate the dust clouds within which the stars are being formed.

8 $\lambda_{max} \sim 10$ μm so observations in the far IR shown a stronger emission if the star is surrounded by such a disc than in its absence.

9 The emission lines in the spectrum of hydrogen are red and blue, so the appearance of the H1 regions is a glowing pink (magenta).

10 The hot spot emits X-rays with a wavelength of around 300 pm (see Q3). Hence, a burst of X-rays will be observed once in each revolution (i.e. ~ once per second). If the observer is directly in the plane of the accretion disc, this might not be observed because of absorption within the disc.

11 (a) A black body is a body or surface that absorbs all the electromagnetic radiation that falls upon it. (It also emits more radiation per second within any wavelength range than any non-black body at the same temperature.)

(b) *Wien's law*: The wavelength of peak emission of e-m radiation for a black body is inversely proportional to its temperature.

The Stefan–Boltzmann law: The power of e-m radiation emitted by a black body per unit area of surface is proportional to T^4, in which T is its kelvin temperature.

12 The white light spectrum will be crossed by dark lines due to absorption of specific wavelengths by the hydrogen atoms. Absorption occurs each time a photon raises the electron in a hydrogen atom from a lower to a higher level. The energies of the levels are fixed, so the wavelengths absorbed are fixed.

13 $\frac{\lambda_{peak}(3000K)}{\lambda_{peak}(6000K)} = 2; \frac{P_{3000K}}{P_{6000K}} = \frac{1}{16}$

14 (a) ultraviolet

(b) 400 nm – 700 nm

(c) X will be bluish-white. This is because the peak lies in the u-v, beyond the blue/violet end of the visible spectrum. Therefore in the visible range the spectral intensity will be greater at this end, as the black body spectrum continues to fall from its peak.

(d) 15 000 K

15 (a) 4.45×10^{26} W   (b) 58 nm

(c) 6.32×10^{-10} W m^{-2}

16 The intensity of e–m radiation from the star reaching B is $\frac{1}{4}$ of that reaching A. Assuming each planet reaches thermal equilibrium when radiation power emitted from it as a black body equals radiation power received, $\frac{T_A}{T_B} = 2$

17 (a) visible

(b) (i) 250 nm: ultraviolet

(ii) 1000 nm: infrared

(c) blue giant: $\frac{L}{L_\odot} = 1600$;

red giant: $\frac{L}{L_\odot} = 625$

1. Elementary particles, such as the electron, are not combinations of other particles. Composite particles are combinations of other particles. For example, the proton is a combination of 3 quarks (**uud**).

2. A baryon is a combination of 3 quarks. A meson is a combination of 1 quark and 1 antiquark.

3. Protons have a 3 quark structure (**uud**). They can take part in strong, weak, electromagnetic (and gravitational) interactions. Electrons are structureless in that they are not composed of other particles. They can take part in all types of interaction except the strong.

4. Charge, baryon number, lepton number and (total) quark number are conserved in all interactions. Quark numbers for individual flavours of quark are conserved in all hadron interactions except the weak.

5. Conservation laws for energy, momentum, charge, baryon number, lepton number.

6. The products of the interaction will include neither leptons nor antileptons, as they are not involved in the strong interaction. There will be one more baryon than antibaryon (baryon conservation). The sum of the charges of the products will be zero (charge conservation). Any number of mesons could be produced. The sum of quark numbers for each individual quark flavour will be the same as for the sum for the original baryon and meson.

7. Neither leptons nor antileptons will be produced, as they are not involved in the strong interaction. The sum of the products' charges will be -1 (charge conservation). The number of baryons will be equal to the number of antibaryons (possibly both zero). There may be meson(s). The quark structure of the reactants is **udd** and **ūūd**, so the sum of u quark numbers in the products will be -1; that of d quark numbers will be $+1$, as these individual quark flavour numbers are conserved in a strong interaction.

8 u$\bar{d}$

9 ddd

10 (a) ν_e: lepton; n: baryon

(b) Baryon and (first generation) lepton number conservation show that the products must be 1 baryon and 1 lepton. Since we are told that the products are different from the reactants, the lepton must be an electron. The reactants' charges are both zero, so to conserve charge the product baryon must be a proton (or a Δ^+).

(c) The weak force. The neutrino cannot participate in the strong or e-m interactions. Only the weak interaction can affect the quark flavour change, from d in the neutron to u in the proton (or. Δ^+).

11 (a) Photon involvement indicates an e-m interaction.

(b) Before the collision the total momentum is zero. Therefore, afterwards it must be zero. But a single photon can only be moving in one direction at any time, so must have a momentum (h/λ) in that direction.

12 (a) p: 500 eV, $_2^4$He: 1000 eV

(b) p: 8.0×10^{-17} J, $_2^4$He: 16×10^{-17} J

13 (a) X is an electron neutrino, ν_e. Applying lepton conservation, X must have $L = +1$ because for p, $L = 0$; for n, $L = 0$; for e$^+$, $L = -1$. Applying charge conservation, X must have $Q = 0$ because for p, $Q = +1$; for n, $Q = 0$; for e+, $Q = +1$. The only first-generation uncharged lepton is the ν_e.

(b) Neutrino involvement, change of flavour of a quark (from u to d) and the long half-life all indicate a weak interaction.

14. (a) $n \rightarrow p + e^- + \bar{\nu}_e$

(b) Lepton number, L, is defined as number of leptons – number of antileptons. L does not change in any interaction. This is the case for neutron decay because $L : 0 \rightarrow 0 + 1 + (-1)$.

(c) Baryon number, B, is conserved as $1 \rightarrow 1 + 0 + 0$.

Charge, Q, is conserved as $0 \rightarrow 1 + (-1) + 0$.

(d) To conserve L, B, and Q, proton decay must be p $\rightarrow$ n + e$^+$ + ν.

Total mass-energy of n and e$^+$ is 940.1 MeV, which is > mass-energy of p.

This is impossible for a free proton; proton must be given extra energy.

For neutron decay, mass-energy account is:

939.6 MeV $\rightarrow$ 938.8 MeV

this need not violate energy conservation as the reaction products can have KE.

15 $\Delta^{++} \rightarrow p + \pi^+$ uuu $\rightarrow$ uud + u$\bar{d}$

$\Delta^+ \rightarrow p + \pi^0$ uud $\rightarrow$ uud + u$\bar{u}$ or
uuu $\rightarrow$ uud + d$\bar{d}$ or

$\Delta^+ \rightarrow n + \pi^+$ uud $\rightarrow$ udd + u$\bar{d}$

$\Delta^0 \rightarrow n + \pi^0$ udd $\rightarrow$ udd + u$\bar{u}$ or
udd $\rightarrow$ udd + d$\bar{d}$ or

$\Delta^0 \rightarrow p + \pi^-$ udd $\rightarrow$ uud + d$\bar{u}$

The half-life is far shorter than for e-m or weak decays. There are no changes of quark flavour, confirming the weak interaction, nor is there neutrino involvement.

Unit 2

2.1

1 $(+)1.47 \times 10^{-17}$ C [14.7 aC]

2 (a) (i) 6.25×10^{13} (ii) 1.3×10^{13}

(b) The number of electrons is equal to the number of protons in the atomic nuclei.

3 1.0×10^{-10} A towards the support.

4 (a) 110 C

(b) 100 mA h = 360 C, so the claim is greatly exaggerated.

S&C 1 392 kC

S&C 2 (a) Draw a graph of current [= pd/resistance] against time. The area underneath gives the total charge of 47 mC

(b) 3.25 s

(c) 0.5 (!)

5 (a) 1.3 mA

(b) ~2.0 mA [from the gradient at $t = 0$]

6 20 mC $\rightarrow$ 10 mC = 8.3 s; 24 mC $\rightarrow$ 12 mC = 9.0 s

16 mC $\rightarrow$ 8 mC = 8.4 s; All very close, so exponential.

Mean value of half life ~ 8.6 s

7 (a) n = free electron concentration; v = drift velocity of free electrons

(b) (i) n, e, I (ii) Av = constant. $A_P = 9 \times A_Q$, $\therefore v_Q = 9 \times v_P$

8 (a) Ionic volume = 1.047×10^{-30} m^3

$\therefore N \sim \dfrac{3}{1.047 \times 10^{-30}} = 2.9 \times 10^{30}$ m^{-3}

(b) 2.7×10^{-5} m s^{-1}

9 0.1 m s^{-1}

10 (a) $I_{max} = 7.5$ A $\rightarrow v_{max} = 0.70$ mm s^{-1}

$\therefore$ Graph as shown

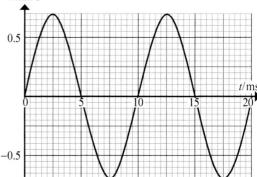

v/mm s^{-1}

(b) Distance = area under graph between 0 and 5 ms = 2.2×10^{-6} m. This small distance is the furthest an electron drifts. In the next half cycle it drifts 'backwards' by the same distance, so, for a.c., the electron doesn't progress. The individual electrons each travel much further than this by their own random thermal motion.

2.2

1 (a) 1.6 J of electrical energy are transferred to other forms in the LED per coulomb of charge which passes through it.

(b) 35 J (2 sf)

2 (a) 2500 J of energy are transferred per second [when the heater is on].

(b) 10.9 A (3 sf)

(c) 21.2 Ω (3 sf) [21.1 Ω if working from 10.9 A]

(d) (i) 2.7 kW (ii) 2.3 kW

(e) In Germany, less energy is transferred per second so it would take longer to bring the water to the boil (~19% longer)

3 (a) The current through a conductor is directly proportional to the pd across its ends.

(b) (i) 3.3 Ω (ii) 20 Ω

(c) Ohm's law only applies to the part of the characteristic where V and I are proportional, i.e. up to 0.7 V, 0.2 A.

In this region the temperature rise of the filament is not great enough to have a significant effect on the resistance.

4 (a) 0.76 mA (b) 2.0 W (c) 4.0×10^{-16} J (d) 3.0×10^{7} m s⁻¹

(e) 2.7×10^{-23} N s

5 (a) 0.78 Ω

(b) Doubling the length doubles the resistance because $R \propto \ell$. Doubling the width and height multiples the csa, A, by 4 hence divides the resistance by 4: $R \propto A^{-1}$. The combined effect is to halve the resistance.

6 (a) (i) 23.9 Ω (ii) 3.4 Ω (iii) Multiply the csa by 7.

(b) (i) $P = I^2 R = I^2 \times \dfrac{\rho\ell}{A}$ $\therefore$ rearranging $\dfrac{P}{\ell} = \dfrac{\rho I^2}{A}$

(ii) $\left(\dfrac{P}{\ell}\right)_X = \dfrac{4\rho I^2}{\pi d^2}$; $\left(\dfrac{P}{\ell}\right)_Y = \dfrac{4\rho(2I)^2}{\pi(2d)^2} = \dfrac{4\rho I^2}{\pi d^2}$. So same $\dfrac{P}{\ell}$.

7 (a) At low voltages the power will be very low and so the temperature will be approximately constant. Thus Ohm's law will be obeyed. At higher voltages the temperature will increase, raising the resistance, so Ohm's law will not be obeyed. The operating temperature is expected to be ~ 2000 K or more (to give a whitish emission).

(b) The series variable resistor is only needed in the low voltage range. Below 2 V, the resistor is adjusted: increasing its resistance decreases the current in the circuit and therefore the pd across the bulb, allowing pairs of values of V and I for $V < 2$ V to be obtained.

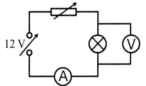

(c) From the graph of I against V (below), the behaviour changes at ~ 1.2 V.

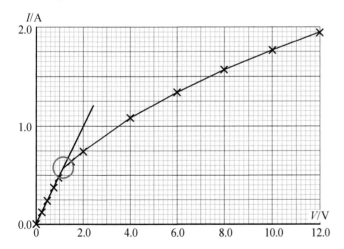

(d) $R = \dfrac{2.0 \text{ V}}{0.99 \text{ A}} = 2.0$ Ω at low temperatures

$R = \dfrac{12.0 \text{ V}}{1.95 \text{ A}} = 6.2$ Ω at operating voltage.

(e) Transition voltage ~ 1.03 V. Above this the graph (below) is almost a straight line (which doesn't extrapolate back to the origin) with a slight downward curve. Suggesting that the value of n is slightly less than 0.6.

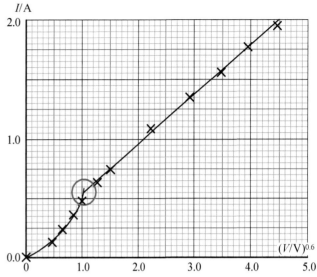

(f) Table of ln values of I and V.

ln (V/V)	ln (I/A)
−1.38	−2.14
−0.69	−1.45
−0.28	−1.04
0.00	−0.76
0.41	−0.45
0.69	−0.30
1.38	0.08
1.79	0.29
2.08	0.45
2.30	0.57
2.48	0.69

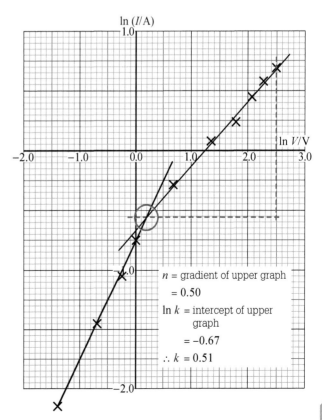

n = gradient of upper graph
= 0.50

ln k = intercept of upper graph
= −0.67

$\therefore k = 0.51$

2.3

1 (a) (i) 4.5 V   (ii) 0.45 A   (iii) 0.75 A

(b) $V_{AB} = 0.75$ A $\times 24$ Ω $= 18$ V. ∴ $V_{AC} = 18$ V $+ 4.5$ V $= 22.5$ V

(c) (i) 30 Ω

(ii) $R_{AC} = \dfrac{V_{AC}}{I} = \dfrac{22.5\text{ V}}{0.75\text{ A}} = 30.0$ Ω

2 160 Ω (2 sf)

3 (a) (i) 6 Ω

(ii) A series combination would have a resistance greater than 10 Ω.

(b) (i)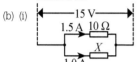

(ii) $X = \dfrac{15\text{ V}}{1.0\text{ A}} = 15$ Ω

(iii) $R_{tot} = \dfrac{10 \times 15}{10 + 15} = 6$ Ω ✓

4 200 Ω

5 (a) (i) 70 Ω   (ii) 3.6 V

(b) (i) Increases: because the total resistance of the circuit decreases

(ii) Increases: because the current through AS increases

(iii) Decreases: because the pd across AS has increased (and the sum makes 12 V)

6 (a) (i) A

(ii) The energy transfer from chemical energy per coulomb of charge leaving (or entering) the cell.

(iii) 1.60 V

(b) (i) 1.20 V   (ii) 0.33 Ω

(c) (i) 0.33 Ω   (ii) the internal resistance

7 (a) $E = I(R + r)$, so $\dfrac{E}{I} = R + r$. Hence $R = \dfrac{E}{I} - r$

(b)

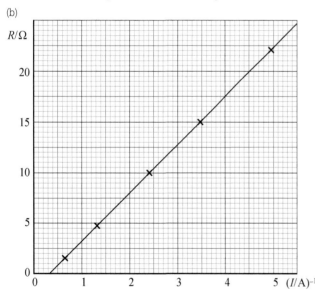

(c) The points lie on a straight line with a positive gradient in line with the equation. There is a negative intercept on the R axis, again as predicted. There is very little scatter in the points about the best-fit line, giving good support to the equation.

(d) $E = 4.76$ V; $r = -1.5$ Ω

(e) To preserve the energy store in the battery (stop it going flat!)

8 (a) (i) 100 Ω   (ii) 44 Ω

(b) (i) 6.4 V   (ii) 106 mA

(c) pd (for both) $= 2.8$ V; I(bulb) $= 51.5$ mA; I(resistor) $= 28.5$ mA

(d) current (for both) $= 43$ mA; V(bulb) $= 1.7$ V; V(resistor) $= 4.3$ V

2.4

1 'Moving' and 'travelling' mean the same, so the answer is confused. It should also mention the particles. Better: The particles in a transverse wave oscillate at right angles to the direction of travel of the wave. The direction of oscillation of the particles in a longitudinal wave is parallel to the direction of travel.

2. (a) With separations of 2λ, 3λ etc. the particles will also be in phase.

(b) If the distance is measured at an angle to the direction of propagation it will be larger than the wavelength.

3 With answer 2, the separation might be 2λ, 3λ etc. or the distance between any two points with a max (or min) midway between.

4. (a) (i) 1.4 mm   (ii) 0.40 m

(b) (i) For the lowest speed, the wave has moved 0.08 m

So lowest speed $= \dfrac{0.08\text{ m}}{0.00025\text{ s}} = 320$ m s^{-1}

(ii) For the next speed, the wave has moved 0.48 m → 1.92 km s^{-1}

(c) (i) 1.25 ms   (ii) 800 Hz

5 (a) (i) 0.100 m, 0.300 m, 0.500 m, 0.700 m, 0.900 m;

(ii) 0.200 m, 0.600 m;

(iii) 0.000 m, 0.400 m, 0.800 m.

(b) Max vertical speed = max gradient × horizontal speed

= 0.022 (est.) × 320 m s−1 = 7 m s^{-1} (1 sf).

6 (i) 0.55 m, 0.95 m, 1.35 m, 1.75 m

(ii) 0.35 m, 0.75 m, 1.15 m, 1.55 m, 1.95 m

7 (a) (i) 0.04 m   (ii) 5.0 Hz

(b) (i) 0.050 s   (ii) With this: speed $= \dfrac{0.30\text{ m}}{0.050\text{ s}} = 6.0$ m s^{-1}

(iii) The time between A and B could be 0.25 s (i.e. 1.25T) which gives 0.2 × speed.

(c) (i) The minimum separation between two points oscillating in phase measured along the direction of propagation.

(ii) 1.2 m.

8 (a) From the information, $f = 2.5$ Hz and $T = 0.4$ s.

(b)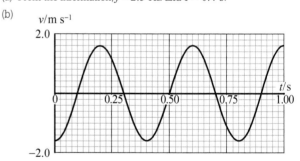

9

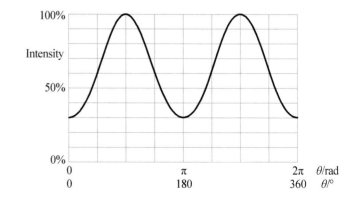

10 (a) The wave speed is lower in the shallow water above the lens shape. The closer to the middle the more the wavefront is held back.

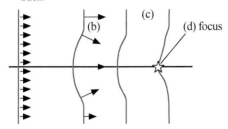

11 (a) The light in between the polaroids will be 100% polarised at $90°$ to the transmission direction for the second polaroid – so no transmission.

(b) The light hitting the second polaroid is polarised at $45°$ to the transmission direction, so some is transmitted – with a polarisation direction at $45°$ to the third polaroid, so again some is transmitted.

12. $A = 10$ cm

$6.28\lambda = 2\pi$, so $\lambda = 1.00$ m. So $\omega = 2\pi f = 2\pi \dfrac{v}{\lambda} = 1600\pi$ (rad) s^{-1}

At $x = 2.5$ m and $t = 0$, $y = 10$ cm $\times \cos(6.28 \times 2.5) = -10.0$ cm

$\therefore$ It is in antiphase to a cos function and $\phi = \pm\pi$ [or $\pm3\pi$, $\pm5\pi$ ....]

2.5

1 (a) The total displacement from two (or more) waves at any point is the vector sum of their individual displacements at that point.

(b) In some places the waves will be in phase, giving an amplitude of 1.5 cm $+ 2.0$ cm $= 3.5$ cm. If the waves are in antiphase, the amplitude will be the difference in the amplitudes, i.e. 0.5 cm. This is the minimum amplitude.

2. B and E must be correct.

3. (a) Yes because their frequencies are identical – they are driven by the same signal generator.

(b) Path difference $= \sqrt{120^2 + 50^2} - 120 = 130 - 120 = 10$ cm

(c) 20 cm

(d) The sound level would be low because the waves arrive in antiphase: the path difference is $\lambda/2$. The principle of superposition.

(e) The level of sound increases to a maximum on the line of symmetry. The phase difference between the waves from A and B decreases towards zero so the waves reinforce each other.

4. (a) So that the frequencies of the two waves are identical.

(b) 336 Hz

(c) The stationary waves are formed by the superposition of the waves travelling in opposite directions. Near a speaker, the amplitude of one wave is much greater than that of the other, so there is incomplete cancellation at a node.

5 (a)

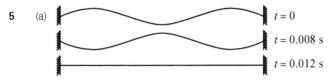

$t = 0$

$t = 0.008$ s

$t = 0.012$ s

(b) 1.65 m $= 1.5\lambda$, so $\lambda = 1.10$ m.

$v = \dfrac{\lambda}{T} = \dfrac{1.10 \text{ m}}{0.016 \text{ s}} = 68.8$ m s^{-1} (3 sf)

(c) All the points between 0 and 0.55 m from either end oscillate in phase. The points in the middle third of the string all oscillate in phase with one another and in antiphase with those in the outer thirds.

(d) 3.30 m; 20.8 Hz

6 $\dfrac{\lambda}{4} = 188 \pm 3$ mm (uncertainty 1.4%)

$\therefore v = 331 \pm 5$ m s^{-1}

7 The equations are: $188 \pm 3 = \dfrac{\lambda}{4} - e$ and $576 \pm 3 = \dfrac{3\lambda}{4} - e$

Subtracting and multiplying by $2 \rightarrow \lambda = 776 \pm 12$ mm $= 0.776 \pm 0.012$ m. This gives $v = 341 \pm 5$ m s^{-1}

And $e = 6 \pm 6$ mm!

8 (a) Diffraction

(b) The point at O receives diffracted waves from both slits. The waves add by superposition. The waves are in phase because the path lengths from the two slits are equal. Because of this the waves interfere constructively, i.e. the resultant amplitude is twice the amplitude in part (a) – so the intensity is 4 × the intensity in part (a).

(c) As the probe moves towards P the probe continues to receive waves from both slits but the waves from the slits are increasingly out of phase because the path length from the bottom slit is greater than from the top slit. When this path difference is $\lambda/2$ the waves are exactly out of phase so (almost) cancel each other out. Further motion in the same direction results in the waves approaching λ path difference at which point they fully reinforce again. This effect is called interference.

(d) (i) (I) 2.94 cm   (II) 2.76 cm.

(ii) The Young's formula which is derived for $d \gg D$ gives quite an accurate answer even when $d = 8$ cm and $D = 50$ cm.

9 (a) The slit separation, $x = 0.64$ mm.

(b) ~ 10 fringes. The interference pattern is mainly visible in the central diffraction maximum. 10×2 mm gives 2 cm.

(c) (i) The central maximum would be ~ 1.6 cm wide and would contain 10 fringes 1.6 mm apart. Both the diffraction and interference pattern spreads are proportional to the wavelength.

(ii) No interference pattern is seen but the diffraction patterns overlap to give a mixture of red and green (appearing as yellow) in the centre shading to red at the edges where the green diffraction pattern central fringe finishes.

(iii) The interference pattern is not affected but is half the intensity because of the absorption of the light which is polarised at right angles.

(iv) No interference pattern is seen. Two light beams at right angles cannot superpose to give a zero resultant. [NB. It is instructive to investigate a photon explanation for this effect.]

10 $\lambda = 2\ell$, so $f = \dfrac{v}{2\ell} = \dfrac{1}{2\ell} \times \sqrt{\dfrac{mg}{\mu}}$, i.e. $f = k\sqrt{m}$ where $k = \dfrac{1}{2}\sqrt{\dfrac{g}{M\ell}}$

11 Angles for the known wavelengths are consistent with a grating slit separation of 2.02×10^{-6} m. Using this value, the wavelength for the mystery line is 546 nm, in good agreement.

12 (a), (b) Diffraction pattern for the shorter wavelength light is more compressed. The extent of the pattern is proportional to the wavelength. The intensity scale for the 450 nm pattern is not to the same scale as the 650 nm.

(c) Slit width ~ 0.5 mm.

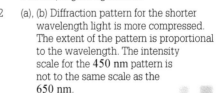

2.6

1 (a) $f = 0.6$ Hz

(b) $v = 3$ m s^{-1}; $\lambda = 5$ m

(c) 22.5° to normal

(d) 11.0° to normal

2 The speed of light in water is $\dfrac{c}{1.33}$ where c is the speed of light in a vacuum.

3 $n_1 > n_2$ and the angle of incidence, $\theta \geq \sin^{-1}\left(\dfrac{n_2}{n_1}\right)$

4 (a) 21.7°

(b) 62.8°

5 (a) 1.41

(b) 67°

(c) 50°

6 (a)

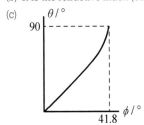

(b) It is the refractive index (1.50)

(c)

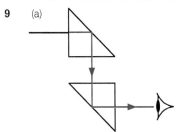

(d) 1.50

7 (a) 1.91×10^8 m s^{-1}

(b) 32.7°

(c) 48.8°

(d) The light ray, travelling in perspex is incident on a boundary with water with an angle of incidence greater than 63.2°.

(e) 0.0293%

(f) 7(.3) m.

8 (a) 1.56

(b) 1.90×10^8 m s^{-1}

(c) 1.05×10^{-4} s

(d) 1.02×10^{-6} s. If data bits are 1 μs apart they will overlap after 20 km. So bit rates must be much less than 1 Mbit s^{-1}

9 (a)

(b) 1.41 (3 s.f.)

10 Ray emerges from middle of opposite face (4.02 cm from top) at 45° below the normal.

11 Angles of incidence on the horizontal boundaries are respectively: 54.7°, 61.0°, 70.4°. TIR occurs at the boundary between the 1.30 and 1.20 layers, with an angle of incidence of 70.4°.

12 See graph below. With these max/min lines

$m_{max} = 1.586$

$m_{min} = 1.435$

$\therefore n = 1.51 \pm 0.08$

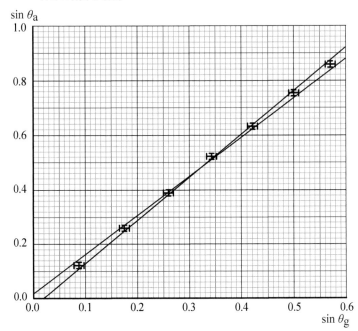

13 (a) 28.4°

(b) [Stretch & challenge]
Emerges from the long face anti-parallel to the incident ray

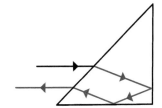

14 (a) 8.3 μm

(b) 0.72 ms

(c) 2.1 mm

15 (a) (i) 86.0°, (ii) 0.12 μs

(b) (i) There is a range of arrival times between that for light rays travelling parallel to the axis and those travelling at the maximum angle of 4° to the axis. Hence the pulse will be spread out. The peak is lower because the total energy in the pulse is the same.

(ii)

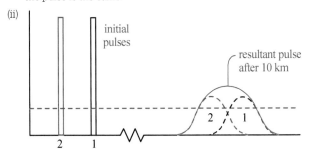

After 10 km both pulses have spread out and overlap as shown in the diagram. The resultant pulse is a merged one, which doesn't drop below the detection threshold, so the two would not be separated by the detection circuits.

16 (a) λ_{red} = 464 nm

λ_{violet} = 261 nm

(b) angle between red and violet rays = 1.6°

2.7

1 1.0×10^{17} s^{-1}

2 About 2 million

3 (a) 1.5 eV
(b) 1.0 eV

(c) The photon energy of the 3×10^{14} Hz radiation is 2.0×10^{-19} J (= 1.2 eV). This is less 2.5 eV; photon energies do not combine to cause photo emission.

(d) 1.0 V. The electrons would lose 1.0 eV with this pd.

4 Assuming 400 nm – 700 nm wavelength range: 2.8×10^{-19} J (1.8 eV) – 5.0×10^{-19} J (3.1 eV).

5 The radiation from other stars will cause promotion to all higher levels because all the energy differences are less than 4 eV. The 4s level is 2.1 eV above the 2s so there will be a 2.1 eV line in the absorption spectrum. Transitions to the 2p and 3s energy levels need photons of less than 1.8 eV (infra-red), and to the 5s needs 3.4 eV photons (ultra-violet), so no more visible absorption lines are produced.

The following downward transitions produce visible photons:

5s → 3s (2.1 eV), 5s → 2p (2.8 eV) and 4s → 2s (2.1 eV)

So the visible emission spectrum has these lines. All other transitions produce either uv or ir photons.

6 (a) The electrons behave as waves. These waves are diffracted by the gaps in the planes and the diffracted waves interfere.

(b) 5.5×10^{-11} m.

(c) First order: 9.4°, 0.165 rad; 5.0 cm from midpoint
Second order: 19.2°, 0.334 rad 10.4 cm from midpoint

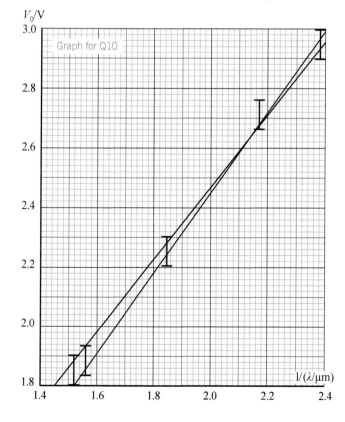

7 $\lambda = \dfrac{h}{p} = \sqrt{\dfrac{h}{2mE_k}}$. The values of E_k are the same:

$\therefore \dfrac{\lambda_e}{\lambda_p} = \sqrt{\dfrac{m_p}{m_e}} = 42.8$

8 3.3 m s^{-1}.

9 4.8×10^5 m^2. [= area of a square of side ~700 m]

10 gradient = $(1.28 \pm 0.07) \times 10^{-6}$ V m

→ $h = (6.8 \pm 0.4) \times 10^{-34}$ J s.

11 Wave: single slit diffraction or Young's slits
Particle: photoelectric effect

12 $\left[\dfrac{p^2}{2m}\right] = \dfrac{(kg\ m\ s^{-1})^2}{kg} = (kg\ m\ s^{-2}) \times m = N\ m = J = [E_k]$

Or: Show that $\left[\dfrac{p^2}{2m}\right]$ has base units of kg m^2 s^{-2} and say that this is the same as energy.

13 The higher the voltage is, the greater the KE and so the greater the momentum of the electrons. So from the de Broglie equation, the higher the voltage is, the smaller the wavelength and hence the smaller the diffraction angles – leading to smaller circles.

14 (a) IR (b) X-ray (c) visible (d) UV

15 3.9×10^{-11} m; X-ray

16. The polarity of the power supply needs swapping [or the photocell] and the power supply needs to be variable.

17 (a) Straight line of positive gradient with a negative intercept on $E_{k\,max}$ axis, i.e. the graph is of the form $y = mx + c$. The gradient of the graph h, the Planck constant; ϕ, the work function of the metal is the negative of the intercept on the $E_{k\,max}$ axis.

(b) The stopping voltage, V_s, is measured for each frequency. $E_{k\,max} = eV_s$.

(c) The values of $E_{k\,max}$ will all be higher by the same amount, so the graph will be parallel and higher. The intercept is less negative because the work function, ϕ, is smaller.

2.8

1 (a) Energy can be conserved if the atom de-excites from state 2 to state 1 if a photon is emitted, which carries away the excess energy. In order to conserve energy an external source of energy is required to excite the atom from state 1 to state 2.

(b) In the absence of an external energy source, all the atoms will eventually de-excite to state 1. With an external energy source, e.g. radiation, atoms can move up to state 2 by absorption of photons. Once there are atoms in state 2 the process of stimulated emission will ensure that as many atoms de-excite as excite and so there will always be more in state 1.

(c) Population inversion; pumping

2 (a) **Ground state**: the lowest available energy state. For a single electron it is the lowest energy level – level 1.

Population inversion: a situation in which more atoms are in a higher state than a lower. For the 4-level laser, more atoms would be in level 3 (4.5 eV) than in level 2 (2.5 eV).

Laser transition: the movement downwards between two levels as a result of stimulated emission; in this case the transition from level 3 to level 2.

Pumping: raising atoms to a high energy level to establish a population inversion; in this case pumping takes atoms from level 1 to level 4.

(b) Level 3 is the upper energy level of the laser transition. Level 2 is the lower level, which must have a short lifetime (compared to level 3) to allow an accumulation of atoms in level 3 and the establishment of a population inversion between these levels. Level 4 (the pumped state) must be short-lived so that level 3 is continuously filled, and so that level 4 itself stays empty enough to accommodate electrons pumped at a high rate.

(c) 0.33 (33%) assuming that the transition from 3 to 2 is by stimulated emission.

3 The two processes absorption and stimulated emission can both occur. In order for stimulated emission to dominate, more atoms need to be in the upper state, i.e. there must be a population inversion.

4 The two photons have the same frequency (wavelength), phase, direction of propagation and direction of polarisation.

5 If fewer than half the atoms are pumped then over **50%** of the atoms must still be in the ground state. Hence U must have <u>fewer</u> than 50% of the atoms, which is less than that of G.

6 (a) An inelastic collision is one in which kinetic energy is lost.

(b) A metastable state is one which lasts a relatively long time before decaying into a lower energy state.

7 (a)

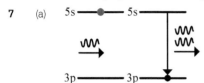

(b) The second photon has the same frequency and polarisation as the first, is in phase with it and travelling in the same direction.

8 The 3p state needs to be shorter lived than the 5s because it needs to empty quickly, so that its population is always less than that of the 5s.

9 633nm: 5s → 3p; 1.15 μm: 4s → 3p; 3.39 μm: 5s → 4p (shorter wavelength = greater energy difference)

Regions: 633 nm: visible (red); 1.15 μm and 3.39 μm: IR

10 5s 20.65 eV $= 3.30 \times 10^{-18}$ J

4p 20.28 eV $= 3.25 \times 10^{-18}$ J

4s 19.77 eV $= 3.16 \times 10^{-18}$ J

3p 18.69 eV $= 2.99 \times 10^{-18}$ J

3s 16.62 eV $= 2.66 \times 10^{-18}$ J

11 339 nm, near UV

12 A helium atom in the excited state collides with a neon atom in the ground state. The excited electron in the helium atom transfers energy to a ground state electron in the neon, promoting it to the 5s state.

13 (a) 33 V

(b) 33 eV

(c) (see sketch)

(d) 20.65 V

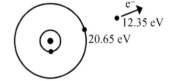

14 58 (max)

Chapter 3

1 (a) (i) length (ii) resistance

(iii) diameter and material of wire

(b) Resistance of leads of the multimeter. Connect the two leads together and measure their resistance. Subtract this value from all the readings.

2 For each sphere:

- Determine its mass using the electronic balance – use the resolution to estimate the percentage uncertainty in the mass.

- Measure the diameter across several diameters using the digital callipers.

- Determine the mean of the diameter of the readings and the percentage uncertainty in the diameter using the spread of the readings.

- Calculate the density, using $\rho = \dfrac{m}{\frac{4}{3}\pi\left(\frac{D}{2}\right)^2}$ along with its absolute uncertainty.

- Compare the density values and check that the ranges of the uncertainties overlap.

3 (a) $p(\ell) = \dfrac{0.1 \text{ cm}}{75.3 \text{ cm}} \times 100\% = 0.132\%$;

$p(w) = \dfrac{0.1 \text{ cm}}{24.6 \text{ cm}} \times 100\% = 0.407\%$;

$p(t) = \dfrac{0.1 \text{ cm}}{1.113 \text{ cm}} \times 100\% = 0.090\%$,

∴ width had the greatest **%** uncertainty.

(b) (2.06 ± 0.01) cm³.

(c) The total width would be halved, i.e. ~ 12.3 cm.

So $p(w) = \dfrac{0.001 \text{ cm}}{12.3 \text{ cm}} \times 100\% = 0.008\%$.

This reduces p(V) to **0.23%**, i.e. less than half of the original, so correct.

4 (a) **Plan**:

- Mark a series of lines at e.g. 10 cm, 20 cm, 40 cm, 60 cm, 90 cm and 120 cm across the inclined plane.

- Release the ball and use the stopwatch to measure the time, t, taken to roll to each of the lines. Repeat the readings to identify any rogue readings.

- Plot a graph of x against t^2, using the mean values of t.

- As $u = 0$, the graph should be a straight line through the origin with gradient $\frac{1}{2}a$, so check that the form of the graph is correct.

- If the graph is a straight line through the origin, determine the gradient and double it to determine the acceleration.

(b) Using $a_{max} = g \sin 8.5° = 1.45$ m s⁻²;

$a_{min} = g \sin 7.5° = 1.28$ m s⁻².

$u = 0$,   so $a = \dfrac{2x}{t^2}$.

Ignoring the uncertainty in x (very small), from Paul's result

$a_{max} = \dfrac{2 \times 1.000 \text{ m}}{(1.45 \text{ s})^2} = 0.95$ m s⁻².

This is lower than the minimum found from $a = g \sin \theta$, so the results are inconsistent with the equation.

5 (a) It is possible to draw a straight line through all the error bars. See the graph. Hence there is a linear relationship between v and t. The max/min lines both have a positive gradient and a positive intercept on the v axis, so the results are consistent with $v = u + at$ where u and a are both positive.

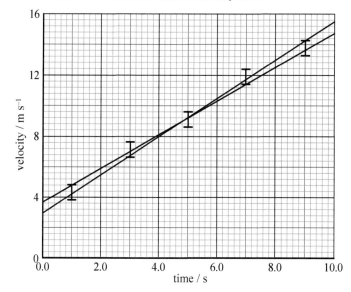

(b) $u = 3.4 \pm 0.4$ m s^{-1}; $a = 1.17 \pm 0.07$ m s^{-2}.

(c) $x = \frac{1}{2}(u + v)t$, so, using the plotted data:

$x = 5.0 \times [(3.4 \pm 0.4) + (15.1 \pm 0.4)]$

$\quad = 5.0 \times [18.5 \pm 0.8] = 92.5 \pm 4.0$ m [or 93 ± 4 m]

The value of 95 ± 3 m overlaps with 93 ± 4 m so it is consistent with the plotted data.

[Note: this is not the only way of answering this question.]

6 (a) Volume of shot, V = Increase in volume reading

$= (48.0 \pm 0.5) \text{ cm}^3 - (23.0 \pm 0.5) \text{ cm}^3$

$= (25.0 \pm 1.0) \text{ cm}^3$

$\therefore p_V = \frac{1.0}{25.0} = 0.040 \ (= 4.0\%)$

Similarly, Mass of shot, $M = (180.25 \pm 0.02)$ g;
$p_M = 0.01\%$ (negligible)

$\therefore$ Density $= \frac{M}{V} = \left(\frac{180.25}{25.0} \pm 4.0\%\right)$

$\quad = (7.2 \pm 0.3) \text{ g cm}^{-3}$

(b) Yes, the air would increase the volume taken up, without increasing the mass significantly (as the density of air is less than one thousandth that of the shot).

The difficulty could be overcome by stirring the shot gently, watching for bubbles rising through the water.

Chapter 4

1 (a) 72 (b) 206 (c) 1.125 (d) 0.4
(e) 16 (f) 216 (g) 4 (h) 160 000
(i) 125 (j) 1 000 000 (k) 100 000
(l) 1100 (m) 8 (n) 0.001 (o) 64
(p) 2.5 (q) 6.0×10^{-7} (r) 11
(s) 3.3×10^{-18} (t) 2.0×10^{-27}
(u) 1.1×10^{-27}

2 (a) 4.44 (b) 9.13 (c) 4.92 (d) −2.86
(e) 4.17 (f) 12.8 (g) 1.19×10^{57}
(h) 2.01×10^{-37}

3 (a) 71.4% (b) 4.03% (c) 64%
(d) 0.5% (e) 4.81×10^8% (f) 89.4%
(g) 0.0714%

4 (a) $\frac{\pi}{2} = 1.57$ rad (b) 0.61 rad
(c) 2.62×10^{-3} rad (d) 7.24 rad
(e) 4.36×10^{-8} rad

5 (a) 5.73° (b) 7.5° (c) 18°
(d) $(4.8 \times 10^{-7})°$ (e) 9.00°

6 (a) $x > 5$ (b) $x < -2.5$ (c) $x < 2$
(d) $x < 3$ (e) $x > 0.5$ (f) $1 < x < 1.5$
(g) $x > 5$ (h) $\frac{5}{3} < x < \frac{7}{3}$ (i) $0 < x < 1$
(j) $x < 0$ or $x > \frac{5}{6}$

7 (a) If $x \sim v$ then $\frac{x}{v} =$ constant.

Checking this: $\frac{25}{5} = \frac{75}{15} = 5$.

So consistent.

(b) z

(c) y, because $75 = 3 \times 25$ and $4 = \frac{1}{3} \times 12$

(d) $w \propto x^2$ and $u \propto x^{-2}$.

(e) With only two data points there are endless functions which can be found. Consider x and y. The two data points are $(x, y) = (25, 12)$ and $(75, 4)$. These points both lie on graphs of each of the following functions: $4x + 25y = 400$; $xy = 300$; $x(y^2 + 48) = 4800$.

[It is suggested that you verify that the two points satisfy each of these function and then find constants c and k for which the two points satisfy the function $x^2(y + c) = k$]

(f) $a = \frac{5}{8}$ and $b = \frac{15}{4}$.

When $v = 10$, $w = 10$.

8 (a) $\theta = 49°, \phi = 41°, x = 18.5$
(b) $\theta = 50°, x = 11.5, y = 9.6$
(c) $\theta = 15°, \phi = 75°$
(d) $x = 34$ m, $y = 94$ m
(e) $\theta = 52°, x = 8.8$ m
(f) $F_h = 5.9$ N, $F_v = 2.7$ N

9 (a) 37°, 143° (b) 37°, 323°
(c) 197°, 343° (d) 134°, 226°

10 (a) (i) $\cos\theta = 0.917$; $\tan\theta = 0.436$
(ii) $\sin\theta = 0.954$; $\tan\theta = 3.18$
(iii) $\sin\theta = 0.287$; $\cos\theta = 0.958$

(b) The same answers as (a) but with $\pm$ sign before each.

11 (a) The gradient.

(b) The gradient becomes 0 at $t = 7.5$ s, so this is where the velocity becomes zero.

(c) 18.6 m s^{-1} [allow yourself ± 1 m s^{-1} tolerance]

(d) Gradient of tangent at 3.5 s = 19 m s^{-1} [allow yourself ± 2.0 m s^{-1}]

(e) Gradient of tangent at origin = 38 m s^{-1} [allow yourself ± 3.0 m s^{-1}]

(f) Method 1: measure the velocity at a series of times up to 7.5 s and plot a velocity–time graph. A straight line indicates a steady deceleration.

Method 2: Plot of a graph of displacement against the (time before 7.5 s)^2. Again, a straight line would indicate a steady deceleration.

12 (a) The acceleration is the gradient; the displacement is the area under the graph.

(b) Before: 60 m s^{-1}; after 7 m s^{-1}.

(c) −18 m s^{-2} [allow ± 1 m s^{-2}]

(d) Max deceleration (at ~ 1.0 s) = − gradient of tangent = 52 m s^{-2} [allow $50 - 55$ m s^{-2}]

(e) 85 m [allow yourself ± 5 m]

13 (a) Equation of graph is $v = 5.6 + 1.26t$

(b) Initial velocity = 5.6 m s^{-1}

14 (a) $a = -0.203$ V mA$^{-1} = -0.203$ kΩ
$b = 9.97$ V

(b) $E = b = 9.96$ V; $r = -a = 203$ Ω

Exam practice answers

Unit 1

1 (a) Moment = Force × perpendicular distance from the point to the line of action of the force

   (b) (i) Clockwise moment = $52 \text{ N} \times 0.15 \text{ m} = 7.8 \text{ N m}$

      (ii) $F \times 0.58 = 7.8$, $\therefore F = 13.4 \text{ N}$ [Note: 8.0 N m can be used → $F = 13.8 \text{ N}$]

   (c) In position 2, the perpendicular distance of the line of action of the weight from the hinge is smaller, so the clockwise moment is decreased. This means that the anticlockwise moment of the force in the bar is less. So the force in the bar must be less and Bethan is correct. [Note: the converse argument about position 1 can be used.]

      Alternative argument: The perpendicular distance between the metal bar and the hinge is greater in position 2 so the force is less [even] for the same moment.

2 (a) The rubber band is suspended from the bar of a clamp stand as shown in the diagram. The rule is placed close to the mass hanger.

      The position of the base of the hanger is measured using the mm scale, with the eye at the same level, when no other masses are added. 100 g masses are added one after the other and the position of the hanger read each time. The extensions are calculated by subtracting the initial reading. The force is calculated using $F = mg$.

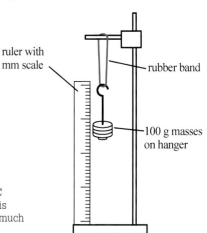

ruler with mm scale — rubber band — 100 g masses on hanger

   (b) (i) $\varepsilon = \dfrac{6.0 \text{ cm}}{8.0 \text{ cm}} = 0.75$

      (ii) At B $\sigma = \dfrac{7.0 \text{ N}}{0.050 \times 10^{-4} \text{ m}^2} = 1.4 \times 10^6 \text{ Pa}$. Note: 140 N cm^{-2} is acceptable.

         $E = \dfrac{\sigma}{\varepsilon}, \therefore \dfrac{1.4 \times 10^6 \text{ Pa}}{0.75} = 1.87 \times 10^6 \text{ Pa}$.   Note: 187 N cm^{-2} is acceptable

   (c) At C the increasing force causes the rubber molecules to straighten out [by the rotation of the C–C bonds] which requires only a small additional force. Hence the gradient is small. At D this process is complete and additional extension requires the lengthening of the covalent bonds, which needs a much larger force and so the gradient is large.

3 (a) (i)

Delta particle:	Δ++	Charge $+2$		Baryon number 1
Electron:	e−	Charge -1		Baryon number 0
Pion:	π−	Quark combination: $\mathrm{d\bar{u}}$		Baryon number 0

      (ii) Electron

   (b) **Charge**: LHS, $Q = (-1) + 1 = 0$; RHS, $Q = (-1) + 2 + (-1) = 0$ so conserved.

      Lepton number: LHS, $L = 1 + 0 = 1$; RHS, $L = 1 + 0 + 0 = 1$ so conserved.

   (c) (i) $\Delta^{++} = \mathrm{uuu}$, $\mathrm{p} = \mathrm{uud}$, $\pi^{+} = \mathrm{u\bar{d}}$

        So U = up quark number = 3 on both sides of the equation, so conserved.

        Down-quark number: LHS, $D = 0$; RHS $D = 1 + (-1) = 0$, so conserved.

      (ii) The very short lifetime (10^{-24} s) is typical of strong decays.

        Only quarks are involved – with no change of quark flavour.

   (d) When the proton and electron were discovered, the discoveries had no practical application. Now there are many uses involving the knowledge of these particles, such as electronic devices (e.g. computers) and proton beam therapy.

      [Note: for this 'issues' question, many acceptable answers are possible.]

4. (a) (i) Work done = $65 \text{ kW} \times 32 \text{ s} = 2080 \text{ kJ} = 2.08 \text{ MJ}$ [2.08×10^6 J]

      (ii) Efficiency $= \dfrac{\text{Work out}}{\text{Energy in}} = \dfrac{mgh}{W_{\text{in}}} = \dfrac{2600 \text{ kg} \times 9.81 \text{ N kg}^{-1} \times 42 \text{ m}}{2.08 \times 10^6 \text{ J}} = 0.52 \text{ or } 52\%$

   (b) Loss in gravitational PE = $2600 \times 9.81 \times 30 = 7.65 \times 10^5$ J

      Work done against friction = $2.8 \text{ kN} \times 36 \text{ m} = 1.01 \times 10^5$ J

      $\therefore$ KE gain = $(7.65 - 1.01) \times 10^5 \text{ J} = 6.64 \times 10^5$ J.

      $\therefore E_k = \dfrac{1}{2}mv^2$, $\therefore$ Speed, $v = \sqrt{\dfrac{2E_k}{m}} = \sqrt{\dfrac{2 \times 6.64 \times 10^5}{2600}}$. The values of $= 22.6 \text{ m s}^{-1}$

5. (a) The wavelength of the maximum intensity of the continuous spectrum can be used to calculate the temperature of surface [photosphere] of the star, using Wien's law: $\lambda_{max} = \dfrac{W}{T}$. The area under the graph gives the intensity, I, of the radiation received from the star, so this can be used, together with the distance, d, of the star to calculate the luminosity, L, of the star using $L = I \times 4\pi d^2$. Together with Stefan's law: $L = A\sigma T^4$, the diameter of the star can be estimated.

The detail shows us the line absorption spectrum. This arises because the radiation from the star's surface passes through the stellar atmosphere. Individual atoms in this gas absorb specific wavelengths of light which are characteristic of the element – hence the chemical composition of the star can be investigated. The strengths of these absorption lines also give information about the temperature and the stage of the star in its life.

(b) (i) Received power per unit area, $I = \dfrac{L}{4\pi d^2}$, where L is the luminosity and d the distance.

$\therefore L = 4\pi \times (1.58 \times 10^{17} \text{ m})^2 \times 1.32 \times 10^{-8} \text{ W m}^{-2}$

$= 4.14 \times 10^{27} \text{ W}$

(ii) $L = A\sigma T^4$

$\therefore 4.14 \times 10^{27} \text{ W} = A \times 5.67 \times 10^{-8} \text{ W m}^{-2} \text{ K}^{-4} \times (7700 \text{ K})^4$

$\therefore A = 2.07 \times 10^{19} \text{ m}^2 = 4\pi \left(\dfrac{D}{2}\right)^2$, where D is the diameter.

$\therefore D = 2.6 \times 10^9 \text{ m}$

Unit 2

1 (a) (i) The emf of a battery is the energy transfer from chemical energy within the cell per unit charge passing through the battery.

(ii) Total external resistance = 6.60 Ω. $\therefore$ Current, $I = \dfrac{4.33 \text{ V}}{6.60 \text{ } \Omega} = 0.656 \text{ A}$.

$\therefore r = \dfrac{4.80 \text{ V} - 4.33 \text{ V}}{0.656 \text{ A}} = 0.72 \text{ } \Omega \text{ (2 sf)}$

[Note: other methods of calculating r are possible, e.g. the potential divider equation can be used instead.]

(iii) (I) The resistance of the parallel combination is less than that of the resistors in series, so the total resistance of the circuit is less. Hence the current is greater. Since r is a constant in $V = E - Ir$, the greater the current, the lower the terminal pd, V.

[Note: again a potential divider argument could be used, e.g. the external resistance is a smaller fraction of the total resistance so has a smaller fraction of the total pd.]

(II) Current through either resistor = $\dfrac{3.35 \text{ V}}{3.30 \text{ } \Omega} = 1.015 \text{ A}$

$\therefore$ Number of electrons per minute = $\dfrac{1.015 \text{ A}}{1.60 \times 10^{-19} \text{ C}} \times 60 \text{ s} = 3.8 \times 10^{20}$

(b) (i) $P = \dfrac{V^2}{R}$, so $R = \dfrac{(4.33 \text{ V})^2}{1000 \text{ W}} = 52.9 \text{ } \Omega$

(ii) $E = Pt = 1000 \text{ W} \times 3600 \text{ s} = 3\,600\,000 \text{ J} = 3.6 \text{ MJ}$

(c) If electrical heating is used, more CO_2 is produced for the same heating effect, if the electricity is generated by a gas-fired power station. This makes a greater contribution to the greenhouse effect / global warming and should be discouraged. However, electricity is increasingly generated using renewable / non-greenhouse installations, e.g. wind or solar power, or nuclear power stations, and the commitment to a zero net carbon economy means that gas heaters will need phasing out.

2. (a)

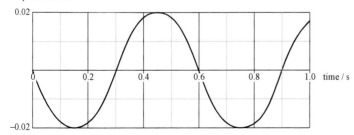

(b) (i) $\lambda = d\sin\theta = 1500\text{ nm} \times \sin 24.9° = 632\text{ nm}$

(ii) For the second order spectrum,

Path difference $AC = 2\lambda = 1264\text{ nm}$

$$\therefore \theta = \sin^{-1}\left(\frac{1284}{1500}\right) = 57.4°$$

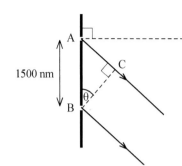

3. (a) (i) $\sin\theta, = 1.60 \times \sin 30°$

$\therefore\ = 53.1°$

(ii) $\theta_C = \text{critical angle} = \sin^{-1}\left(\frac{1}{1.60}\right)$

$$\sin\theta_C = \frac{x}{100\text{ mm}}$$

$$\therefore \frac{x}{100\text{ mm}} = \frac{1}{1.60}$$

$$\therefore x = \frac{100}{1.60}\text{ mm} = 62.5\text{ mm}$$

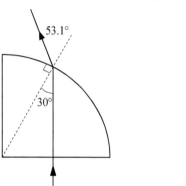

(b) The fibre consists of a low-impurity glass core with a cladding which has a lower refractive index. Light travels through the core at small angles to the axis, so that the angle of incidence with the cladding is greater than the critical angle for these materials. The light is totally internally reflected at the core-cladding boundary and is able to propagate for long distances with very low loss.

The light travels at a range of angles to the axis. These paths have different lengths between two widely separated positions along the fibre, so that for the same starting time, the arrival times are slightly different and a pulse of light becomes spread out. Successive pulses overlap, increasingly as the distance increases, until eventually they cannot be distinguished.

4. (a) (i) To eject an electron, a photon must have at least energy ϕ. The energy of a photon of frequency f is hf, where h is Planck's constant. Therefore, if the frequency is lower than ϕ/h, the photon will have insufficient energy. Increasing the intensity only increases the number of photons per second – the photon energy is unchanged.

(ii) The photon energy $E_{ph} = hf = 6.63 \times 10^{-34} \times 6.59 \times 10^{14} = 4.37 \times 10^{-19}\text{ J}$.

This is less than the work functions for calcium and zinc – so these are ruled out!

0.35 V corresponds to an energy of $0.35 \times 1.6 \times 10^{-19}\text{ J} = 0.56 \times 10^{-19}\text{ J}$. Hence the work function of the metal must be between 4.37×10^{-19} and $(4.37 - 0.56) \times 10^{-19}\text{ J}$, i.e. $3.81 \times 10^{-19}\text{ J}$: Barium is the only one which is possible.

(b) (i) See graph

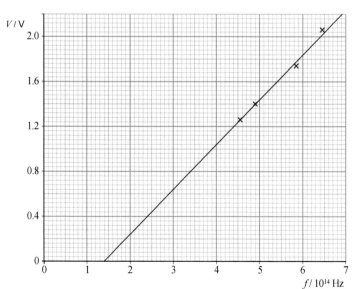

(ii) The equation predicts a straight line through the origin. The line is straight, with only a small scatter of points on either side. But it does not pass through the origin, so is only partly in agreement with the equation.

(iii) Gradient $= \dfrac{2.20\text{ V}}{(6.90 - 1.40) \times 10^{14}\text{ Hz}} = 4.0 \times 10^{-15}\text{ V s}$.

$h = \text{gradient} \times e = 4.0 \times 10^{-15} \times 1.60 \times 10^{-19} = 6.4 \times 10^{-34}\text{ J s}$.

Glossary

Absolute (or Kelvin) temperature scale   A scale whose zero (0 K) is the temperature (absolute zero) at which atoms have their lowest possible random energy. The temperature, T, in kelvin is related to the temperature, θ, in °C by $T/\text{K} = \theta/°\text{C} + 273.15$.

Absorption spectrum   The drop in intensity of radiation at certain wavelengths due to absorption by a material.

Acceleration (a)   If the velocity of a body is changing it is said to be accelerating, i.e. acceleration is defined by the change in velocity per unit time (units are m s^{-2}).

Acceleration (mean)   Change in velocity/time taken ($\Delta v/\Delta t$) (units are m s^{-2}).

Acceleration due to gravity (g)   Free objects, in the absence of air resistance, fall to Earth with the same acceleration. Close to the surface of the Earth this acceleration is almost constant at $g = 9.81$ m s^{-2}.

[Air] resistance (drag)   Force that opposes relative motion between an object and a fluid [= liquid or gas] through which the object is moving.

Amorphous   Literally, without form. The atoms of an amorphous substance are not arranged in a crystal lattice nor in parallel chains.

Antibaryon   Hadron composed of 3 antiquarks, e.g. antiproton.

Antiparticles   For each of the particles there is a corresponding antiparticle with an identical mass and (opposite) charge, if a charge is present.

Atom   The smallest particle of an element, consisting of a nucleus surrounded by electrons.

Baryon   Hadron composed of 3 quarks, e.g. proton, neutron.

Black body   A body (or surface) which absorbs all the electromagnetic radiation that falls upon it. It also emits more radiation at any wavelength than a non-black body in the continuous spectrum.

Brittle fracture   The failure, by crack propagation, of a brittle substance under tension.

Brittle substance   A substance that reaches its breaking point under tension without undergoing plastic extension.

Celsius temperature (θ) is defined by:
$\theta/°\text{C} = T/\text{K} - 273.15$

Centre of gravity (C of G)   The point at which we consider all of an object's weight to act.

Coherent sources   Two (or more) sources of waves with a constant phase difference.

Continuous spectrum   Consists of all wavelengths within a range.

Crystalline   Materials that have a periodic structure called a lattice.

Delta (Δ)   A useful symbol to represent the change of some quantity. So Δv = final velocity – initial velocity.

Density (ρ)   The ratio mass/volume for a substance (units are kg m^{-3} or g cm^{-3}).

Diffraction   The spreading of a wave when it meets an obstacle, such as the edges of a slit. Some of the wave's energy travels into the geometrical shadow of the obstacle.

Diffraction grating   An opaque plate ruled with parallel, equally spaced slits that produces beams of light at angles that depend on the wavelength of the light shone on to it.

Displacement (x)   A quantity which includes direction as well as distance (vector).

Ductile   Able to be drawn into wires. Ductile materials are also malleable.

Dynamic friction   The force which opposes the relative motion when one surface is sliding over another.

Dynamics   Concerned with the causes of motion and its changes. Why things move, not how.

Edge dislocation   Where an additional part of a plane of ions is present within the crystal lattice.

Efficiency   The fraction of the energy input which is transferred usefully by the system. Useful work out/work in.

Elastic collision   A collision where there is no change in the total kinetic energy.

Elastic limit   The point at which deformation ceases to be elastic.

Elastic potential energy   The energy stored in a body by virtue of its deformation.

Elastic strain   The strain that disappears when the stress is removed, i.e. the specimen returns to its original size and shape.

Electric current (I)   The rate of flow of electric charge (units are A)

Electromagnetic (e-m) interactions (forces)   Affect all charged particles and also affect neutral hadrons because quarks have charges. Infinite range.

Electron volt (eV)   The energy transferred when an electron moves through a potential difference of 1 V (1.00 eV $= 1.60 \times 10^{-19}$ J).

Elementary charge (e)   The charge on the proton. Its value is 1.602×10^{-19} C (4 s.f.). The charge on the electron is $-e$.

Elementary (or fundamental) particle   A particle that is not a combination of other particles.

Emission spectrum   The particular combination of wavelengths emitted by a substance. Atoms in a gas emit discrete wavelengths, so these 'lines' act as a 'fingerprint' that we can use to identify the gas.

Energy   The ability to do work.

Equilibrium   A body is in equilibrium if it is moving and rotating at a constant rate. In some cases this means it is not moving at all.

Extension ($\Delta \ell$)   The increase in length of an object in tension.

Flavour   The different types of quark, up, down, etc., are said to possess different flavours.

Fraunhofer lines   The dark lines in the solar spectrum are called Fraunhofer lines after the German scientist who noticed them in 1814.

Friction   A force that acts between two surfaces in contact, parallel to the surfaces, opposing their sliding over each other.

Gravitational interactions (forces)   Attractive forces acting between every particle of matter and every other particle. Affect all matter but negligible for subatomic particles. Infinite range.

Gravitational potential energy   The energy possessed by an object by virtue of its position or the positions of its particles.

Hadron   High mass particle consisting of quarks and/or antiquarks.

Homogeneous equation   Quantities can be added together or subtracted or equated only if they have the same units, and then the answer has the same units.

Hooke's law The tension (in a spring or wire) is proportional to the extension, provided that the extension is not too great.

Hysteresis Literal meaning: lag. For elastic substances its presence means that the graph of force against extension (or stress against strain) follows a different path when unloading from that when the load is increasing.

In phase Oscillations of the same frequency are in phase if they are at the same point in their cycles at the same time.

Inelastic collision A collision where kinetic energy is lost.

Intensity (I) The intensity of electromagnetic radiation is the power per unit area crossing a surface at right angles to the radiation.

Internal energy (U) Sum of the potential and kinetic energies of the particles in an object. It is often incorrectly called heat.

Ionisation The removal of one or more electrons from an atom.

Ionisation energy of an atom The minimum energy needed to remove an electron from the atom in its ground state.

Isolated system A system on which no external forces act and no energy or particles enter or leave.

Kelvin scale (K) On the kelvin scale, ice melts at 273.15 K, water boils at 373.15 K and absolute zero is 0 K.

Kinematics The science of motion and its mathematical description. How things move, not why.

Kinetic energy The energy of a body by virtue of its motion.

Lepton Low mass, elementary particles, e.g. electron, neutrino.

Limiting friction The maximum value of static friction (FR) between surfaces when a given normal contact force acts between them. If another force parallel to the surfaces and greater than FR is applied, the surfaces will start to slide over each other.

Line spectrum Consists of a series of individual wavelengths (or, more accurately, a series of very narrow wavelength bands).

Longitudinal wave A wave where the particle oscillations are in line with (parallel to) the direction of travel (or propagation) of the wave.

Luminosity (L) The luminosity of a star is the total power it emits in the form of electromagnetic radiation. (units are W m^{-2})

Malleable Able to be hammered into shape.

Mechanical energy Kinetic and gravitational potential energies are often grouped together under this title.

Meson Hadron composed of a quark and an antiquark, e.g. pion.

Moment (torque) (τ) The mathematical expression of turning effect of a force about a point and is the product of the force and the perpendicular distance from the point to the line of action of the force (units are N m)

Momentum (p) A vector quantity. $p = mv$, where v is the velocity of the body and m is the mass. Momenta (plural). (units are kg m s^{-1}).

Monomer A molecule with a double bond which is broken open to form the repeat unit of a polymer.

Monomode optical fibre: Fibre with a very thin core in which light can only propagate in only one direction (parallel to the axis).

Multimode optical fibre One in which light rays with a range of directions can propagate (by total internal reflection).

Multimode dispersion Data degradation because each pulse (in a multimode fibre) travels by a range of different paths, so arrives spread out over time and may overlap neighbouring pulses.

Neutrinos Neutral leptons of very low mass which only interact via the weak force.

Newton's 1st Law of Motion N1. A body's velocity will be constant unless a force acts upon it.

Newton's 2nd Law of Motion N2. The rate of change of momentum of a body is directly proportional to the resultant force acting upon it and takes place in the direction of the resultant force.

Newton's 3rd Law of Motion N3. If a body **A** exerts a force on body **B**, then **B** exerts an equal and opposite force on **A**.

Normal force (F_N) If an object rests against a surface, the surface exerts a force on the object. The word 'normal' is used to mean 'at $90°$'.

Perfectly elastic See *Elastic collision*.

Photoelectric effect The emission of electrons from a surface when light or ultraviolet radiation of short enough wavelength falls upon it.

Pion First generation meson. There are three types: π^+, π^-, π^0.

Plastic (or inelastic) strain The strain that decreases only slightly when the stress is removed, i.e. the specimen does not return to its original size and shape.

Point defect Where a lattice ion is missing or a 'foreign' atom or just an additional ion is present.

Polycrystalline A substance consisting of many small, randomly orientated crystals.

Polymer A substance whose molecules consist of long chains of identical sections called repeat units.

Population inversion A situation in which a higher energy state in an atomic system is more heavily populated than a lower energy state of the same system.

Potential difference (V) The work done between two points, X and Y, that is the loss of electrical potential energy, per unit charge passing between X and Y. (units are volt (V) = J C^{-1}.)

Power Work done per unit time or the energy transferred per unit time (units are W).

Principle of conservation of energy The total energy of an isolated system is constant though it can be transferred within the system and the way it is classified (e.g. kinetic or potential) can change.

Principle of conservation of momentum (CoM) The vector sum of the momenta of the bodies in a system is constant provided there is no resultant external force.

Principle of moments (PoM) For a body to be in equilibrium the sum of the anticlockwise moments about any point is equal to the sum of the clockwise moments about the same point.

Principle of superposition For waves, the resultant displacement at each point is the vector sum of the displacements that each wave passing through the point would produce by itself.

Projectile An object which is thrown/kicked/made to move obliquely upwards and continues on its path under the influence of gravity.

Pumping Feeding energy into the amplifying medium of a laser to achieve a population inversion.

Quantity A physical property of an object or material which can be measured. A quantity is represented by a number multiplied by a unit.

Quark Elementary particle, not found in isolation, which combines with other quarks or with an antiquark to form hadrons. The two first generation quarks are the up quark (u) and the down quark (d).

Refraction The change of direction of travel of light (or other wave) when its speed of travel changes, e.g. when it passes from one material into another.

Refractive index (n) Defined by $n = \dfrac{c}{v}$, where v and c are the speeds of light in the material and a vacuum respectively.

Resolution The smallest measurable change that can be observed using an instrument.

Resolving Finding the components of a force in particular directions.

Scalar quantity Has magnitude only (cf. vector).

Speed (mean) Distance travelled/time taken (units are m s^{-1}).

Spring constant (k) The ratio tension/extension. A constant for the spring provided that the limit of proportionality is not exceeded.

Standard model A unifying theory of fundamental particles.

Static friction (grip) (FR) A frictional force that stops two surfaces sliding over each other.

Stefan constant (σ) The constant in Stefan's law. $\sigma = 5.67 \times 10^{-8}$ W m^{-2} K^{-4}

Stefan's law (or the Stefan–Boltzmann law) The total power, P, of the electromagnetic radiant energy emitted by a black body of surface area A and temperature T is given by $P = \sigma A T^4$

Stimulated emission Emission of a photon from an excited atom, triggered by a passing photon of energy equal to the energy gap between the excited state and a state of lower energy in the atom (or molecule).

Strong interactions (forces) Affects all quarks and also affects interactions between hadrons (e.g. nuclear binding). Effective range of $\sim10^{-15}$ m.

Superconductor Material that loses all its electrical resistance below a certain temperature, the *superconducting transition temperature* (or superconducting critical temperature), θ_c.

Tensile strain (ε) The extension per unit length due to an applied stress. $\varepsilon = \Delta\ell/\ell$, where ℓ is the original length and $\Delta\ell$ the increase in length.

Tensile stress (σ) The tension per unit cross section. $\sigma = F/A$, where F is the tension and A the cross-sectional area.

Tension The force which an object exerts on external objects because it is being stretched. If an object has a tension, it stretches.

Transverse wave A wave in which the oscillations are in a direction at right angles to the direction of travel (or propagation).

Vector quantity Has magnitude and direction (cf. scalar).

Velocity (mean) Displacement/time taken (units are m s^{-1}).

Wavefront A surface on which, at all points, the oscillations are in phase.

Wavelength The wavelength, λ, of a progressive wave is the minimum distance (measured along the direction of propagation) between two points on the wave oscillating in phase. (units are m).

Weak interactions (forces) Affect all particles but are only significant when e-m and strong interactions are not involved. Effective range of $\sim10^{-18}$ m.

Wien constant (W) The constant in Wien's displacement law. $W = 2.90 \times 10^{-3}$ m K.

Wien displacement law The wavelength, λ_{max}, of peak emission from a black body is inversely proportional to the absolute temperature, T, of the body, that is $\lambda_{max} = W/T$.

Work Work is done when a force moves its point of application in the direction of the force. Work = force × displacement of point of application × cos (angle between force and displacement). The unit of work is the joule (J).

Work function The minimum energy required to remove an electron from a metal surface. The work function can be expressed in either joules or **electron volts**.

***xuvat* equations**
x = displacement;
u = initial velocity;
v = final velocity;
a = acceleration;
t = time.

Five equations, each relates four of the variables, so each has one variable missing.

Young modulus (E) For a material that obeys Hooke's law, the Young modulus is: E = tensile stress / tensile strain.

Index